A Practical Guide to SAP Integration Suite

SAP's Cloud Middleware and Integration Solution

Jaspreet Bagga

Apress®

A Practical Guide to SAP Integration Suite: SAP's Cloud Middleware and Integration Solution

Jaspreet Bagga
Texas, TX, USA

ISBN-13 (pbk): 978-1-4842-9336-2   ISBN-13 (electronic): 978-1-4842-9337-9
https://doi.org/10.1007/978-1-4842-9337-9

Managing Director, Apress Media LLC: Welmoed Spahr
Acquisitions Editor: Divya Modi
Development Editor: Laura Berendson
Coordinating Editor: Divya Modi
Copy Editor: Kezia Endsley

Cover designed by eStudioCalamar

Cover image by Luemen Rutkowski on Unsplash (www.unsplash.com)

Distributed to the book trade worldwide by Apress Media, LLC, 1 New York Plaza, New York, NY 10004, U.S.A. Phone 1-800-SPRINGER, fax (201) 348-4505, e-mail orders-ny@springer-sbm.com, or visit www.springeronline.com. Apress Media, LLC is a California LLC and the sole member (owner) is Springer Science + Business Media Finance Inc (SSBM Finance Inc). SSBM Finance Inc is a **Delaware** corporation.

For information on translations, please e-mail booktranslations@springernature.com; for reprint, paperback, or audio rights, please e-mail bookpermissions@springernature.com.

Apress titles may be purchased in bulk for academic, corporate, or promotional use. eBook versions and licenses are also available for most titles. For more information, reference our Print and eBook Bulk Sales web page at http://www.apress.com/bulk-sales.

Any source code or other supplementary material referenced by the author in this book is available to readers on GitHub (https://github.com/Apress). For more detailed information, please visit http://www.apress.com/source-code.

Printed on acid-free paper

I dedicate this book to my valued customers, who have entrusted me with the important task of integrating their SAP systems with other critical applications. It has been my absolute privilege to work with you on hundreds of successful integration projects over the years, and I am grateful for the opportunity to have been a part of your digital transformation journey. Your trust, collaboration, and feedback have been instrumental in helping me grow and improve as a consultant.

Table of Contents

About the Author

Jaspreet Bagga is an executive consultant with expertise in SAP and non-SAP integrations. He is a hands-on SAP architect who does, solution architecture, development work, leads the delivery of complex integration programs, manages global teams, and ensures successful project go live/goals. Jaspreet has made a lasting impact on more than 73 global businesses, delivering more than 200 IT projects for Fortune 500 clients, including Walgreens, McKinsey & Company, the state of Nevada, Discovery Channel, Aflac Insurance, the city of San Diego, Siemens, and more.

Jaspreet graduated from the State University of New York, Buffalo Engineering School and was inducted as an official member of the Forbes Technology Council.

His latest interests are in building and integrating SAP Business Technology Platform applications, developing interface-monitoring software like DOST Add-on®, and mentoring SAP community newcomers regarding integration topics.

About the Technical Reviewer

Miguel Figueiredo is a passionate software professional with more than 30 years of experience in technical solution architecture. He has a degree in information systems and an MBA from Mackenzie University, as well as an international MBA in business administration from the FIA Business School, in partnership with Vanderbilt University.

Miguel gained his experience delivering business intelligence solutions for a number of Fortune 500 companies and multiple global corporations. As the SAP HANA Services Center of Excellence leader, he was responsible for the evangelization and best-practice adoption of data management and business intelligence in his region.

Currently, as a customer success partner, he advises companies to maximize value realization in their digital transformation journeys and move to cloud initiatives.

Miguel is dedicated to supporting his family and encouraging the development of good habits for a healthy body and mind.

Acknowledgments

I want to thank the SAP Integration community for sharing their knowledge, expertise, and passion with me. Your contributions have inspired me to keep learning and pushing the boundaries of what's possible in this exciting field.

I also want to express my deep appreciation to my family and team, who have supported me throughout this journey. Your unwavering encouragement, sacrifice, and hard work have enabled me to deliver the best possible results to our clients. I couldn't have done it without you.

Finally, this book would have not been possible without the help of my editors, technical reviewers, and publishing team at Apress, who worked relentlessly with patience to ensure that the book is well written and on time.

Introduction

Welcome to the world of SAP Integration! Over the past few years, SAP has emerged as one of the leading providers of ERP solutions for businesses of all sizes. With its suite of powerful tools and applications, SAP has become the go-to choice for organizations looking to streamline their operations and improve their bottom line. However, as businesses grow and evolve, they often find themselves working with multiple SAP or non-SAP systems that need to be integrated seamlessly to increase efficiency, improve user experience, avoid duplication, and reduce errors.

That's where the SAP Integration Suite comes in. This suite of tools and technologies enables businesses to connect different SAP and non-SAP systems and applications, allowing for real-time data exchange and streamlined workflows. As an SAP Integration consultant to hundreds of businesses, I have spent years working with different SAP systems and helping businesses set up seamless integrations. I have been fortunate to work with some of the best minds in the industry. However, I also know that mastering the SAP Integration Suite can be a daunting task, especially for those who are new to the field. Therefore, I wrote this book to provide a comprehensive guide that covers all aspects of SAP Integration Suite Cloud Integration, from the basics to advanced concepts.

This book is the culmination of my experience and expertise, providing a comprehensive guide to the SAP Integration Suite and the different integration options available. This book covers the entire spectrum of knowledge required to gain mastery over the subject and get hands-on knowledge to work on the latest SAP BTP-based Integration technology—SAP Cloud Integration.

This book is designed for anyone who wants to understand the SAP Integration Suite, regardless of your level of experience. Whether you are a business owner, IT professional, or consultant, you will find valuable insights and practical advice in these pages. The book is divided into chapters that cover different aspects of the SAP Integration Suite Cloud Integration, and each chapter includes real-world examples that illustrate how to use this technology in practice.

CHAPTER 1

■ ■ ■

Introduction to Integration

This chapter discusses the idea of integration and its significance in modern business operations. To enable data exchange and workflow automation, various systems and applications must be connected through the integration process.

Businesses use a number of systems and applications to handle various parts of their operations in today's globally connected environment. The fact that these systems frequently work in isolation results in inconsistent data, a need for manual intervention, and ineffectiveness. The key to utilizing these technologies and applications to their full capacity is integration, which enables organizations to link, automate, and improve workflows. This chapter explains the fundamentals of integration, together with its advantages, varieties, and challenges. iPaaS, ESBs, and other essential technologies and tools for integration are covered. By the end of this chapter, you will have a clear understanding of integration, as well as understand how it can support your company in the digital era.

To understand what integration is, the next section digs into the concept of system integration.

1.1 What Is System Integration?

System integration, very broadly speaking, is the act of joining various sub-systems (components) into a single bigger system that works as a single unit. System integration is often referred to as the process of connecting disparate IT systems, services, and/or software to make them functionally compatible software solutions.

System integration is used by businesses to increase output and raise the standard of their operations. Through integration, multiple IT systems can speak to each other across the organization, thus accelerating information flow and lowering operational expenses. System integration can also be utilized to link an organization's external business partners, not just its internal systems.

The next section looks at some examples and approaches to integration that you may encounter and use.

1.2 Examples of Integration in Real Life

Examples of integration include ordering something online from Amazon or another e-commerce site, booking a hotel, purchasing a flight ticket, and many more.

© Jaspreet Bagga 2023
J. Bagga, *A Practical Guide to SAP Integration Suite*, https://doi.org/10.1007/978-1-4842-9337-9_1

Let's look at how a flight ticket is purchased to get a better understanding of the complete integration process in practice:

1. Booking a flight ticket from Google.

   You first enter the source and destination, let's say New York (source) and Tokyo (destination), when you purchase a ticket using Google.com (see Figure 1-1). You typically receive the ticket price after entering those details. The integration application's input comes from these sources and destinations. Google supplies you with information from every airline company. See Figure 1-2.

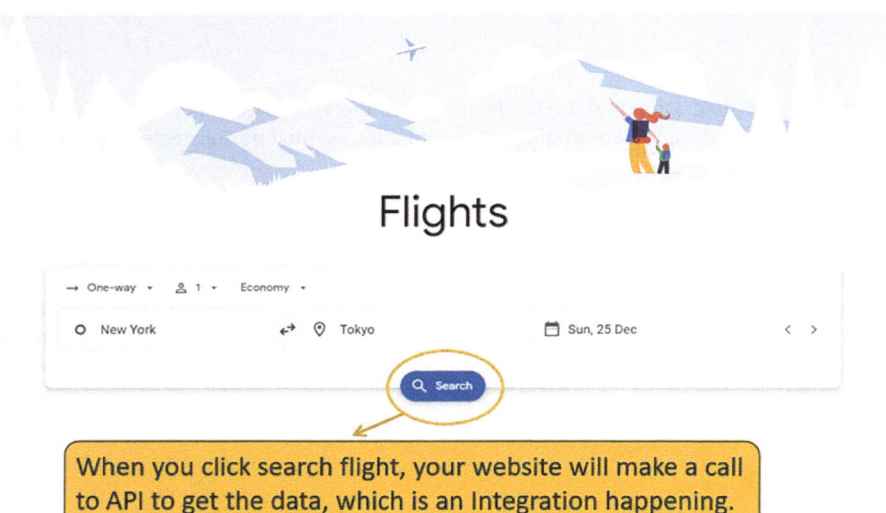

Figure 1-1. *Home page of the Google Flights website*

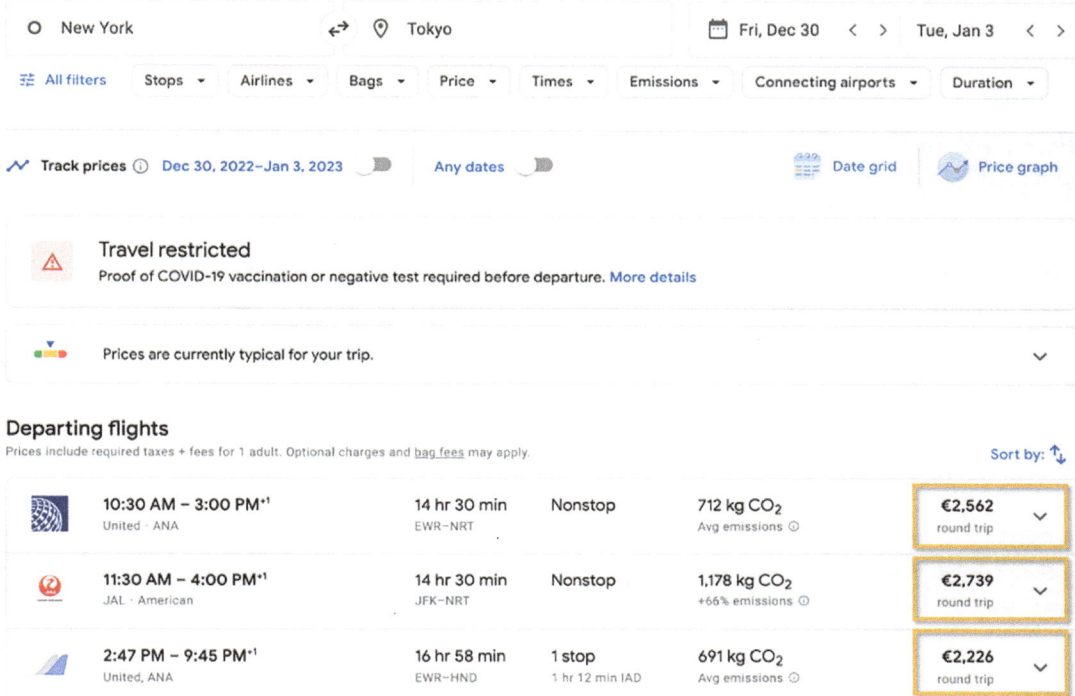

Figure 1-2. *Getting the price of the flights*

Although Google does not operate the airplanes nor sell the airline tickets, it helps you book a ticket from any of the airline's websites due to integration.

On a business website, you can make a reservation just like you would on Google. Additionally, those businesses use integration for their ticketing systems.

When you purchase a flight ticket from any airline company, the airline company integrates using its own system. But when Google integrates, it does so from a variety of airline company that provide the data from their systems. This is known as *layered integration* (integration on integration).

2. Online shopping from an e-commerce website.

    When you shop online, say at Amazon or any other website, you select the required product and add it to the cart, then add your address details, and so on. When you proceed to the payment, there is again an integration happening with your bank or payment provider. Only after the payment is deemed successful is the order shipped, and many integrations comes into play.

    For example, if you search for tires on Amazon.com (as illustrated in Figure 1-3), you get the price of the tires, which means the integration has happened in the backend. You see the price of the tires that you searched for, as shown in Figure 1-4.

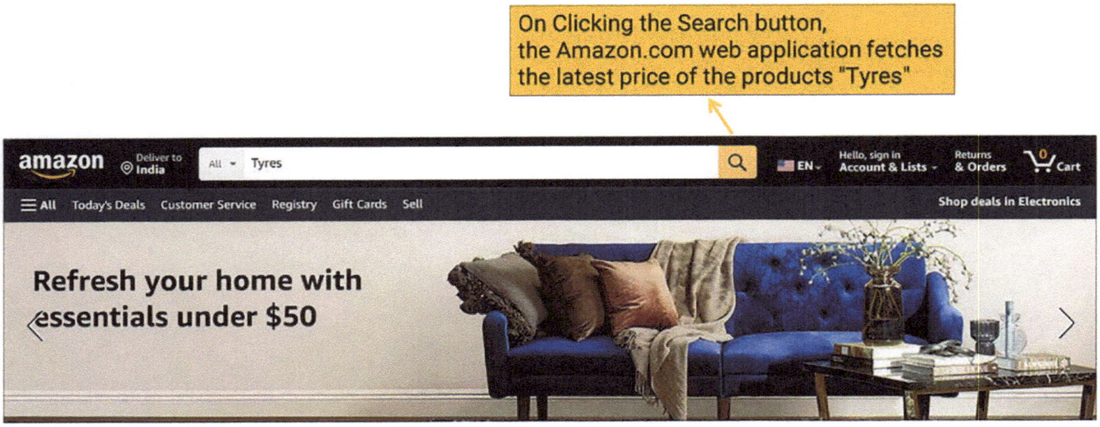

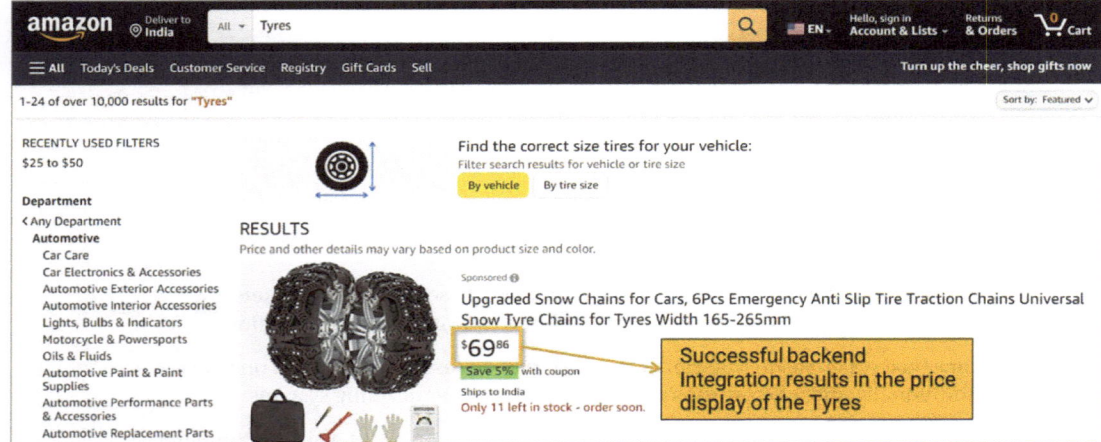

Figure 1-3. *Home page of an e-commerce website*

Figure 1-4. *Amazon.com giving the price of the tires*

1.3 Types of Integration

1.3.1 Process Integration

Process integration refers to the practice of connecting and harmonizing different processes within an organization. It involves the seamless exchange of data, information, and activities across various systems, departments, and stakeholders. To get better results, companies can use PI to coordinate their internal processes as well as those of their partners, clients, and suppliers.

As a business expands, processes become more complicated. For example, one vendor can manage purchase orders without too much difficulty. To address this process, you might add software as you work with more providers. You could then add software for managing your accounts, paying bills, and keeping track of inventories. Sooner or later, a single process may involve several different technologies. Because the data is siloed, this might reduce productivity and obstruct your automation processes.

According to the IT and the engineering leaders, Integration and retrieving external data are the most significant obstacles to process automation. Businesses need a process integration solution to integrate their systems and provide uniform access to reliable, consistent data from all sources. Lack of business integration can result in high production costs.

The P2P cycle is the greatest example of the process integration. The procure-to-pay (P2P) cycle is a business procedure that entails acquiring products and services from suppliers and paying for those products and services respectively. P2P integration is the process of integrating several P2P cycle systems and processes to automate as much of the manual labor as possible.

The process of acquiring products or services from a third-party supplier involves a series of actions that a business needs to follow. These actions include purchasing goods, making payments to the supplier, and considering any available discounts. This cycle outlines the step-by-step procedures that businesses must undertake to successfully obtain the desired goods or services from the supplier while ensuring timely and accurate payments.

In Figure 1-5, each line represents the process integration between a tire manufacturer and its retailers.

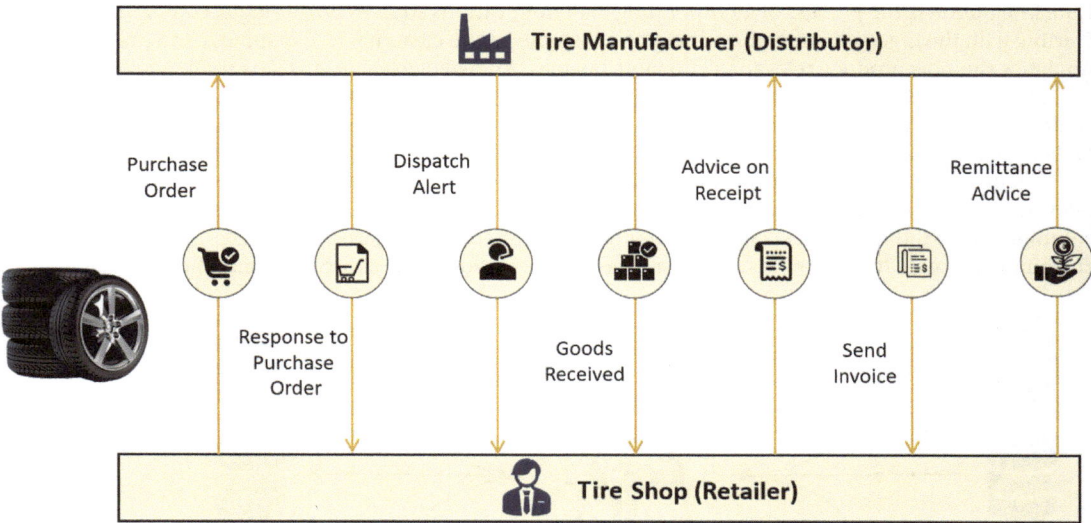

Figure 1-5. *Example of process integration in the P2P cycle*

Three Types of Process Integration

There are generally three types of process integration, as follows:

- **Native integration**—Data is effortlessly transferred between software applications. The program has these integrations built-in, making the setup procedure simpler. All you have to do is provide permission to connect to the applications

  An example of native integration is the SuccessFactors Employee Central Payroll (ECP)/benefits programs that read data files from ECP SFTP or IDOCS being exchanged between two SAP systems, such as SAP CRM and S/4HANA.

- **Application programming interface (API) integration**—Data is exchanged between two or more applications. If there are no native integrations, you might develop a custom API to link your online store to a payment processor to take orders.

For example, you can create a SAP S/4HANA Sales Order and call the PayPal API to process the payment.

- **Third-party integration**—Businesses with specific requirements frequently create their software. However, using an existing tool to add a particular feature is often more helpful. Instead of creating a feature from scratch in these circumstances, you can use third-party connectors. An example of third-party/custom integration is developing adapters or modules to fulfil a specific requirement.

  For example, developing custom AWS S3 buckets or Salesforce adapters to be deployed on SAP Process Orchestration, SAP does not deliver these adapters out of the box.

1.3.2 Data Integration

Data integration is the process of incorporating data from various sources into a single, coherent perspective. Starting with the ingestion procedure, integration encompasses cleaning, ETL mapping, and transformation. Analytics can now produce relevant, actionable business knowledge because of data integration.

Data integration involves the process of combining data from various sources in different ways. It can be achieved through different methods and techniques. Clients accessing data from a central server, a network of data sources, and other shared components form the basis of data integration systems.

Clients typically request data from controller servers as part of the data integration process. The relevant data is then ingested by the controller server from internal and external sources. The information is compiled into a thorough data collection after being acquired from a variety of sources. This is given to the customer, who uses it. Figure 1-6 shows the data transformation of the source data and the target data.

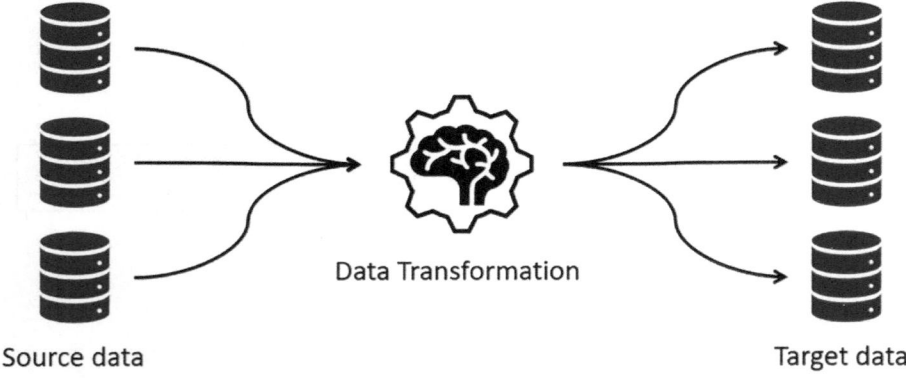

Source data   **Target data**

Figure 1-6. *Data integration architecture*

Companies that develop data integration skills have a considerable competitive advantage, including:

- Reduction of the need to manually alter and merge data sets, which increases operational efficiency.

- Improved data quality by applying business rules to data through automated data transformations.

- A comprehensive perspective of the data and more accessible analysis, which leads to more insightful information.

6

A digital business is based on data and the algorithms that process it. It derives the most value from its information assets by accessing them whenever and wherever needed across the whole business ecosystem. Data and related services move freely yet safely across the IT landscape in a digital organization. Data integration prepares your data and provides a complete view of all the information moving through a business.

Advantages of Data Integration

Companies need access to reliable, current information to stay competitive. Systems that use real-time integration can improve every aspect of their performance. Not only will gathering data and transforming it be more efficient and accurate, but it also provides real-time intelligence, agility, and actionable insights. Advantages include:

- Data quality and data integrity
- Quick, simple, and accessible links between data storage
- Seamless system-to-system knowledge transmission
- Improved cooperation
- Real-time, comprehensive business analytics, intelligence, and insights
- Improved ROI and efficiency

Disadvantages of Data Integration

Numerous products are available on the market to assist you with these issues. Even with so many resources at your disposal, there are still frequent problems to avoid when developing an effective data integration plan, including these:

- Different data sources and formats
- Data not being accessible when it should be
- Poor or out-of-date information
- Use of inappropriate integration software
- Excessive data

1.3.3 Business-to-Business Integration

B2B integration is the process of integrating several corporate applications and systems to allow information and data exchange. This kind of integration can increase productivity, lower errors, and expedite company procedures. It is frequently used to promote communication and collaboration between enterprises. B2B integration frequently includes the use of specialized software and technology to link systems, and its setup and maintenance may call for the assistance of IT specialists, as shown in Figure 1-7.

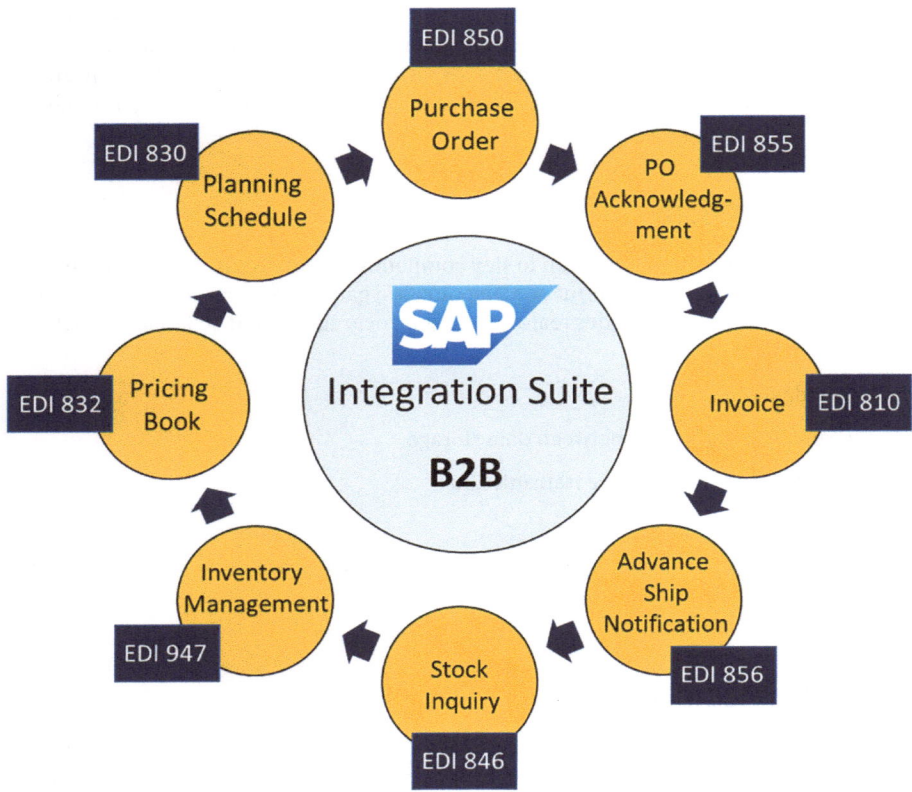

Figure 1-7. *Representation of B2B integration*

Using APIs to allow multiple systems to connect and using electronic data interchange (EDI) systems to exchange invoices, purchase orders, and other business documents are two typical examples of B2B integration.

Components of B2B Integration

There are two types of components of business-to-business integration:

- **Data-level integration**—This entails automating the interchange of documents between trading partners and converting paper documents to digital ones. This kind of B2B integration has its origin in electronic data interchange, or EDI.

- **People-level integration**—This enables trading partners to collaborate and communicate effectively to carry out end-to-end business processes. It include the features like partner onboarding and community management.

Advantages of B2B Integration

The main advantages of B2B integration involve enabling cooperative partnerships and new work practices that support modern supply chains, in addition to the cost, speed, and productivity increase of completing digital transactions.

Every company needs to collaborate with the other enterprises, and B2B integration represents the digital transformation of these external collaborations. A fully integrated end-to-end B2B integration solution offers several advantages, many of which were previously impossible, such as:

- Manual processing time and cost are reduced.

- Make fewer mistakes when processing company documents.

- Automate critical company operations, such as order-to-cash and procure-to-pay, to increase efficiency.

- Increased supply chain automation boosts productivity and lowers inventories.

- Increase your supply chain's visibility to all activity and transactions.

- Improve cooperation among your commercial partners.

- Manage trading partner performance more effectively.

- Promote innovation by developing tighter ties with your trading partners.

- Better visibility and control over your B2B data to provide actionable insights that will enhance decision-making.

Challenges in B2B Integration

The so-called "100 percent trading partner enablement" is the main obstacle to use this integration. Your company will be more efficient and reap more rewards if you can interact with more of your trading partners electronically. The old business strategy, however, has led to businesses concentrating on the 20 percent of partners that produce 80 percent of the income. Businesses must develop methods to collaborate closely with all their partners, as the 80/20 rule becomes less applicable to businesses as they become increasingly digital.

Here are the following challenges of B2B integration:

- There are many types of business documents that can be exchanged, for example, Mapping of EDI 810, 850, 820, 834, 997, and so on, which makes design and development of integration complex.

- Working with many trading partners requires the support of numerous B2B standards, including ANSI, EDIFACT, and XML.

- The B2B standard is likely to be interpreted differently by each business partner.

- There are many communications protocols (AS2, SFTP, FTP, VAN) that can be used.

- Smaller business partners might not have the necessary funds, internal resources, or business needs to develop a complex B2B connectivity solution.

- Smaller trading partners may be reluctant to alter the way things are done.

- Partner onboarding can be laborious and slow, creating backlogs that hinder company productivity.

- EDI messages are not easy to read and interpret by business users unless they are transformed into readable, XML, file, or database tables.

1.4 Patterns of Integration

System integration involves combining different systems, parts, or subsystems to create a larger and more complex system. This process, known as patterns of system integration, focuses on the design and integration of multiple systems into a cohesive whole. These patterns encompass the techniques, approaches, and communication methods used to integrate system components, as well as the overall design of the integrated system. By following these patterns, organizations can successfully integrate diverse components into a unified and functional system.

While there can be many patterns of integration, the following sections describe some of the most commonly used patterns in the business world.

1.4.1 Star Integration

A collection of point-to-point system connections make up a *star integration*. In other words, when multiple basic connections are joined into one, a star connection is created. As the number of connected subsystems rises, so do the number of points at the beginning and the lines that follow. Figure 1-8 illustrates this.

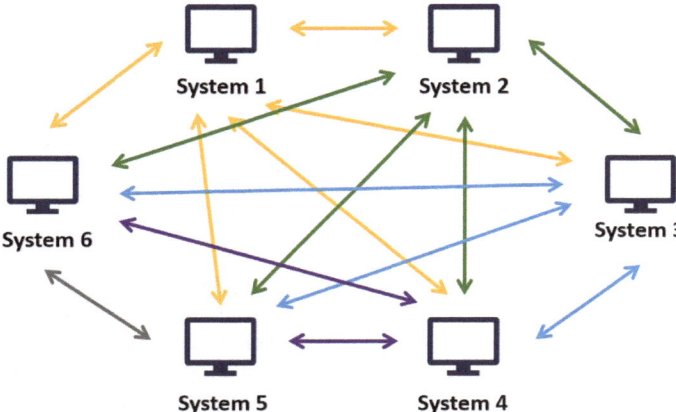

Figure 1-8. *Representation of the star integration pattern*

The network of links created by this system integration method may under ideal circumstances resemble a star polyhedron when integrating each system into the remaining subsystems. The basic system integration diagram is more likely to resemble a plate of spaghetti, which is why star integration is sometimes called spaghetti integration. When a company focuses on system integration in this way, the initially well-structured IT architecture can quickly become complex and difficult to navigate. This complexity arises from the need to manage multiple integrations, resulting in greater capability compared to a single point-to-point connection. However, the management of these integrations becomes more challenging due to the increased complexity of the overall system. An effective Enterprise Integration Strategy can prevent the systems integrations of an enterprise from taking a giant Star Integration pattern which can become complex to operate.

Advantages of Star Integration

- It is incredibly dependable; even if one interface breaks, the others will still function.

- Since there are no data collisions and middleware resource sharing, it performs well.

- Simpler to integrate and robust in nature.

- Simple defect identification because the relationship is frequently simple to identify.

- There are no network hiccups when adding, unplugging, or disconnecting systems.

- Each system needs one port to connect to the hub.

- Best suitable for scenarios where there are minimum integration points or systems to be connected.

Disadvantages of Star Integration

- More connection is needed than with a enterprise service bus (ESB).

- The singular interface monitoring becomes challenging and tedious at application level.

- It poses security risks to open individual system connections to the internet.

- Scaling the architecture becomes challenging or impossible due to vast number of distributed connections.

- Because the interfaces typically do not use middleware, it needs more resources and ongoing upkeep.

- The need for additional software, like adapters or certificates, raises the price of implementation.

- Performance is dependent on a single concentrator or hub.

1.4.2 Horizontal (ESB) Integration

The process of horizontal integration involves connecting all of the subsystems together by employing one specialized subsystem as the common user interface layer. In other words, because the subsystems are connected indirectly rather than directly through the main system, fewer connections are needed for system integration. If there are five subsystems, there will only be five connections. If there are ten subsystems, there will only be ten connections. Because fewer connections are required to maintain operation, this approach has the major benefit of requiring less time, effort, and money to develop the system. This sort of system integration employs an enterprise service bus (ESB) as an intermediary layer or subsystem, as shown in Figure 1-9.

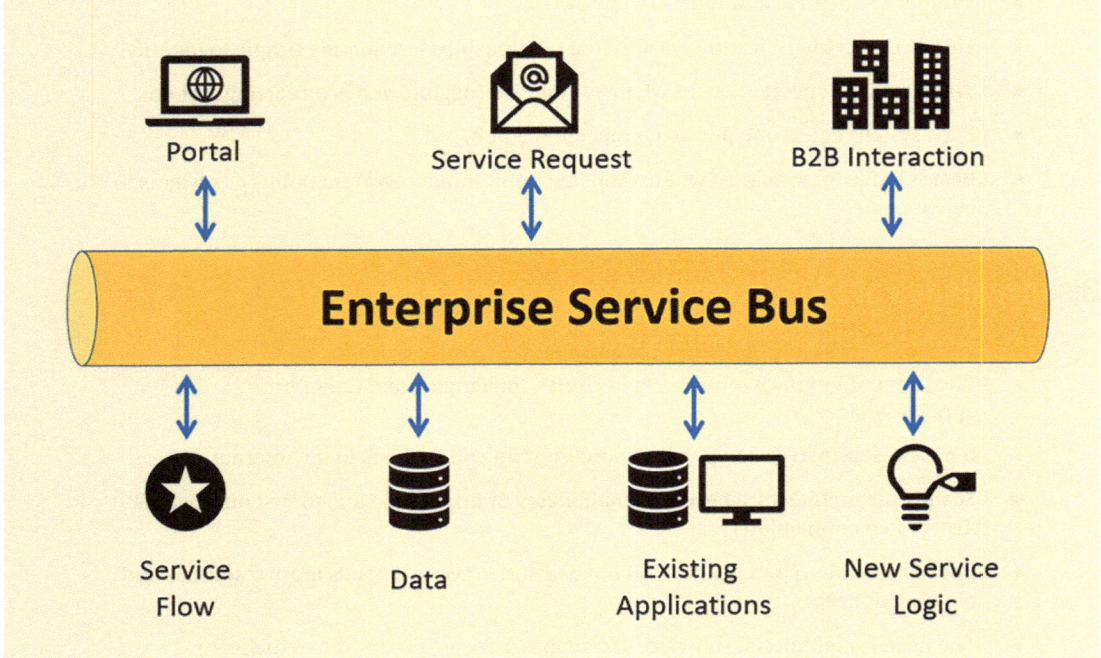

Figure 1-9. *Representation of the horizontal integration pattern*

- **Pros and cons**—One of the best things about ESBs is that since one subsystem is separated from the others by a "message bus," it is possible to replace or modify one subsystem without affecting the operation of the others. This supports highly scalable architecture. Such projects are also dependable and straightforward to develop. The dispersion of integration duties across the systems increases the complexity of maintenance and troubleshooting. However, well-orchestrated middleware products by vendors like SAP Integration Suite, Process Orchestration, Dell Boomi, and so on, take care of the scalability and stability of the software components and can be effectively utilized as an Enterprise level Service Bus (ESB).

- **Usage circumstances**—Large projects like enterprise application integration (EAI) are best implemented using an ESB model since it enables them to scale as necessary. If a business has to put it together onsite, it fits well.

1.4.3 Point-to-Point Integration

Point-to-point integration/connectivity is not system integration in its purest form. Despite the system functioning, complexity is limited. A 1:1 link, or one system to another system, is best served by this type of system integration, because it typically handles one business function at a time. Figure 1-10 shows how point-to-point system integration quickly becomes unmanageable as more systems are used and there are more links between them.

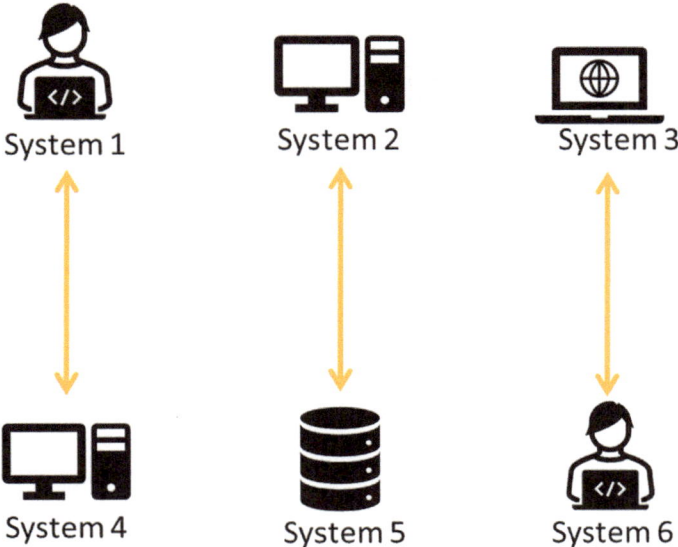

Figure 1-10. *Representation of a point-to-point integration pattern*

In point-to-point integration, data is extracted from one system, changed or formatted, and then transferred to another system using a point-to-point link. The logic for data translation, transformation, and routing is implemented by each application while considering the protocols and supporting data models of other linked components.

- **Pros and cons**—One of the key benefits of point-to-point integration is that the IT team may swiftly construct a small-scale integrated system. On the other hand, the model is challenging to scale, and managing all the integrations becomes extremely difficult as the number of applications increases. Say you need to run 15 integrations to connect six modules. In this case the so-called star/spaghetti integration comes into play.

- **Usage circumstances**—This strategy is appropriate for businesses with straightforward business logic that rely primarily on a small number of software components. Additionally, it is the best choice for companies connecting to SaaS services.

1.4.4 Hub and Spoke Integration

Hub and spoke integration involves connecting multiple systems to a central hub. The hub then manages the flow of information between the different systems. This type of integration is typically used when many systems need to be connected. Figure 1-11 shows the different number of systems connected to a central hub, which is called the *integration hub*.

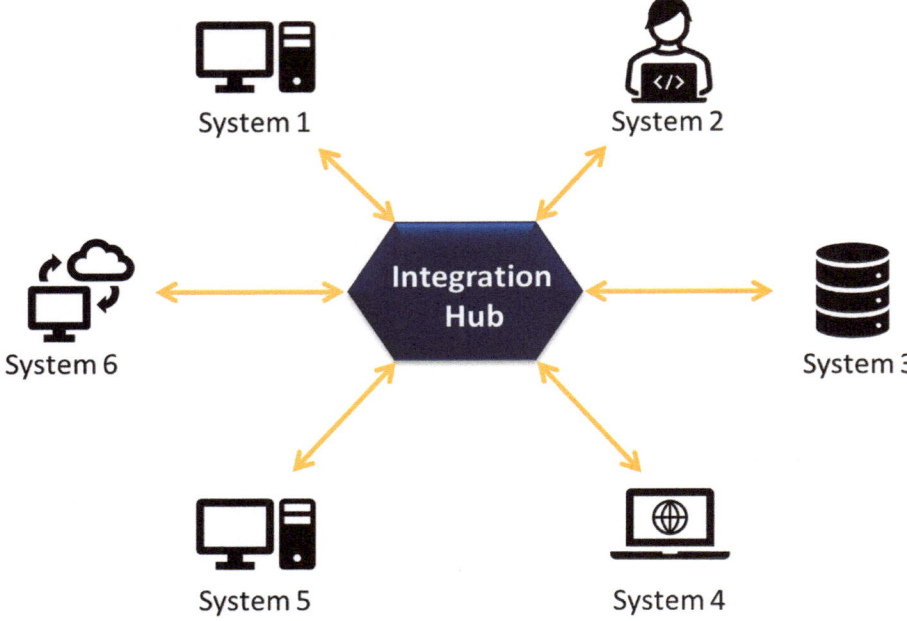

Figure 1-11. *Representation of a hub and spoke integration pattern*

- **Pros and cons** —The model offers several advantages over the point-to-point pattern, including greater scalability. Things improve in terms of security and design simplicity because each system has one connection to the central hub. Meanwhile, a drawback of this strategy can be the hub's concentration. The only integration engine that powers the entire infrastructure is susceptible to become the main bottleneck as the workload increases.

- **Usage circumstance**—The hub and spoke model is frequently used in payment processing, financial operations, and e-commerce. Additionally, it is the preferred architecture for heavily regulated industries with high-security risks.

1.5 Advantages of Integration

Any new software system must be implemented slowly throughout the business. Why would you want to disrupt the post-implementation balance that your company has established? Running several software programs simultaneously without integrating them has many disadvantages. This section covers the advantages of system integration and explains how it helps resolve issues brought on by disorganized systems.

1.5.1 Automation and Simplification

Gathering relevant and connected data is one of the process's most important results. This translates to considerably simpler and more straightforward retrieval and processing for all subsystems. If the development company you choose performs a fantastic job with your integration process, you will have an integrated system with correctly designed data flow channels. That will enable you to automate and streamline your business activities.

1.5.2 Availability, Correctness, and Coordination

The integration process often involves collecting data from numerous sources and storing it in one place. As a result, there is no longer a need to manually wait for data to sync across several systems over extended periods. Instead, the data is immediately updated for all other systems when one subsystem updates it. As a result, the real-time availability and accessibility of data is one of the essential benefits of system integration.

As consumers are less likely to access out-of-date data, automatic updating and synchronization of data results in higher accuracy.

With simple access to the most recent data, all subsystems can engage with one another and improve overall teamwork. It is also simpler to train users on a more straightforward system. Additionally, fast data accessibility speeds up decision-making across all departments, enhancing the entire company process. Employees will be constantly informed of previous events and will act appropriately in the future.

1.5.3 Effectiveness and Productivity

Low-quality performance might result from divisions between an organization with multiple departments and segmented roles (functions). If one subsystem alters its piece of the data, the other subsystems may need to catch up by manually inputting the revised data into their respective data stores.

By eliminating time-consuming manual data entry, an integrated system improves efficiency. As a result of the system's automatic update of the main database, employees can receive accurate data for further processing. There is a significant time-saving result. Each department can concentrate on its responsibilities without having to worry about keeping up or being in sync with other departments' operations. The resources that were saved might subsequently be used for crucial business processes. System integration, therefore, results in two important benefits: it boosts worker productivity and efficiency.

1.5.4 Cost-Effectiveness

The necessity for repetitive tasks is eliminated by an integrated system, as I talked about in the last section. As a result, crucial jobs can now be finished more quickly while utilizing the same resources, cutting down on needless expenses.

Central storage eliminates the necessity for many data stores to store the same data. Each subsystem's specific data can be stored there, and any data overlaps can be categorized appropriately. As a result, the price of underutilized data storage space is less.

When you realize the need for a better coordinated organization-wide software system, system integration enables you to avoid building a new, expensive, and complicated system from the beginning. Alternatively, you might employ an expert to combine the current systems to operate in concert. In addition to saving you money, doing this also saves the time and effort that would be used to train staff members on a new system.

1.5.5 Scalability

Integrated systems increasingly rely on the cloud due to technological developments in data storage. Massive resources are needed if each subsystem has its own storage or processing platform. Each subsystem's capacity must be raised independently as data volume increases.

When a system is integrated, this repetition is eliminated. All subsystems can use shared resources as necessary. You can quickly ask the cloud provider for more resources as your company's computing or storage needs increase. Thus, scalability is one of the main advantages of integration.

1.5.6 Availability of Performance Insights

When data is dispersed over numerous departments or subsystems, the ever-growing volume of data makes it challenging to monitor the organization's overall operation closely. It takes a lot of work to gather and combine data and produce the analytics report. You might need to carry out several imports and exports to ensure there are no inconsistency.

With an integrated system, you always have access to all essential data and can determine the effectiveness of each department using a central dashboard. An integrated dashboard can access the core data repository. Your ability to examine relevant data as required will enhance the clarity of your analytical and performance reports. By choosing the proper parameters, you can evaluate departmental progress or conduct a comprehensive analysis of your company all in one spot.

1.5.7 Security

In isolated systems, it is challenging for attackers to simultaneously compromise all subsystems, especially when each system use a different level of security. It could be challenging for you to manage security for various platforms.

With a centralized system, your data is equally vulnerable to theft. However, by using a more potent security tool or algorithm, this problem can be resolved. Thus, it becomes considerably simpler to handle the security of a single platform.

1.6 Disadvantages of Integration

Even though you might want to believe that system integration services are an ideal choice for any business, there are a few disadvantages to consider.

1.6.1 Security Issues

It's preferable to have multiple programs than a single integrated system regarding security issues. After completing system integration, any hack or fraud could access all your data rather than just a portion of it.

Because of the pathways through which data moves from one program to another during system integration, your information is more exposed than it formerly was.

1.6.2 Issues in Upgrades or Maintenance Updates

Upgrading can be a time-consuming procedure if your company uses multiple integrated systems. Each module refreshes while you wait. Additionally, you need to test, upgrade, and maintain separate patches for each program. The more space they occupy, the more time and effort it takes.

1.6.3 High Cost

The cost to integrate and update the company's systems will vary depending on several variables. These factors include the scope and complexity of the enterprise, the number of programs that must be integrated, and the degree to which each program needs support. Additionally, businesses must consider the costs of switching programs following integration.

The upfront cost of IT system integration services may be significantly more than your typical monthly operational expenses.

1.7 The Role of Integration in Enterprise

Enterprise integration connects different systems and applications to streamline business processes. It allows for data exchange between systems, which can then be used to make better-informed decisions.

Enterprise integration has many benefits, including increased efficiency, improved decision-making, and reduced costs. However, it is essential to note that enterprise integration is not a one-size-fits-all solution. Each organization is unique and will have different requirements for their integration solution.

Enterprise integration is essential for the design, execution, and distribution of crucial applications, as well as the improvement of internal business processes and activities. Companies may improve their operational scalability, broaden their customer base, and boost their revenue by sharing crucial information, streamlining procedures, and seizing chances. The Role of Integration in Enterprise are as follows:

1.7.1 Sharing Important Information

Enterprise integration facilitates the transfer of data between operating systems and complex information by providing a middleware layer to serve as the common interface between each application, design, and service. Application developers can easily share data and expose interfaces without the need to understand other applications, their locations, or anticipate future issues. Furthermore, it simplifies the exchange of data between multiple programs and the various consumers who rely on that data.

1.7.2 Streamlining IT Procedures

Enterprise integration combines functionality and information transmission across several applications to facilitate fluid cooperation. Their interconnectedness makes it easier for individuals and organizations to use IT processes by helping to streamline them. It speeds up user access to data and aids IT businesses in streamlining data integration and services. This simplification modernizes the development and use of enterprise integration patterns such gateway services, message queues, file transfer, and enterprise service bus by making it possible for them to be created, deployed, operated, and maintained through Agile and automated processes (ESB).

1.7.3 Expanding Possibilities

Additionally, enterprise integration enables teams to act proactively in order to seize opportunities and address new and evolving company needs. Teams can immediately recognize and respond to time-sensitive events, such as unanticipated policy changes or new application management procedures, by taking control of all data access points without having to modify the apps themselves. In the end, teams are given the tools they need to design, develop, and streamline various integration solutions by utilizing a standard method for communication and collaboration.

1.7.4 Benefits of Enterprise Integration

- **Automating and streamlining corporate procedures**—Customers and staff get a seamless, personalized experience across many digital touchpoints. Businesses are concurrently delivering exciting consumer experiences by streamlining and automating dispersed internal business operations.

- **Providing client insight**—Today's market leaders must use data from inside and outside their enterprises to predict client preferences and demands. To foster loyalty and expand their competitive edge, businesses can develop a 360-degree understanding of their audiences and customers.

- **Creating a future-proof IT environment**—Often, a successful digital strategy necessitates adapting existing systems to accommodate new business models. With an API-first strategy, cohesive enterprise integration solutions can maximize your current investments.

- **API economy facilitation**—Businesses are unlocking exclusive services and developing fresh company ideas for a competitive edge. By utilizing digital channels to create a nearby ecosystem of customers, partners, and suppliers, businesses can propel these economies within and beyond their immediate organizational bounds.

1.7.5 Enterprise Integration Scenarios

An organization can benefit from integrating various critical systems, processes, data, and applications from all business lines. Figures 1-12 shows some of the commonly used integration types.

Consider an example of a company called Abusiness, as shown in Figure 1-12. This company sells groceries using integration scenarios. Process integration occurs to support its purchase ordering to the supplier. The AWS data lake, which receives the bulk records, demonstrates data integration. Following the sale of the goods to the customers, Salesforce records the customer reviews, demonstrating cloud integration. The POSDM database is application integration; it sends the daily sales receipts to Abusiness. A separate company provides the API Integrations and tweets about grocery store discounts. Finally, by connecting to the warehouse management system, another company obtains the cold storage temperature and modifies the warehouse temperature, demonstrating device integration.

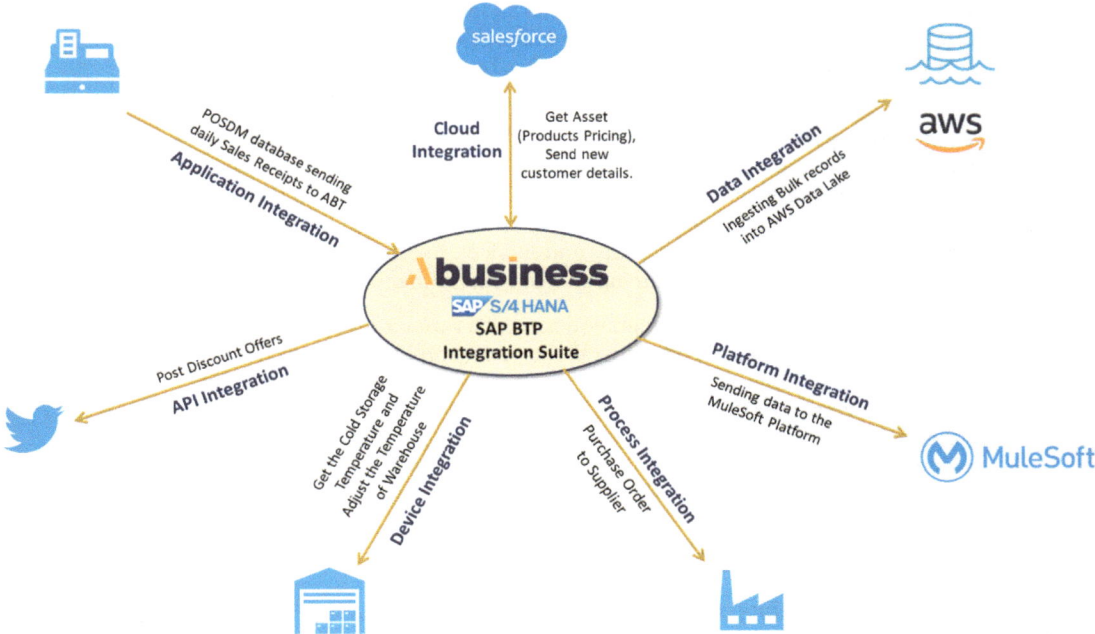

Figure 1-12. *Abuisness is using all the enterprise integration scenarios*

Some of the commonly used enterprise integration scenarios include:

- **Application integration**—Using enterprise application integration (EAI), processes and data can be upgraded, integrated, and exchanged among different software programs in real time to increase visibility, productivity, and insights across the organization. For example, integration of a CRM system (SAP CRM, Salesforce, Sugar CRM) with an ERP financial system (Oracle, S/4HANA).

- **Data Integration**—The process of merging data from various sources into a single, cohesive view is known as data integration. This enables firms to understand their data more thoroughly and precisely. It can be applied to many tasks, including analysis, reporting, and decision-making. An organization needs to build its data warehouse cloud by integrating daily sales data into a data lake (Amazon S3, Azure).

- **Cloud integration**—To create a unified IT infrastructure that handles data, processes, system architectures, and enterprise applications, numerous hybrid cloud environments (public and private clouds) are brought together through cloud integration. For example, an enterprise needs to integrate its Amazon Marketplace. Customer review data is integrated with the Salesforce cloud.

- **API integration**—Application programming interface (API) integration connects two (or more) enterprise applications through their respective APIs, enabling the exchange of data sources between the systems. These vital connections enable workflows and processes across the organization, syncing data to improve productivity and spur growth. For example, a website can use an API to connect to a service that handles payments, or a mobile app might use an API to connect to a service that provides weather forecasts.

- **Platform integration**—Platform integration allows IT professionals to create secure integration flows that connect and control many cloud-based applications, systems, services, and data sources using various software solutions. Integrated platform as a service (iPaaS) and platform integration go hand in hand.

- **Process integration**—Process integration is the process of integrating various company operations, alignment of business processes across departments/teams, and integration of various systems, technologies, and data sources. Process integration seeks to increase productivity, decrease waste, and help businesses provide their clients with better goods and services. An organization needs to integrate its purchasing and sourcing system with its financials.

- **Device integration**—Device integration connects various devices so they can interact, communicate, and work together to meet corporate goals and boost productivity. For example, an enterprise has the requirement to integrate its warehouse sensors or RFID devices with its enterprise warehouse management system or an organization using thermo sensors needs to control the temperature of a warehouse.

1.8 Legacy System Integration

Linking outdated software with modern applications is known as legacy system integration. When properly implemented, integrated applications work in harmony with one another and benefit your company. Here are a few benefits of this process:

- **Few manual tasks**—Repeatedly entering the same data lowers performance and increases the chance of human error. You only need to enter the data once when all your crucial systems are in sync. You become much quicker and much more accurate as a result.

- **Enhanced decision making**—By integrating legacy systems with modern solutions, you can access the priceless data you've been gathering for years, spot trends, and use that information to make better strategic choices. For example, you may use your past CRM data to identify more effective strategies for luring new clients and keeping current ones.

- **Access to new features and technologies**—Without starting from scratch, integration is the ideal way to upgrade your legacy system with new features and cutting-edge technology.

- **Low learning curve for old workers**—Whenever you replace a legacy system with new software, you must train your employees on how to use it, which can be difficult, mainly if you use the old system for your core business operations.

- **Low learning curve for new workers**—Many legacy systems are less user-friendly than contemporary ones, making it difficult for new staff to learn how to use them. Integration helps your business become less dependent on these technologies, reducing new hires' learning curve and streamlining their onboarding.

1.8.1 Key Challenges of Legacy System Integration

Integrating legacy software with modern IT infrastructure may seem like a simple task at first, especially when compared to the challenges of updating or creating a new system from scratch.

On the other hand, the nuances of legacy system integration reveal that it is by no means smooth sailing. According to Accenture's research, 44 percent of financial sector organizations believe that the biggest challenge to a successful digital transformation is the inability to link new technologies to current IT architecture. Here are some of the challenges:

- **Lack of knowledge**—Integration of legacy systems will require experts in outmoded technologies. You're stuck without them if you don't have any on staff. For example, in the 1980s, many colleges stopped offering courses in COBOL, one of the most popular legacy coding languages. This indicates that it will be challenging to locate COBOL experts.

- **Inadequate documentation**—Legacy apps frequently have weak documentation. If the programmers who created your legacy system are currently retired, its inner workings and years of patching are shrouded in mystery (called "spaghetti code"). This increases the difficulty of the integration process.

- **Poor data quality**—The legacy data is structure and format are outdated. You must first organize your data to perform integration and then figure out how to give it a "modern look."

- **Lack of upgrades, falling behind**—Despite the complexity of your existing legacy system, your staff, who have extensive experience and familiarity with its intricacies, may feel reluctant to embrace something new. In such cases, it becomes your responsibility to demonstrate the benefits of integration and ensure that the transition is as smooth and user-friendly as possible.

- **Security concerns**—Cyber risks might affect older apps. According to one poll, 74 percent of healthcare companies experienced at least one severe cyber attack in 2018, while about 70 percent of those organizations still used legacy systems. Your legacy system will be even more vulnerable due to the integration. Therefore, you need a solid plan to protect your legacy data during and after integration.

1.8.2 Legacy Integration Methods

In the past, system integration often involved complex and time-consuming manual processes. This included manually exporting data from one system and importing it into another or writing custom code to connect different systems. These methods were often error-prone and required a lot of technical expertise, making it difficult for many organizations to effectively integrate their systems. Additionally, these methods were not scalable and did not support real-time data exchange, making it challenging for organizations to keep their systems up-to-date and in sync. As a result, many organizations struggled to fully leverage the benefits of system integration and were unable to fully automate and optimize their business processes. The different methods for legacy integrations are as follows:

1. Files

    In an ideal world, your company would run on a single integrated piece of software that was created from the ground up to function uniformly and cogently. Naturally, even the tiniest procedures don't function in this way. Several pieces of software handle different sides of the business. Numerous factors account for this are:

    - People purchase products created by external groups.

    - Systems are developed at various eras, which influenced the technological choices.

- People construct various systems, and depending on their backgrounds and interests, they use multiple methods for creating applications.

- Even more so when integration doesn't offer value to the application being developed, getting an application out there and providing value is more critical than ensuring that integration is addressed.

Any company must be concerned about information exchange between highly diverse applications. These can be created using many platforms, in various languages, and with different operating system presumptions.

Such apps must be tied together, which necessitates a deep understanding of how to do so at both the business and technological levels. The less you need to understand how each application functions, the simpler this will be.

Enterprise applications can be created in a variety of languages, run on a variety of hardware, and use a variety of operating systems. As a result, it's crucial to have a standard method of data transfer that a range of programming languages and operating systems may utilize. The most straightforward strategy is to incorporate the apps using files and a shared storage system in all workplace operating systems.

Make each application create files containing the data that the others must consume. Integrators oversee converting files into various formats. Depending on the nature of the business, you need to create the files on a regular basis, as shown in Figure 1-13.

Figure 1-13. *Exchange of files from one application to another*

One of the key benefits of using files for integration is that integrators typically only need a file and minimal understanding of how the application functions internally. The structure and content of the file were discussed with integrators, although the options were frequently constrained if a package was utilized. Then, integrators handled the modifications needed for other programs or let consuming applications choose how to read and manage the file.

2. Database Connectivity

The process of gathering data from various sources—including social media, IoT sensor data, data warehouses, consumer transactions, and more—and sharing an up-to-date, accurate version of it across an organization is known as database integration. The home base, to and from which all shared information will flow, is provided through database integration, as shown in Figure 1-14.

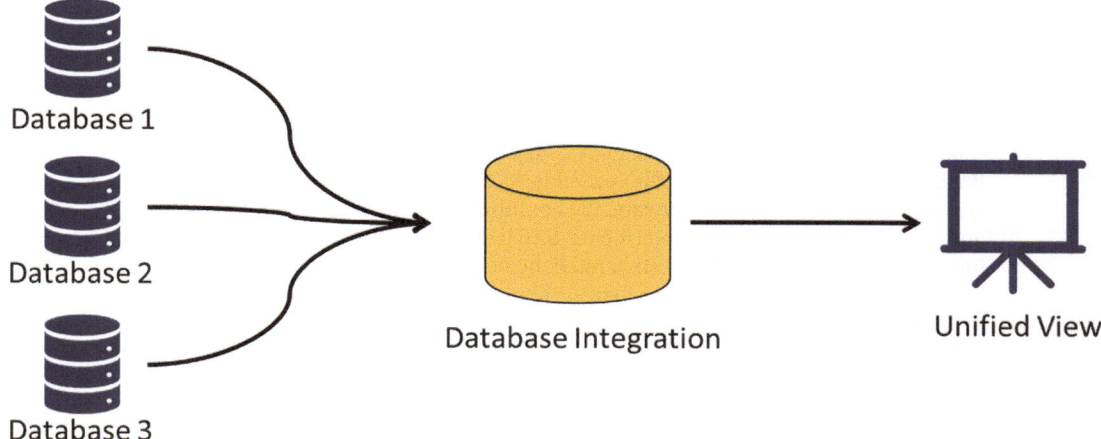

Figure 1-14. *Representation of a database integration pattern*

For example, when two companies merge, the information in their formerly separate databases is crucial to run the new combined company. Data deduplication, rule-based storage, cleansing, and secure stakeholder sharing are all possible with database integration.

As big data continues to drive business intelligence and the capacity to compete in an increasingly fast-paced digital marketplace and the cloud becomes the new standard for operations, database integration plays a crucial role in ensuring that businesses are effectively utilizing their data rather than becoming overwhelmed by it.

Benefits of Database Integration

Data is the foundation of contemporary business, which has replaced physical infrastructure like servers, routers, and other office buildings with digital exchanges.

These difficulties are transformed into quantifiable operational improvements by properly managed database processes. Benefits of database integration include:

- Global enterprises may retain a single source of business truth by ingesting, cleaning, protecting, and resharing data from a limitless number of heterogeneous sources.

- A valuable tool for finding bottlenecks, enhancing user and customer experiences, and cutting down on delivery times and other things is provided by holistic operations supervision, which involves managing business-wide intelligence from a central, visible operations screen.

- Security is easier for businesses due to the prevalence of high-profile hacks, revealing that isolated, on-premises network infrastructures now face more ports of entry and higher security dangers than ever before. A central database integration deployment makes it much easier to secure sensitive data because the final copies of the data entry are from a single source.

- Modern, digital businesses have a more outstanding obligation to adhere to national and international operating requirements, such as HIPAA, PCI, and GDPR. Database integration offers centralized administration to guarantee compliance inside the business.

3.   RFCs

The Remote Function Call protocol is exclusive to SAP.

Enterprise resource planning (ERP) software leader SAP (Systems, Applications, and Products) helps businesses manage a variety of business activities, including finance, procurement, production, sales, and distribution, among others. Business organizations can handle several tasks and operations in a single system with the help of SAP software, thus optimizing business procedures and lowering human labor requirements. SAP is widely utilized in a variety of industries, including manufacturing, retail, healthcare, and financial services, by companies of all sizes, from small firms to major organizations. More than 440,000 customers in 180 countries use SAP software, according to the SAP website, to manage their businesses.

A communications interface called Remote Function Call (RFC) is based on CPIC (Common Programming Interface for Communication) but has more features and is more straightforward for application programmers to utilize. Data is sent/retrieved across systems via the CPIC based RFC Protocol by SAP.

As shown in Figure 1-15, the RFC protocol is used to invoke special function modules across a network. Function modules can be compared to C and C++ procedures. They have a specified interface that enables the interchange of data, tables, and return codes. The R/3 or ECC system's Function Builder function library is where function modules are controlled.

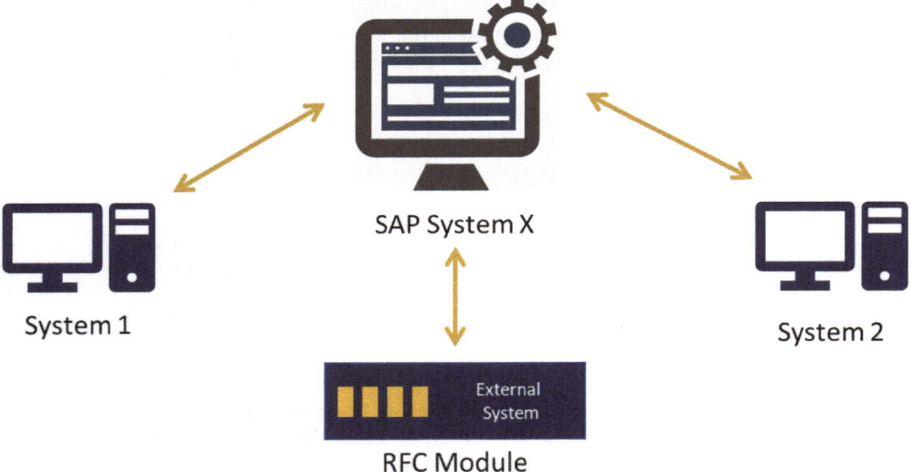

Figure 1-15. *Representation of the RFC integration pattern*

Application programmers have access to a suitable environment for writing, documenting, and testing function modules that may be called locally and remotely through the Function Builder (transaction SE37). The SAP system automatically creates the extra code (RFC stub) for remote calls.

In latest versions of S/4HANA, RFC-based integration is limited or discouraged and API based designs are gaining popularity for their resilience and easy maintenance.

Advantages of RFC

RFC enables programmers to save time by avoiding the need to re-create modules and methods at remote systems. It has the capacity to:

- Easily expose or consume data as RFCs in SAP.

- Transform the data so the remote (target) system can understand it.

- Activate specific procedures required to begin communication with the remote system.

Disadvantages of RFC

- RFC calls are processed slowly by the registered server program on the gateway (transaction SMGW).

- An ABAP dump occurs while calling a Function Module that is RFC enabled: OPEN TASK LIMIT EXCEEDED.

- Error messages appearing in the system log include "No WP block received," "No APPC block received," and "No free block discovered in the WP Communication Area" (TA SM21).

- With asynchronous RFC, no resources were made available for parallelization (dump RESOURCE FAILURE).

- Exceeding the RFC call's execution time limit.

- Limited maximum number of concurrent RFC connections are supported: Examine the system variable rdisp/rfc max login.

- Data volume limitation is returned to the caller app by the RFC-enabled Function Module.

1.8.3 Modern Methods of Integration

1. Application Programming Interface (API)

   An API is a set of guidelines that specify how an app should "behave" to access data or the functionality of another app. It essentially functions as a "messenger" that takes requests and replies in a language that both systems can understand. For example, to determine the ideal room rates for tomorrow, a hotel management system (App A) needs data from a weather app (App B). When App A Contacts App B's API, App B responds with the desired data if the request is handled correctly and by protocol.

   When one application needs to use the features of another program, the same process is followed. Therefore, App A might use App C's API to access predictive analytics whenever it needs to determine appropriate hotel room pricing depending on variables (including the weather).

As you can see, App A and App B, as well as App A and App C, only ever share a minimal amount of data or functionality. Of course, ESBs and APIs have similar functions. Application programming interfaces perform better than the latter in many ways, however. For example:

- A single system can accommodate hundreds of applications while remaining manageable, thanks to APIs. Your ecosystem is much more unstable when using an ESB, and you may link fewer applications.

- APIs are compatible with internal and external programs and may be included in microservices and cloud systems.

- There is no need to fiddle with other integration techniques because most contemporary programs provide API access.

- APIs can be reused. This means you don't have to create a new API each time you want to provide information or service features to a new system.

- APIs can be used to access the functionalities of API gateways, including tools for monitoring and analytics to access the performance of your APIs, limiting the number of requests to prevent crashes, allowing authentication to prevent illegal access, and more.

- It's a beautiful technique to monetize the data or features of your service because you may make your API available to clients of other companies.

The most popular way to connect technologies right now is through an API. This indicates a strong likelihood that all the services you want to integrate with your legacy environment support API connectivity and that adopting alternative integration techniques will provide a different set of difficulties. In some circumstances, APIs are your only option. For example, in some countries, banks uses APIs to share data with financial service providers.

2. Enterprise service bus (ESB)

An ESB can be the answer if you want to link your legacy system with several other apps. It is a centralized software that unifies several programs into a single environment. An ESB serves as a "translator" that enables communication between various systems. It comprises message transformers, a message queue, and applications that are "wired" into it.

How does it work? Let's look at a hospital management system consisting of several apps that require data in XML, CSV, JSON, and TXT forms. If an ESB is in existence, the information flow in the system looks like this:

- The doctor's application receives a prescription from a healthcare professional and records it. This application needs to be capable of exchanging information with various other applications such as laboratories, nutrition services, pharmacies, and billing systems. It should do so in a standardized format that each application can understand. The exchanged data includes details about necessary lab tests, dietary restrictions, prescribed medications, and the cost of medical services.

- Rather than sending data directly to every app, it sends all the prescription information to a message queue, after which the doctor's app is no longer in charge.

- Applications require prescription information to sign up with the message queue.

26

- The message queue sends the requested information to the message transformers after verifying that each app is active and "eligible" to receive the information (converters).

- Each prescription component is formatted according to expectations and provided to the recipient app. The ESB saves the requested message and transmits it to the recipient as soon as it becomes available if any apps are down.

While an ESB appears to provide everything that P2P integration lacks, this approach still has substantial drawbacks. For example, an ESB doesn't provide load balancing or a limit on the number of message calls; hence crashes frequently happen during busy scenarios.

In addition, most ESBs run locally and are not compatible with microservices architectures. Finally, they can store enormous amounts of data and are more suited for internal systems than third-party apps. ESBs are viewed as legacy integration techniques that hardly ever function with contemporary applications for these and other reasons.

1.9 Integration Technologies (Middleware)

Middleware, usually referred to as integration technologies, is a class of software tools and platforms that permit data sharing and communication between various applications and systems. By serving as a "bridge" between several systems, middleware enables information sharing and data processing between them. This can come in handy in several circumstances, such as when many applications need to exchange data in real-time or when a business needs to link its numerous systems and databases to improve operations. Basically, there are three types of middleware, discussed next.

1.9.1 On-Premises Middleware

Middleware that is installed and operated on a company's own servers as opposed to being accessed through the cloud is known as on-premise middleware. This indicates that, as opposed to being hosted by a third-party source, the middleware and the systems it connects are situated on the company's actual property, as shown in Figure 1-16.

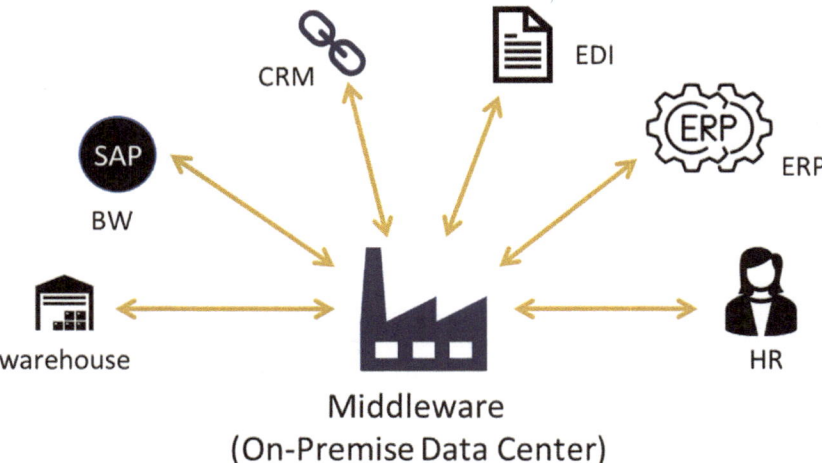

Figure 1-16. *Representation of on-premises middleware*

The runtime and design time are part of the customer's infrastructure. Examples of on-premises middleware include:

- SAP Process Orchestration (SAP PO)
- Oracle WebLogic
- Informatica
- IBM WebSphere
- Microsoft BizTalk Server
- TIBCO Connected Intelligence

1.9.2 Hybrid Middleware

Middleware that blends on-premises and cloud-based technology is known as hybrid middleware. This indicates that while the corporation hosts some middleware and the systems it connects to its own servers, other middleware is hosted by a cloud provider. Companies that want to profit from both on-premises and cloud-based systems or that want to progressively transition their current on-premises systems to the cloud may find hybrid middleware to be helpful.

This is an environment where customers want more control and a regulated environment, as shown in Figure 1-17.

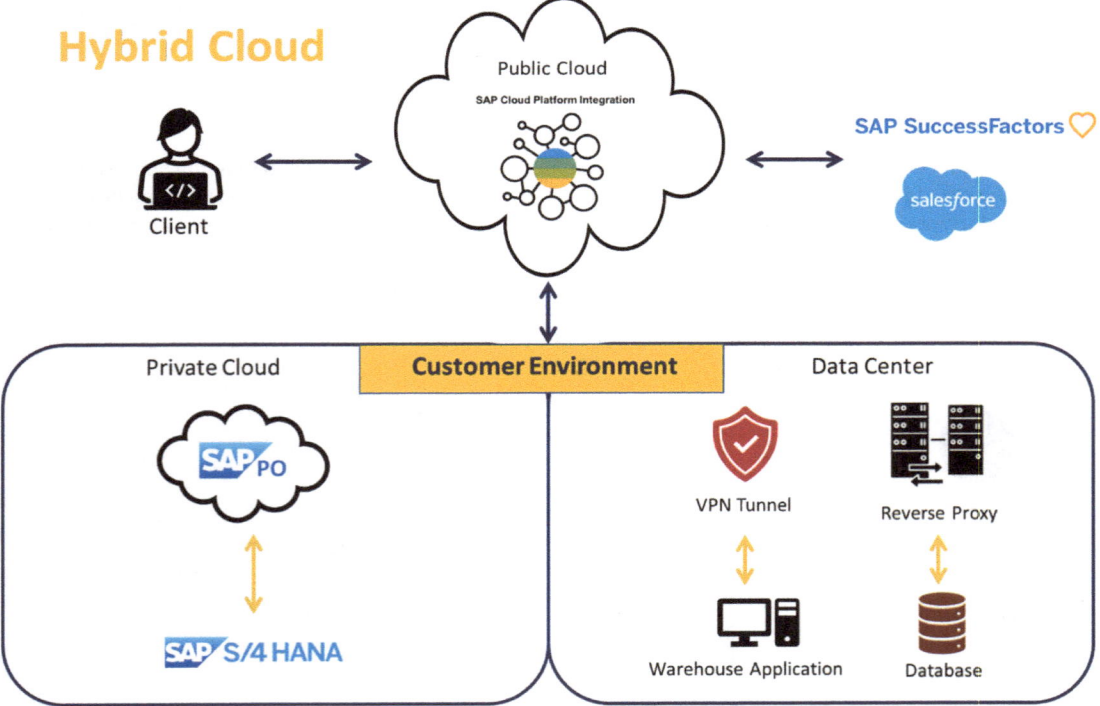

Figure 1-17. *Representation of hybrid middleware*

Examples of hybrid middleware include:

- Running SAP Integration Suite (Cloud Integration) interfaces on Process Orchestration runtime

- Dell Boomi

- MuleSoft

- Jitterbit

1.9.3 Cloud Middleware

Cloud middleware is a type of middleware that is hosted and accessed through a cloud computing platform. This means that rather than being installed and hosted on the business's own premises, the middleware and the systems it connects are located on servers that are owned and managed by a third-party cloud provider. Companies who want to lower the expense and complexity of managing their own on-premises infrastructure or businesses that need to quickly and cheaply link a large number of distributed systems and applications can benefit from using cloud middleware. Figure 1-18 shows a cloud middleware example, where all the infrastructure, platforms, and systems as services are connected to the cloud middleware.

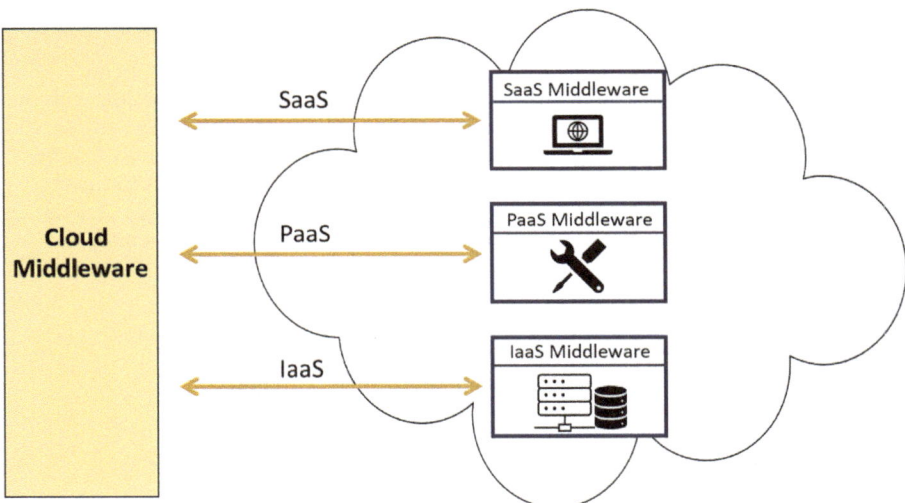

Figure 1-18. *Representation of cloud middleware*

Runtime and design time are 100 percent part of the public vendor-managed infrastructure. Examples of cloud middleware include:

- SAP Integration Suite

- Dell Boomi

- MuleSoft

- Apigee X

- Microsoft Azure Integration Services

A true integration platform as a service (iPaaS) has the characteristics of SaaS, PaaS, and IaaS. An integration platform as a service (iPaaS) can be your best option if you lack experience creating APIs for legacy or cloud systems from scratch and feel you need more time/money to maintain an on-premises infrastructure.

An iPaaS is a collection of SaaS products created to connect applications with different levels of integration capability. The abundance of preconfigured adapters, connectors, APIs, and even ESBs makes it possible for companies to create and maintain a variety of interfaces quickly. An iPaaS provides the advantages listed here when it comes to integrating legacy systems:

- The vendor handles the maintenance of your integrations, so you don't need to worry about it.

- X12, EDIFACT, and COBOL are examples of old programming languages and data formats used with iPaaS solutions to convert them into forms that modern systems can understand.

- Most of the time, integration platforms handle your compliance requirements, giving you everything you require to comply with various standards, including GDPR, HIPAA, SO2, and more.

- Platforms for integration let you connect applications hosted on-premises and in the cloud. Additionally, the connectivity between on-premises and cloud systems is provided by best-in-class technologies.

There is, however, another side to this story—the iPaaS provider dependency. It might offer you something other than the connectors or ESBs you require. For example, it can also be incompatible with the legacy system protocols by not supporting file formats or database drivers that your older system supports. Whatever the cause, you won't be able to take advantage of an integration platform's user-friendly features if at enterprise level you are running applications that are outdated or out of support as per the industry standards. For example, many iPaaS middleware do not support outdated security protocols like TLS 1.0, 1.1, and so on, which is a good thing and forces your enterprise to adopt the latest secure infrastructure.

Types of iPaaS

1. **Simple Automation iPaaS**

   Simple automation iPaaS are characterized by their simple, intuitive, and user-friendly approach to connecting apps and establishing seamless workflows across them. An example of a simple automation iPaaS tool is SAP Process Automation.

   Characteristics of simple automation iPaaS:

   - Operating requires little to no technological expertise.

   - There are a lot of prebuilt connectors to pick from.

   - Utilizing prebuilt connectors, web-based and cloud-based apps can be integrated with ease.

   - Point-and-click workflows can be made using an intuitive UI.

   - You can create independent connectors to link more applications to your current platforms.

   Limitations of simple automation iPaaS:

- When increasing the complexity of a business requirement, the out-of-the-box features, workflows, and integrations can be a limiting factor.

- The use of primarily officially supported and connected apps has a constraint. For example, SAP tools deliver SAP ecosystem connectors and packages first compared to competing application connectivity, like Salesforce.

2. **Hybrid iPaaS**

    The Hybrid/ESB iPaaS offers a range of examples, such as the SAP Integration Suite combined with SAP Process Orchestration runtime, MuleSoft, and Dell Boomi atoms. This hybrid iPaaS enables seamless integration of both on-premises and cloud applications, serving all your integration needs. Additionally, it supports ESB (Enterprise Service Bus) use cases.

    Characteristics of hybrid/ESB iPaaS include:

    - It is equipped to handle all ESB use cases.

    - It possesses the ability to adhere to criteria for the government, insurance, and medical industries by providing on-premise runtime capabilities.

    - It can be applied to sophisticated IT infrastructures.

    - It can link cloud-based applications.

    - It includes simple developer tools for managing, publishing, and creating APIs.

    Limitations of Hybrid/ESB iPaaS include:

    - Costly compared to the True or Simple Automation iPaaS.

    - Compared to True iPaaS and Simple Automation, requires more developer and infrastructure training.

    - Less agility because of the hybrid ESB/iPaaS cloud app integration procedure being too slow.

3. **True iPaaS**

    True iPaaS provides developers with a simple and user-friendly method for creating process/B2B interfaces, APIs, IoT integrations, and on-premises/ cloud integration through a nimble, no-code, and low-code user interface. It also has the capability to extend functionality of the interfaces by using custom code written in JavaScript, Groovy, Data Weave, XSLT, and so on. True iPaaS solutions examples are SAP Integration Suite, Jitterbit, Dell Boomi, and MuleSoft, and so on.

    SAP Integration Suite links and contextualizes processes and data, making it possible to build new content-rich applications more quickly and with less reliance on IT. With the use of prebuilt integration packs and current investments, new results can be produced with less engagement from integration professionals.

As mentioned by Boomi-iPass-infographic, True iPasS has five key characteristics:

- **Cloud-native**—Since True iPaaS is created to operate on cloud infrastructure, it offers scalability, dependability, and flexibility to accommodate changing business needs.

- **Open**—True iPaaS is based on open standards and APIs, making it possible for systems and applications of all types, regardless of vendor or technology, to integrate with it without any issues.

- **Distributed**—True iPaaS can handle distributed integration situations, allowing organizations to integrate systems and applications over various settings, networks, and locations.

- **Low-code**—True iPaaS is low-code, making it possible for non-technical users to create and manage integrations using a visual interface without the need for intricate coding or scripting.

- **Unified**—True iPaaS offers an integrated platform for various integration types, such as application-to-application (A2A), business-to-business (B2B), and data-to-application (D2A) integrations. This enables companies to handle all of their integration requirements in one location with a single user experience and governance model.

Limitations of True iPaaS include:

- True iPaaS requires more technical experts to construct solutions than Simple Automation iPaaS does.

- To prevent an accidental breach of sensitive data, data security and access control measures must be managed carefully by the vendor.

- There is limited visibility on the infrastructure of the iPaaS vendor, which can be challenging for regulated industries like defense, healthcare, government, and utilities.

1.10 Summary

This chapter provide an overview of the concept of integration and its importance in modern business operations. The chapter explained the concept of integration, which is the procedure of integrating several systems and applications to facilitate data interchange and workflow automation. Integration has become a crucial component of managing operations as digital technologies have become more vital to enterprises.

The chapter examined the advantages of integration, including how it can streamline processes, increase effectiveness, and spur creativity. It also drew attention to the difficulties that companies have in managing and implementing integrations. The chapter also presented significant integration platforms and technologies, including iPaaS, APIs, and middleware, and their function in facilitating seamless integration.

The next chapter discusses the overview of the SAP Business Technology platform (BTP), the SAP Integration Suite, and hands-on installation and setup of SAP Integration Suite Cloud Integration (f.k.a CPI or CI) components.

CHAPTER 2

■ ■ ■

SAP Integration Suite

The SAP Integration Suite, a powerful cloud-based platform that enables businesses to effortlessly integrate their SAP and non-SAP systems and applications, is discussed in this chapter. Application-to-application (A2A), business-to-business (B2B), and data-to-application (D2A) integrations are all supported by the SAP Integration Suite's single platform.

The chapter starts by covering the value of integration in the current business environment and explains how the SAP Integration Suite can support businesses in overcoming integration hurdles. It also covers some of the platform's key advantages and features, such as its support for cloud-to-cloud, cloud-to-on-premise, and hybrid integration situations.

The chapter dives into the many integration technologies and tools offered by the SAP Integration Suite, including APIs, pre-built integration flows, and B2B add-ons. It also looks at ways to use these technologies to improve data visibility, expedite digital transformation, and streamline business operations. You also see how to set up the trial account for the SAP Integration Suite in the SPA BTP Cockpit.

You will have a firm grasp of the capabilities of the SAP Integration Suite by the end of this chapter, including understanding how it can assist your organization in achieving its integration objectives, modernizing its IT environments, and generating business benefits.

Before going through the SAP Integration Suite, We will first see What is SAP Business Technology Platform (BTP)?

2.1 What Is SAP BTP?

SAP BTP, or SAP Business Technology Platform, is a cloud-based platform that enables businesses to create, expand, and connect SAP and non-SAP applications. It offers a variety of tools and services that simplify the processes of developing and deploying SAP applications on the cloud, and it is a crucial part of SAP's broader cloud strategy.

The key features and capabilities of SAP BTP include the following:

- A cloud-based development environment that dispenses with local infrastructure and enables businesses to create and test SAP applications in the cloud.

- A variety of integration tools and services that make it simpler to interface SAP applications with other systems and technologies, such as connectors, data mapping tools, and API management tools.

- Support for well-known programming frameworks and languages, like SAPUI5, Node.js, and Java, makes it easier for developers to create and implement SAP apps.

- Organizations can create more comprehensive and integrated cloud-based solutions by integrating other SAP cloud services, such as SAP Analytics Cloud and SAP Cloud Integration.

© Jaspreet Bagga 2023
J. Bagga, *A Practical Guide to SAP Integration Suite*, https://doi.org/10.1007/978-1-4842-9337-9_2

- It is the platform of choice for innovating and extending SAP solutions like S/4HANA, SuccessFactors, Ariba, SAP Analytics Cloud, and Concur.

- Built on the Cloud Foundry platform, SAP BTP offers a flexible and scalable environment for creating and deploying cloud-native applications.

- Including pre-built AI models and interacting with well-known AI frameworks, this is a collection of tools and services for developing and deploying artificial intelligence (AI) and machine learning (ML) applications and models.

Overall, SAP BTP provides a powerful and flexible platform for organizations developing and deploying SAP applications on the cloud.

Companies need access to real-time data to use cutting-edge technologies and industry best practices inside Agile, integrated business processes. Integrating end-to-end processes is a crucial component of SAP's strategy, whether the solutions are from SAP, its partners, or third parties.

Lead to Cash, Source to Pay, Design to Operate, and Hire to Retire are just a few of the business scenarios that SAP actively promotes.

For connected end-to-end business operations spanning SAP and external applications, the SAP Business Technology Platform offers integration options.

Figure 2-1 shows some of the major capabilities of the SAP Business Technology Platform, including automation, analytics, data, and integration.

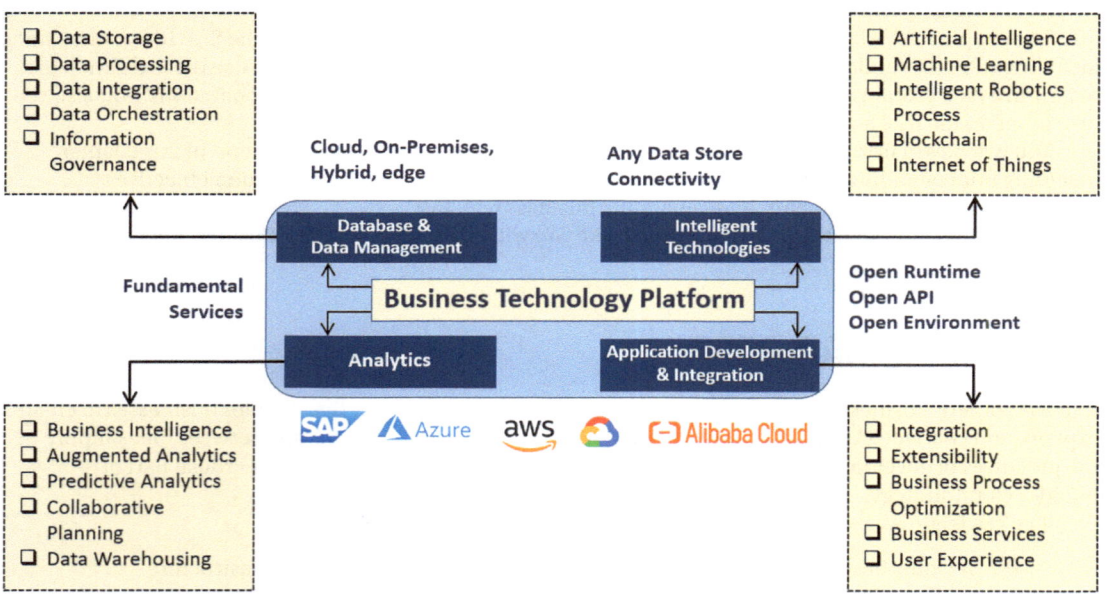

Figure 2-1. *The capabilities of the SAP Business Technology Platform (BTP)*

2.2 Overview of the SAP Integration Suite

The SAP Integration Suite is a collection of tools and services, as shown in Figure 2-2, that SAP offers to businesses so they can interface their SAP systems with other enterprise software systems, including customer relationship management (CRM) and enterprise resource planning (ERP) systems. Because of this, companies can share data and processes throughout their platforms, streamlining workflow and increasing productivity. The SAP Integration Suite offers services for managing and monitoring the integration process and solutions for data, process, and message integration.

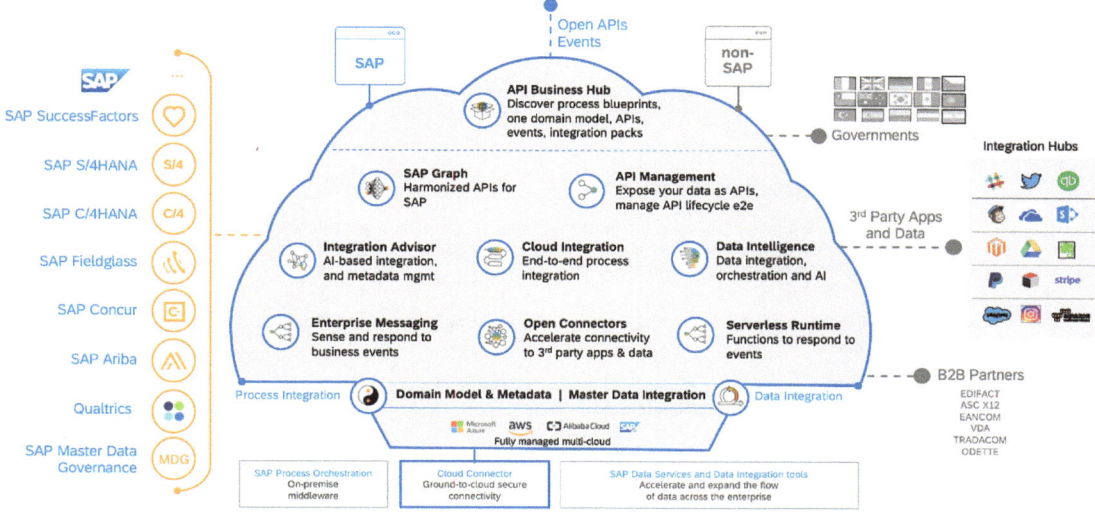

Figure 2-2. *The SAP Integration Suite capabilities, Source: SAP SE*

The SAP Integration Suite links and contextualizes processes and data, making it possible to build new, content-rich applications more quickly and with less reliance on IT. With less engagement from integration professionals, new outputs can be produced using pre-built integration packs and existing investments.

The SAP Integration Suite runs in the SAP BTP, Cloud Foundry Environment.

2.3 Capabilities of the SAP Integration Suite

The major capabilities of the SAP Integration Suite are as follows:

2.3.1 Cloud Integration

It is essential to consider the various components that must be taken care when the businesses migrate to the cloud. Businesses must integrate several systems and applications that are created and maintained using various technology stacks, according to various security standards, and having various business interface requirements. Regardless of the environment, businesses can quickly and easily integrate these numerous applications due to the Cloud Integration feature of the SAP Integration Suite.

You can process communications in real-time situations spanning several businesses, organizations, or departments in one organization by integrating SAP and non-SAP, cloud, and on-premises apps.

Figure 2-3 shows the Discover section of integration, which you can find in the Integration Suite once you get into it. The list in the image shows the pre-built integration packages available in SAP. In order for organizations to rapidly and simply connect their SAP and non-SAP systems and applications, the pre-built integration packages offered by SAP Cloud Integration are extremely important. These pre-built packages include integration information and templates that have already been established; they can simply be changed to match particular business requirements, cutting down on the time and expense of integration projects.

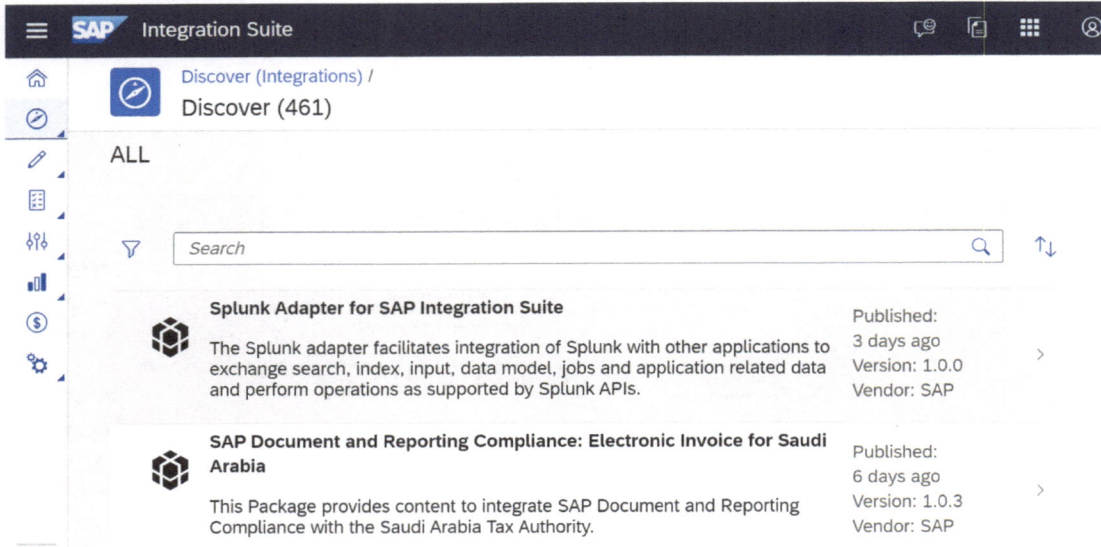

Figure 2-3. *The Discover Integrations section of Cloud Integration*

2.3.2 API Management

The API Management feature of the Integration suite offers a central layer for governing, managing, and metering APIs. It offers a web-based, code-free framework for creating new APIs, managing current APIs, enhancing APIs with security and access restrictions, creating logical groupings using product catalogues, and making APIs available to the community via a developer portal. To commercialize API access, pricing plans can also be connected to APIs. A plug point that can be utilized to enable the development of applications and extensions, as well as the integration of businesses, is the main goal of an API-based integration strategy.

Application programming interfaces (APIs) enable access to simple, scalable, and secure digital assets, which can then be consumed.

Figure 2-4 shows the Discover API section in SAP Integration Suite. Discover APIs offers a wide selection of pre-built API packages. SAP API management is crucial because it enables organizations to quickly and effectively link their systems and applications. These packages offer a wide range of integration scenarios, including integrating with cloud-based applications and services as well as linking SAP systems to non-SAP systems.

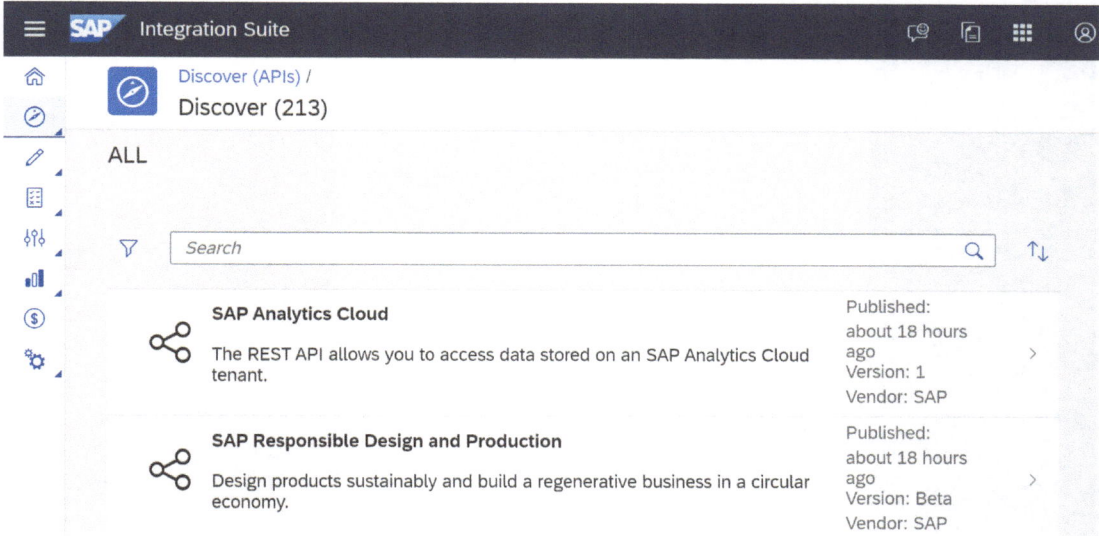

Figure 2-4. *The Discover section of SAP API Management*

Businesses can accelerate their digital transformation by using Discover APIs packages, which can considerably minimize the time and effort needed to design and deploy APIs. These packages are easily adaptable, allowing organizations to customize them to meet their unique needs.

SAP API Management delivers tools and services that make it simple to manage and govern APIs throughout their lifecycle in addition to pre-built packages. This covers attributes that help guarantee the dependability, scalability, and security of APIs, such as API documentation, testing, versioning, and security.

2.3.3 Open Connectors

With standardized authentication, error handling, and connectivity protocols, the Open Connectors feature of the SAP Integration Suite offers third-party connectors. Rather than learning about the technology required to link with third-party systems, this enables developers to concentrate on creating business integrations.

Utilizing pre-built connectors, you can create seamless interfaces with more than 160 non-SAP applications.

Figure 2-5 shows the home page of SAP Open Connectors, which has various features, including Home, Connectors, Instances, Common Resources, Formulas, Activity, Security and Settings. Each option has its own functionalities.

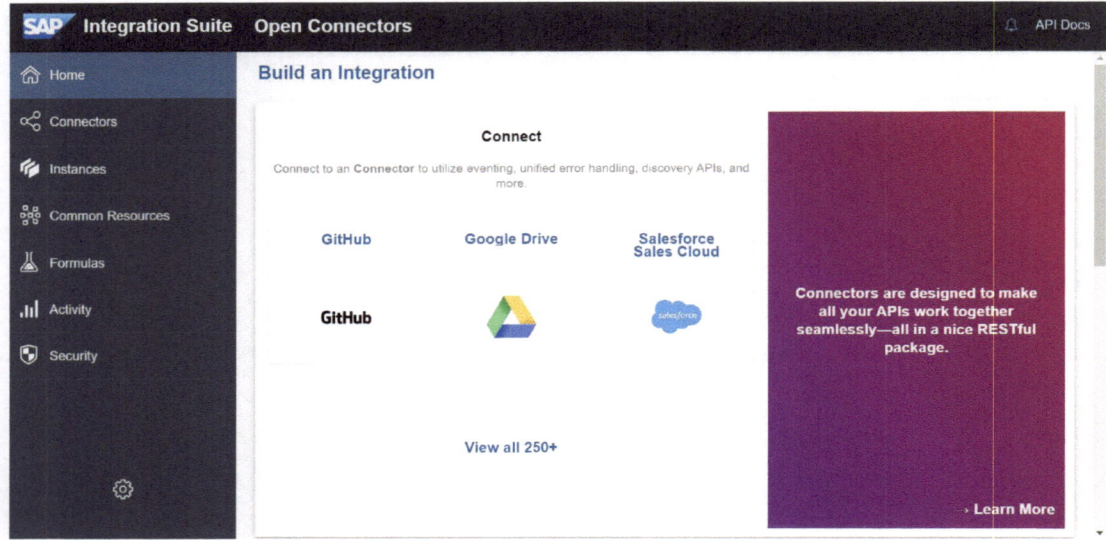

Figure 2-5. *Dashboard of SAP Open Connectors*

2.3.4 Integration Advisor

A project's integration process includes mapping business interfaces, which the technical developer and functional consultant do. But doing so it can be expensive and time-consuming. You can accelerate the creation of your business interfaces and mappings with the assistance of the SAP Integration Suite's Integration Advisor feature. It features pre-built information for EDI industry standards and SAP S/4HANA.

You can create runtime artifacts quickly, construct business-oriented interfaces and mappings more rapidly, and put forth a lot less work.

Figure 2-6 shows the list of pre-built type systems available in the SAP Integration Advisor. The Type Systems refer to a set of predefined structures and data types that are available within the integration platform. The SAP Integration Advisor's pre-built systems are important because they offer quick and simple approaches to begin developing links without the need for major customization or configuration. These pre-built systems are examples of regularly integrated technologies and systems used by businesses, including SAP S/4HANA, Salesforce, and RESTful APIs.

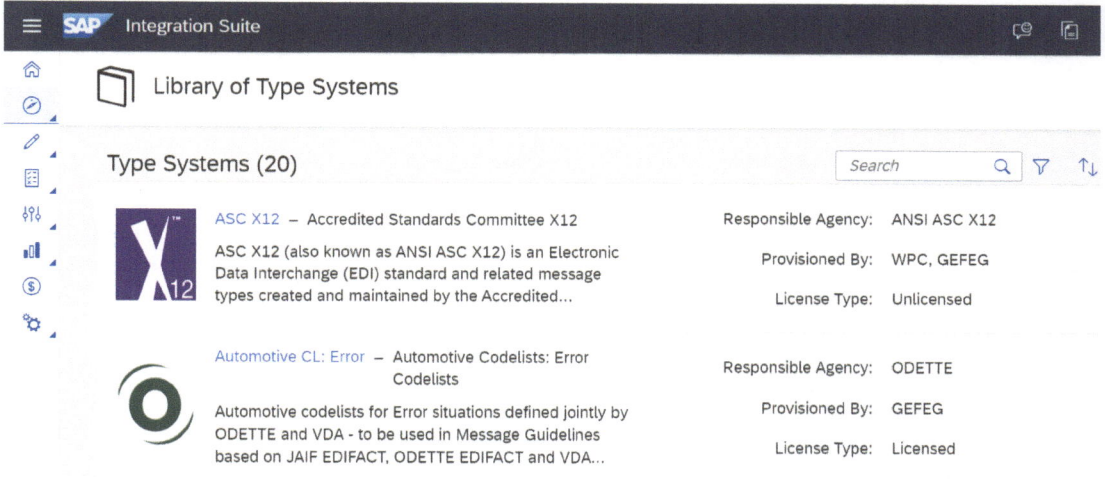

Figure 2-6. *Dashboard of the SAP Integration Advisor*

2.3.5 Trading Partner Management

Trading Partner Management is a collection of products and services offered by SAP that lets organizations use, SAP systems to manage their interactions with trading partners. This consists of tools for finding and integrating new trading partners, creating and administering trading contracts, and keeping track of and controlling trading partners' performance. Businesses can enhance their supply chain processes and maximize their interactions with trading partners using trading partner management. The SAP Integration Suite, which offers several tools and services for linking SAP systems with other enterprise software systems, includes it.

Figure 2-7 shows the home page of Trading Partner Management, where you can create a company profile, trading partners, and agreement templates. Each option has its own functionalities, which are covered in Part 2 of the book.

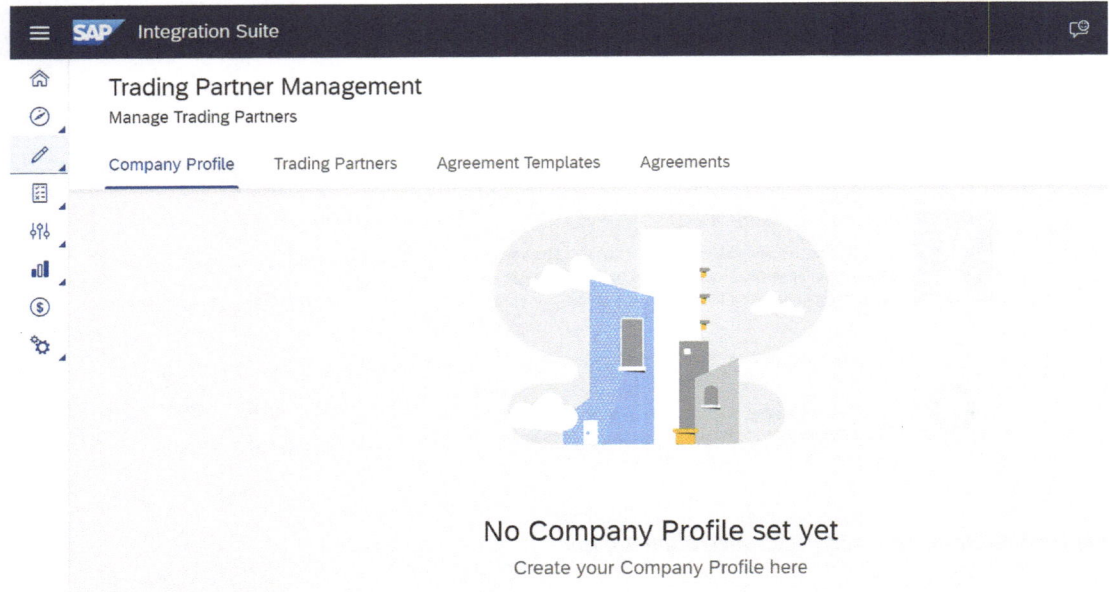

Figure 2-7. Dashboard of the SAP Trading Partner Management feature

2.3.6 Integration Assessment

As a starting point, SAP defined integration advice and templates. Integration Assessment comprises important ISA-M (Integration Solution Advisory-Methodology) master data, such as the definition of integration domains, integration styles, use case patterns, and common key integration technology characteristics.

The SAP Integration Assessment uses ISA-M (Integration Solution Advisory—Methodology) to assist enterprises in assessing and enhancing their integration capabilities. It is a methodical process with a number of steps intended to evaluate an organization's current integration landscape and offer suggestions for streamlining its integration procedures.

Figure 2-8 shows the home page of SAP Integration Assessment, in which you can see the different components of integration assessment, including Request, Analyze, Configure, and Settings.

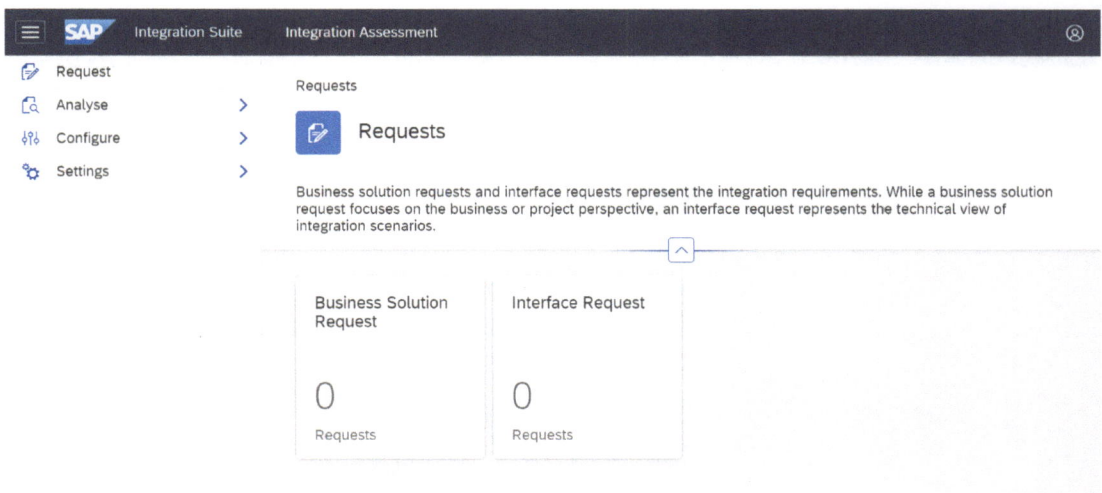

Figure 2-8. *Dashboard of the SAP Integration Assessment*

2.3.7 Migration Assessment

A new feature called Migration Assessment has been recently added to the capabilities of the SAP Integration Suite. This feature helps you estimate the technical work needed for the conversion process by evaluating how various integration scenarios can be moved from the On-Premises SAP Process Orchestration system to SAP Integration Suite. With this version, SAP is attempting to consider switching from its on-premises SAP Process Orchestration system to its SAP BTP-based Integration Suite in order to take advantage of the next generation of integration capabilities.

The Migration Assessment helps in evaluating the technical effort necessary for the migration process by analyzing potential migration paths in various integration scenarios.

Figure 2-9 shows the Request page in Migration Assessment; the two options are Data Extractions and Scenario Evaluations. Data Extractions is a procedure whereby the application collects data from a connected SAP Process Orchestration system, such as integration scenarios, mapping objects, communication channels, and other design time artifacts, and prepares the data for evaluation. After the information has been collected, Scenario Evaluations assesses it using predetermined rules.

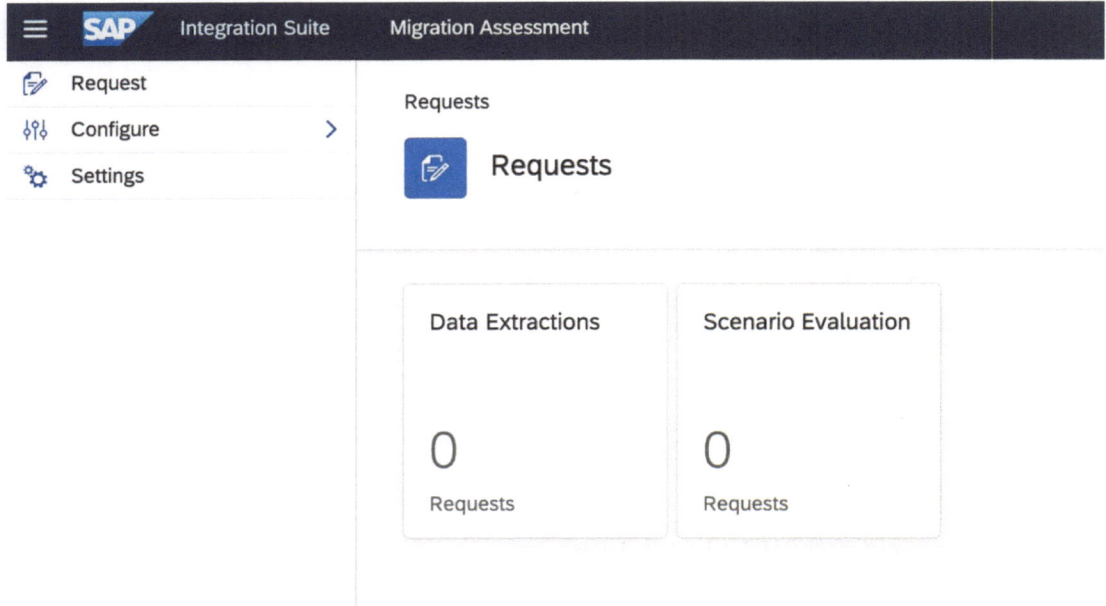

Figure 2-9. Migration Assessment

2.4 Features of the SAP Integration Suite

The SAP Integration Suite includes various features that enable businesses to integrate their SAP systems with other enterprise software systems. Some of the key features of the SAP Integration Suite include:

- **Build integration scenarios** —With cloud integration, you can explore, create, and run scenarios for process integration from beginning to end.

- **Manage APIs**—With API Management, you can find, create, and regulate APIs for API consumers.

- **Enable application connectivity**—To easily combine SAP and non-SAP applications, use open connectors to select pre-built connectors from a catalog.

- **Put integration packs to use**—Cloud integration, as shown in Figure 2-10, uses pre-built integration packs for end-to-end scenarios like hire-to-retire, lead-to-cash, procure-to-pay, SuccessFactors, Jira, ServiceNow, and many more. You can accelerate the construction of integrations. For example, partner-developed integration packages are also available to copy/configure the integration packages.

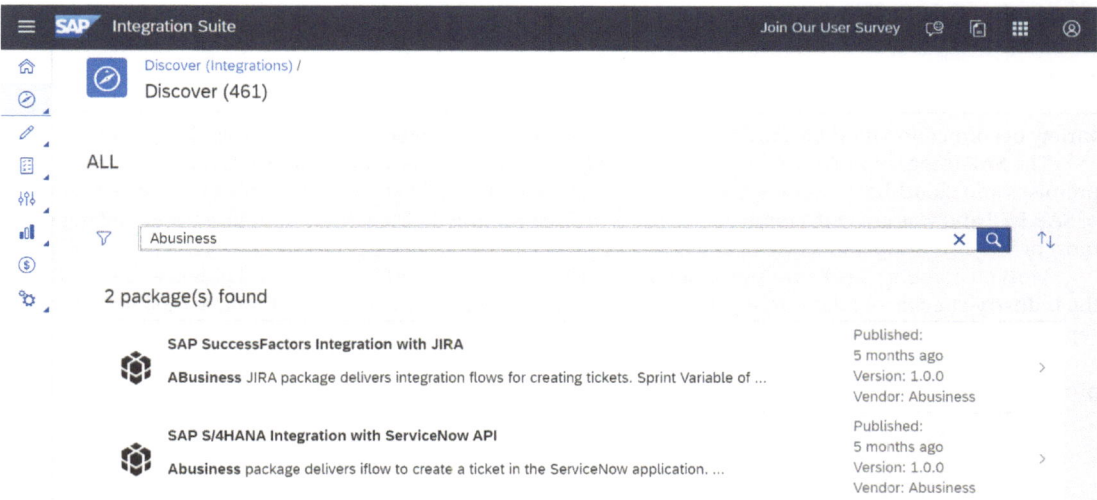

Figure 2-10. *Discover Integrations*

- **Implement mapping more simply**—Integration Advisor creates user interfaces and mappings for commercial applications using crowdsourcing and machine learning.

- **Organize trading partner integrations**—Trading Partner Management constructs, preserves, carries out, and keeps track of business-to-business scenarios.

2.5 Benefits of the SAP Integration Suite

Using the SAP Integration Suite to integrate SAP systems with other enterprise software systems has several benefits. Among the main benefits are:

1. **Improved productivity**—By connecting SAP systems to other enterprise software systems, businesses can simplify operations and do away with the need for manual data input and process management.

2. **Enhanced data accuracy**—Businesses can guarantee that their data is consistent and up to date by sharing data between SAP systems and other enterprise software systems. This increases the correctness of their information and decision-making.

3. **Better customer service**—By connecting SAP systems with customer relationship management (CRM) systems, companies can give their clients more precise and timely information, thus enhancing the customer experience.

4. **Increased flexibility**—By integrating SAP systems with other enterprise software systems, companies can more readily adjust to changing business needs and processes, enabling them to be more flexible and responsive to changing market conditions.

5. **Reduced costs**—By implementing automation in procedures and eliminating the necessity for labor-intensive manual data entry and process management, companies can achieve cost savings in the long run.

2.6 SAP BTP Integration Suite Landscape

The SAP Integration Suite is a collection of tools and services that allow businesses to link their software to SAP BTP and other platforms and systems. To link SAP BTP with other platforms and systems, it offers a variety of connection methods, including messaging, APIs, data integration, and process integration.

The SAP Integration Suite can be used to integrate various systems and platforms, including on-premises and cloud-based systems, and it is intended to be versatile and scalable. It is a crucial element of SAP BTP and was created to assist enterprises in streamlining their business operations and making the most of SAP BTP's capabilities.

Most customers use one non-production subaccount for Dev and QA. As a good practice or based on the industry-specific regulations, if you need to maintain different environments for data/business process separation, you can create N number of subaccounts for Development, Test, and Production usage.

In Figures 2-11 and 2-12, you can see the setup of SAP BTP Integration Suite landscape for non-production and production environments.

- Integration Suite—Development

- Integration Suite—Test

- Integration—Production

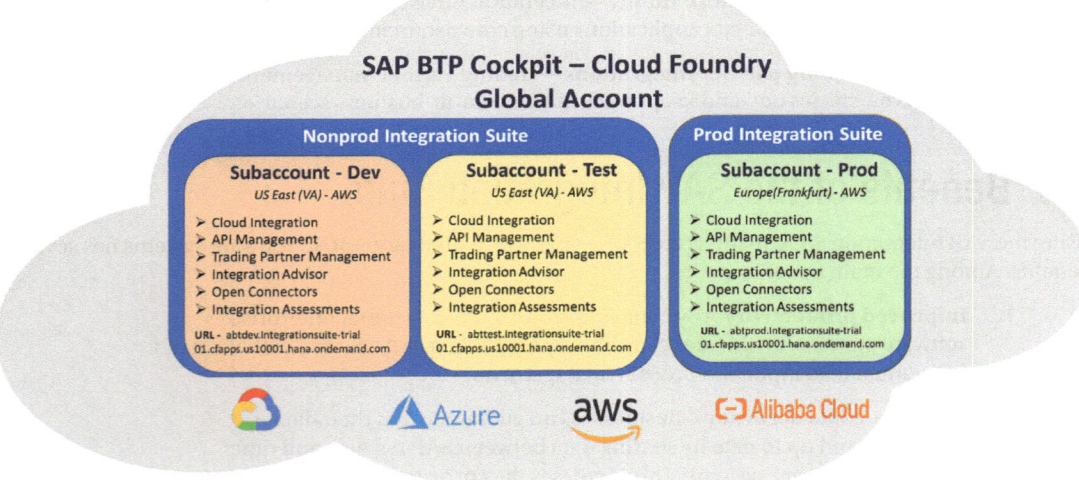

Figure 2-11. *Subaccounts of the SAP BTP Integration Suite can be created in any of the four major cloud providers (AWS, GCP, Azure, and Alibaba Cloud)*

SAP has discontinued issuing New NEO (SAP Data Center) Cloud Integration Enterprise licenses for new customers, and it encourages existing customers to migrate to Cloud Foundry based Cloud Integration. Contact your SAP Account Executive for more details. Existing customers can still continue to use their NEO based licenses.

For example, API Management Service is not available for new customers as a standalone BTP service anymore. It is only available in Cloud Foundry Environment as part of the Integration Suite.

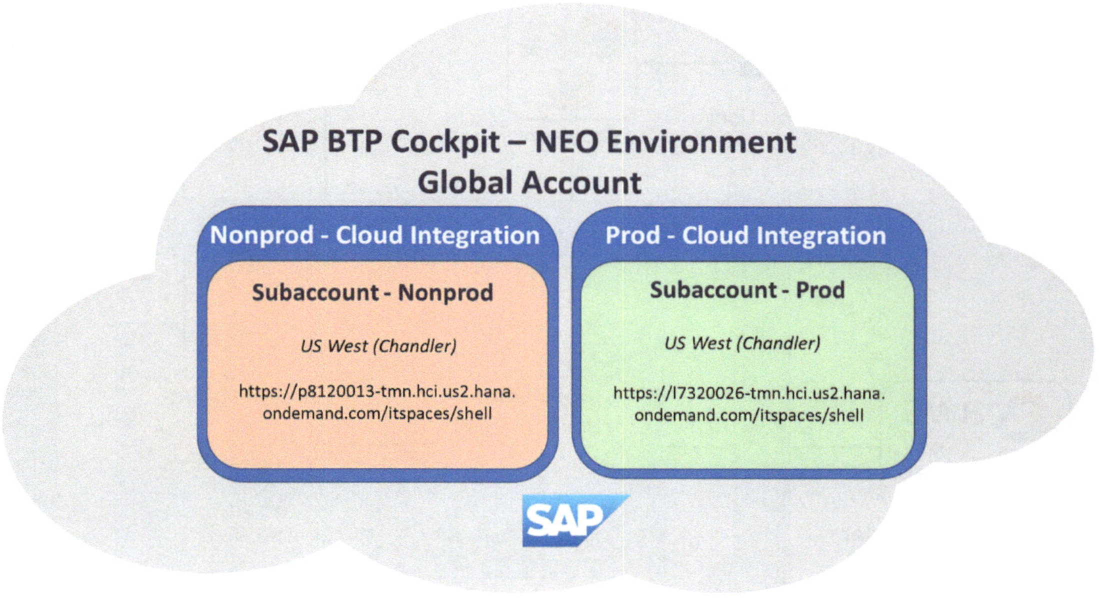

Figure 2-12. *Subaccount created in SAP Data Center NEO*

2.7 The SAP Integration Suite and Security

The security-related features of the SAP Integration Suite are discussed in this section, along with the precautions you can take to safeguard client data when it is transmitted over the SAP Integration Suite during the execution of an integration scenario.

Clients who use SAP Cloud Integration agree that a significant portion of their confidential data, as well as that of their customers, is managed and stored in an infrastructure that does not belong to them. An integration platform's primary function is to act as a hub for messages that might include sensitive client data. These messages must, first and foremost, be shielded from prying eyes and unlawful access.

Technically speaking, the integration platform is created as a cloud-based clustered and containerized integration platform. Different areas of the platform handle messages, handled via integration flow from various clients (referred to as tenants).

In terms of CPU, data storage, and user access, tenants handling integration flows from various clients are carefully isolated from one another.

Figure 2-13 shows the high-level architecture of the SAP Integration Suite key components.

45

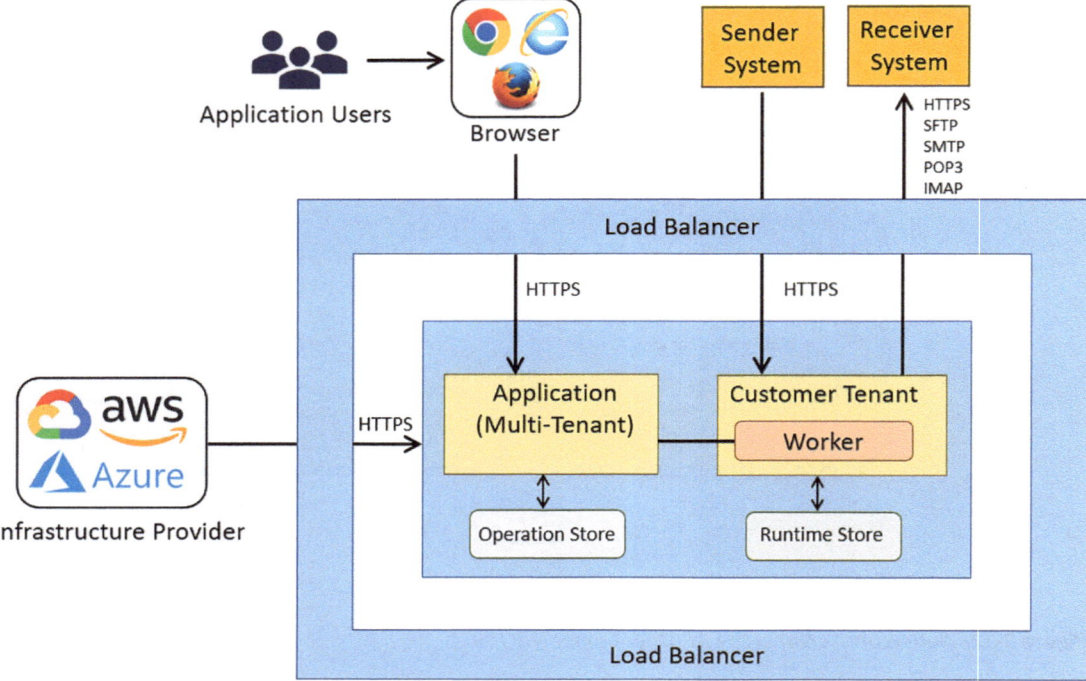

Figure 2-13. *High-level technical architecture of the SAP Integration Suite*

2.7.1 Transport and Message Level Security

The transfer of any data, whether between internal or differet components, can be secured using techniques like encryption.

The connected remote systems communicate with each other in a scenario, depending on the chosen transport protocol. These protocols offer various options to protect the transmitted data from unauthorized access. Additionally, the content of the messages exchanged can be further secured through digital encryption and signatures, in addition to the security measures applied at the transport level. In terms of Security of SAP integration Suite there are two levels of security which are as follows:

- Transport Layer Security

  Transport layer security in the SAP Integration Suite refers to the steps taken to protect communication between two systems or applications at the transport layer. The third layer of the OSI (Open Systems Interconnection) paradigm is the transport layer, which ensures dependable end-to-end communication between the two systems.

  The Secure Sockets Layer (SSL) and Transport Layer Security (TLS) protocols are two examples of how transport level security can be implemented. For communication between the systems, these protocols enable authentication, secrecy, and integrity.

  In the SAP Integration Suite, transport-level security can be configured and enabled for various integration scenarios, such as web service calls, file transfers, and messaging. It is typically used to secure communication between SAP systems and external systems, or between different SAP systems.

Depending on the underlying transport protocol, each adapter enables you to configure a certain security level.

Table 2-1 shows some of the adapter-specific transport-level security options that can be configured as per your integration requirements.

■ **Note**   The SAP Integration Suite does not support unsecure communication at the transport level and follows the security best practices. For example, inbound HTTP messages are not supported, and you must configure client/server SSL certificates for secure end-to-end communications. The SAP Integration Suite supports only HTTPS communications.

Table 2-1. *Adapter-Specific Security Options*

Transport Protocol	Description
SFTP (Secure Shell File Transfer Protocol)	Both the SFTP transmitter and receiver adapters support this protocol.
	In an open network, Secure Shell (SSH) transfers data securely.
	To secure FTP transmission, SSH utilizes a symmetric key length of at least 128 bits. The SAP asymmetric key's default length is 2048 bits.
	Supported forms of authentication:
	• Username/password authentication, in which the SFTP server uses the username and password to verify the calling component.
	• Public key authentication, in which the SFTP server uses a public key to verify the caller component.
	SFTP additionally ensures that the participants are using only permitted public keys when asymmetric key pairs are in use.
HTTP(S) (Hypertext Transfer Protocol Secure)	All adapters that permit communication via HTTPS, such as Soap Adapter, IDoc adapter, and the HTTP adapter support this protocol.
	Utilizing Transport Layer Security, communication can be safeguarded (TLS). In this example, at least 128 bits of symmetric key length are used (which is technically enforced). The default SAP asymmetric key length is 2048 bits.
	Receiver adapters additionally offer SAP Cloud Connector-based principal propagation.
	Numerous possibilities for authenticating depending on the sender or receiver adapter, simple authentication using user credentials, client certificates, or OAuth—are available.
SMTP (Simple Mail Transfer Protocol)	The exchange of emails is supported via these protocols (in combination with the Mail adapter).
POP3 (Post Office Protocol)	The STARTTLS extended operation supports transport encryption.
IMAP (Internet Message Access Protocol)	You can submit a username and password in plain text or in encrypted form to authenticate against the email server; the latter is only an option if the email server supports it.

- Message-Level Security

  Message-level security in the SAP Integration Suite describes the steps taken to protect a message's payload (or message content) at the application layer. The OSI (Open Systems Interconnection) model's seventh layer, the application layer, oversees services that are particular to the program being utilized.

  There are many approaches to establish message-level security, such as employing encryption, digital signatures, or access control systems. These precautions guarantee that the message can only be seen by the intended recipient and that the message contents are not altered during transmission.

  Message-level security can be set up and made available in the SAP Integration Suite for a number of integration situations, including messaging, file transfers, and web service calls. Typically, it is used to protect sensitive or private data being communicated between SAP systems.

  In addition to the choices for transport-level security, you can also secure communication at the message level, where the contents of the sent and received messages can be secured using digital signatures and encryption. This can be accomplished using a variety of security standards, as listed in Table 2-2.

  Utilizing certain integration flow security features, you can set message-level security choices. Table 2-2 shows the supported standards and algorithms.

Table 2-2. *Supported Standards and Algorithms*

Standards	Security Features
PKCS#7/CMS Enveloped Data and Signed Data	Message content encryption and decryption. Payload verification and signing.
PKCS#7/CMS Enveloped and Signed Data	Payload encryption, decryption, signature, and verification.
Pretty Good Privacy (PGP)	Message content encryption and decryption. Encryption, decryption, message signing, and verification.
XML Signature	Payload verification and signing.
WS-Security	Signing and validating the SOAP body.

2.7.2 Access SAP BTP Cockpit

As a license owner of Enterprise accounts, your S-user ID (SAP ID) can access multiple global accounts. This access is provided by SAP during the license purchase to the Account Owners S-user ID. The account owner must create the access for administrators/developers to continue the configuration process by adding the right role collections/roles/authorizations as per the desired personas of BTP Integration Suite users. SAP BTP also offers the trial accounts for (90-days) in which the users have access to the global account.

Role Collection for Global Account

There are two key predefined Role Collections that are assigned to the user when managing the SAP BTP Cockpit. These Role Collections are connected to the account administrator. Personas are useful for managing the BTP global account.

To view the Role collections in SAP BTP Cockpit, Navigate to the Security ➤ Role Collections in the SAP BTP Cockpit, as shown in Figure 2-14. The Role Collections available are:

- Global Account Administrator

- Global Account Viewer

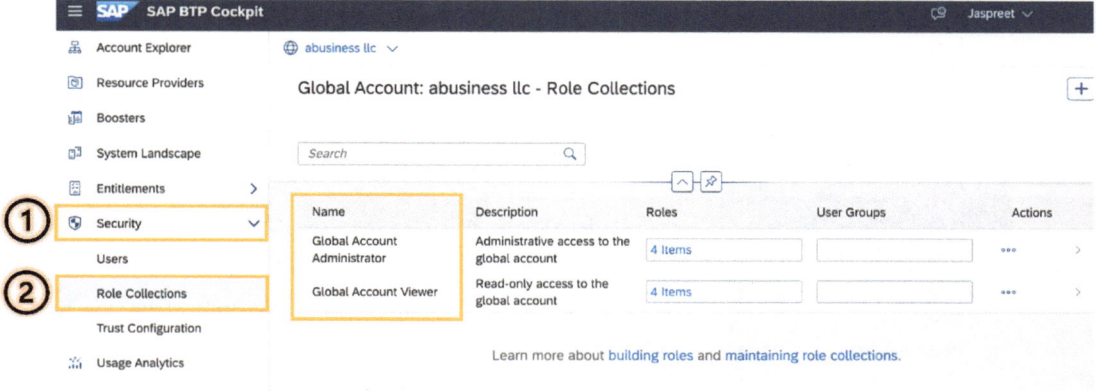

Figure 2-14. *Role collection for the global account*

After you have been assigned the Global BTP Account Administrator Role Collections, you should be able to see all the global accounts for which you have assigned access, as shown in Figure 2-15.

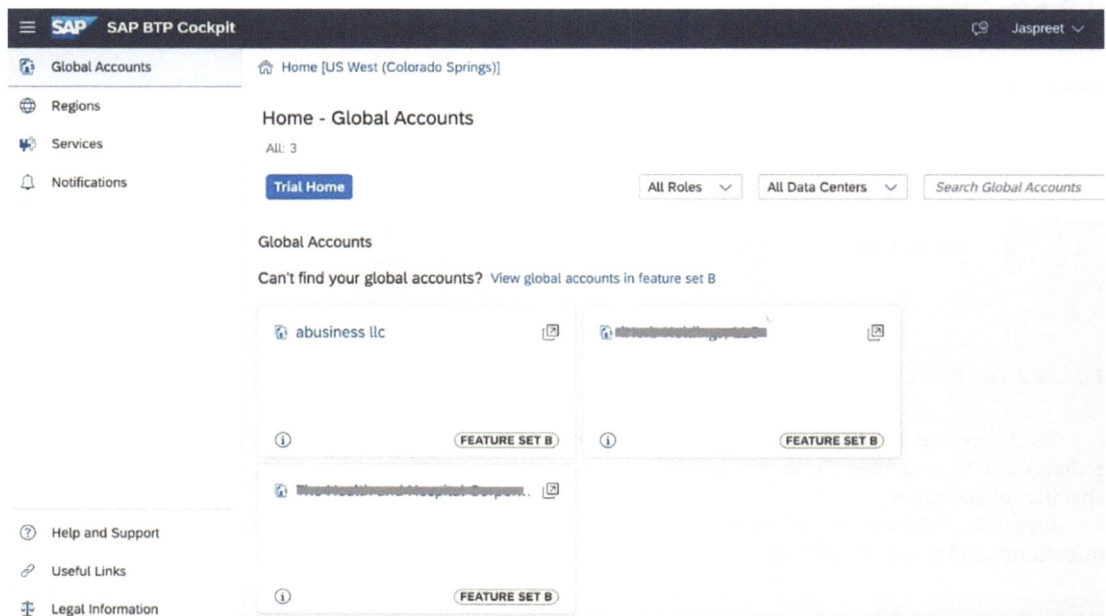

Figure 2-15. *The SAP BTP landing page*

Role Collections for the Subaccount

There are three key predefined Role Collections that are assigned to you when managing SAP BTP subaccounts. These Role Collections are connected to subaccount Administrator personas and they are useful for managing the BTP subaccount.

Navigate to Security ➤ Role Collections in the SAP BTP Cockpit, as shown in Figure 2-16. Search for *subaccount* to see the Role Collections associated with the Subaccount, which are as follows:

- Subaccount Administrator

- Subaccount Service Administrator

- Subaccount Viewer

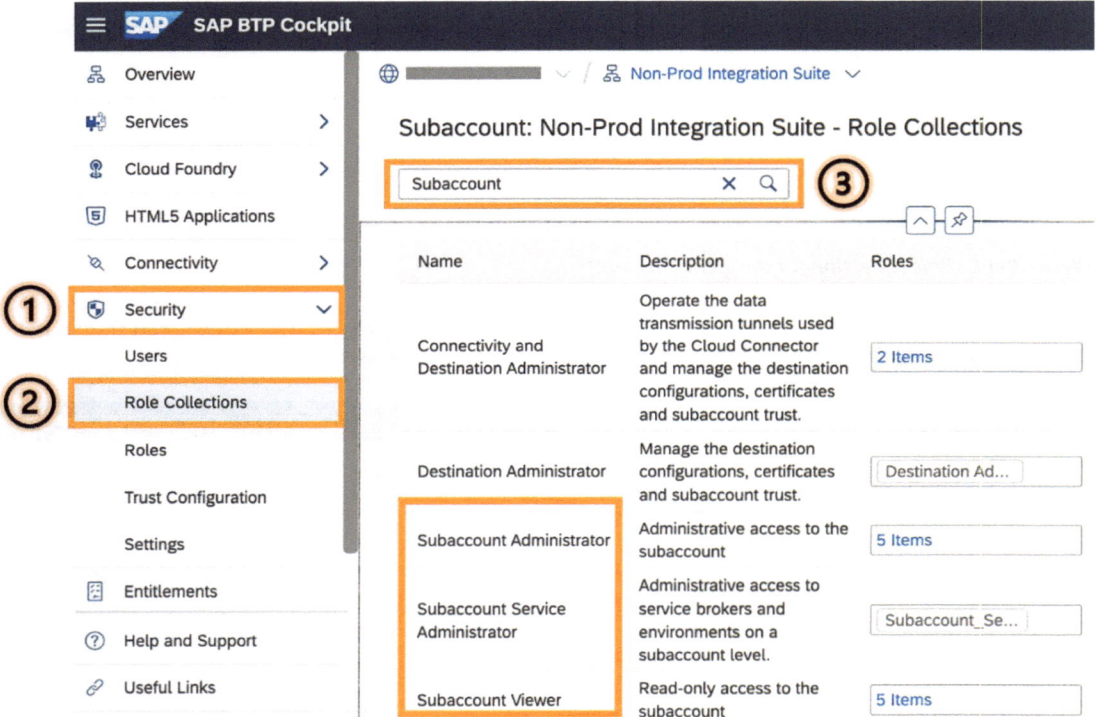

Figure 2-16. *Role Collections for Subaccount*

For Enterprise License users, once you have access to the global account and have assigned the Subaccount Administrator Role to yourself or the administrators, you can then create BTP Integration Suite-specific subaccounts.

Figure 2-17 shows two NEO subaccounts created inside the abusiness llc global account. This setup is mandatory and is the same for Cloud Foundry subaccounts as well.

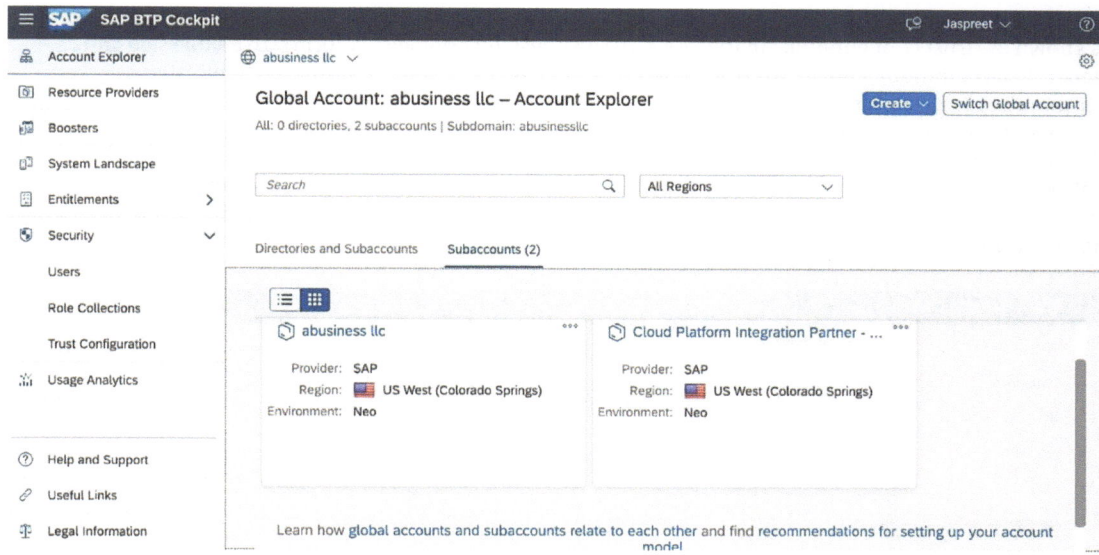

Figure 2-17. Navigation from the Global Account to the Subaccount

2.7.3 Access SAP BTP Integration Suite

You can access the SAP Integration Suite if you have a SAP S-User ID/email address/universal ID-based account.

Accessing the SAP Business Technology Platform (BTP) Integration Suite is possible in several ways:

1. **Using the SAP Integration Suite browser-based application:** The SAP BTP Integration tool, a graphical user interface (GUI) for creating and deploying integrations, can be used to access the Integration Suite. Various Integrated Development Environments (IDEs), including Eclipse and Visual Studio, used to be compatible with the Integration Tool as a standalone application or as a plug-in.

   SAP does not support Eclipse-based cloud integration plugins anymore. Developing integration flows using Eclipse Web-ID is strongly discouraged.

2. **From the SAP BTP Cockpit:** You can access the Integration Suite through the SAP BTP Cockpit if you have an account with the platform. Log in to the Cockpit and search for the integration tools to accomplish this.

3. **Using SAP BTP API:** Using the SAP BTP API, you can programmatically access the Integration Suite as well. Using RESTful APIs, this API enables you to monitor, automate, set up, and control interfaces between SAP and non-SAP services.

■ **Note**   Access to SAP BTP Integration Suite capabilities (Cloud Integration, Integration Advisor, Integration Assessment, etc.) can also be federated through the customer's own custom identity provider (Microsoft AD/ Azure, Okta, etc.) for increased enterprise-level security by using SSO/MFA.

Establishing and monitoring trust relationships between various distributed system components is known as "trust configuration" in the SAP BTP (Business Technology Platform). To enable safe and dependable communication across these elements, which can include cloud-based applications, on-premises systems, and third-party services, trust configuration must be in place.

The Trust Configuration in SAP BTP (see Figure 2-18) is used to enable secure and dependable communication between various components of a distributed system. SAP BTP can guarantee that data is sent securely and that only authorized parties have access to it by creating and monitoring trust relationships. This maintains the confidentiality, integrity, and accessibility of data in the system and helps defend against cyberattacks, data breaches, and other security threats.

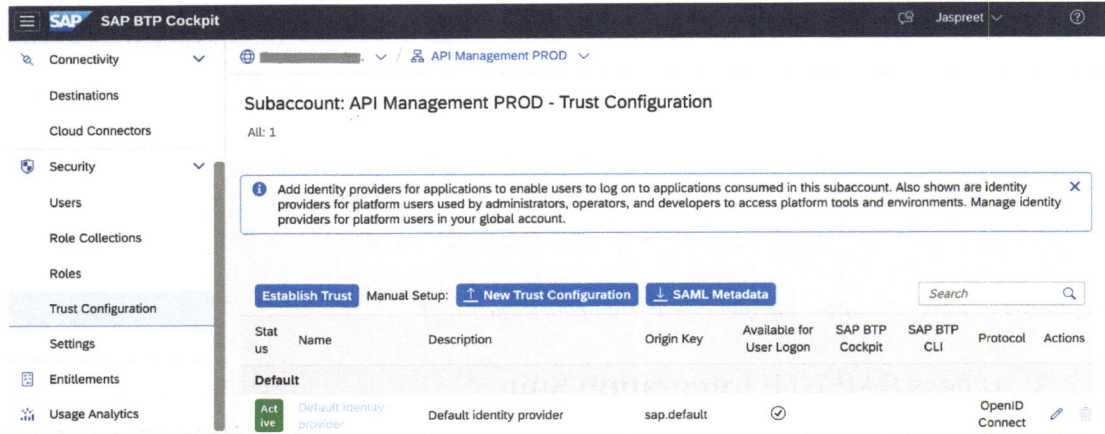

Figure 2-18. *Trust configuration of the SAP Integration Suite capabilities*

If you have configured security trust configuration to use a custom identity provider in place of the default SAP Identity provider, you will get the screen shown in Figure 2-19 to access the Integration Suite capabilities. You must log in with your enterprise credentials.

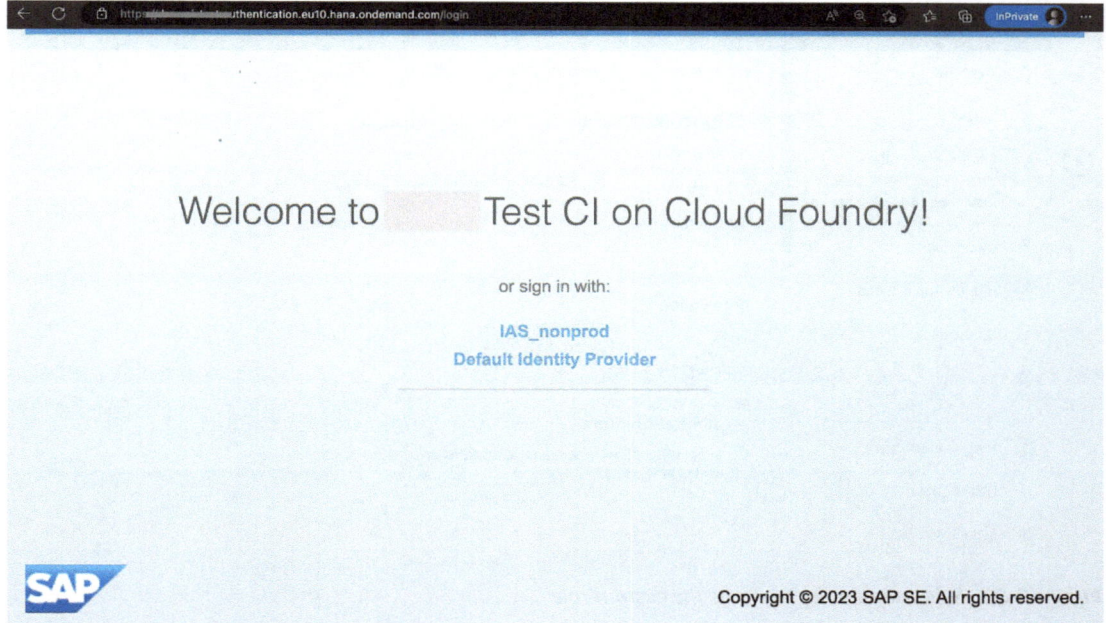

Figure 2-19. *The SAP Integration Suite login screen when a custom identity provider is enabled*

2.7.4 Creating Custom Roles

To give users fine-grained access to different features and functionality within the platform, you create custom roles. For example, you should create a role that allows users to access certain API endpoints but not others.

Procedure

1. Go to the Subaccount for your Cloud Foundry environment in the SAP BTP Cockpit.

2. In the left window, select the Service Marketplace. To create a custom role, select Integration Suite, or you can also choose any other service (see Figure 2-20).

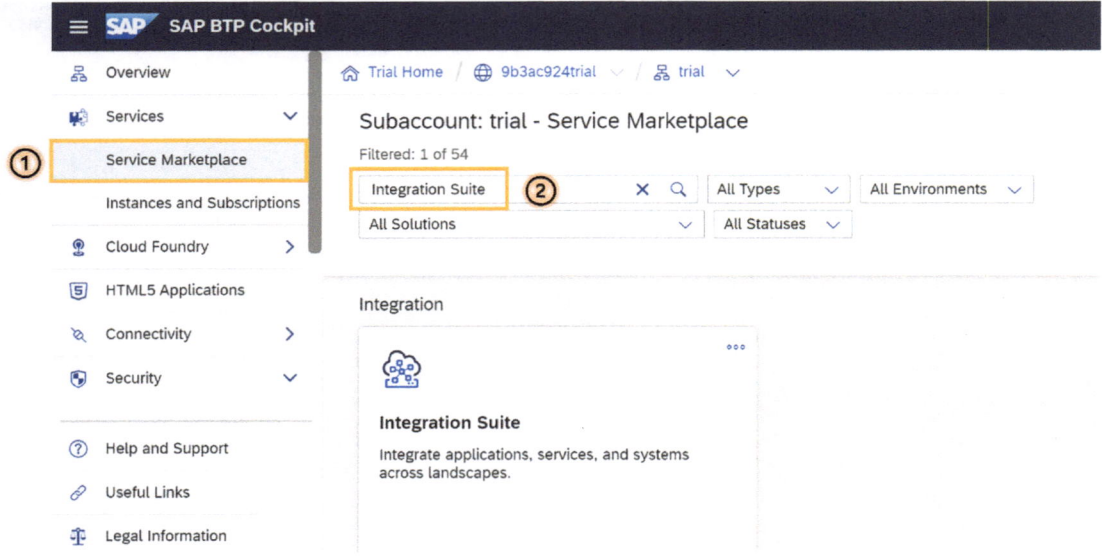

Figure 2-20. *Select the service to create the custom role*

3. Select Manage Roles from the Action icon under Application Plans for Integration Suite (see Figure 2-21).

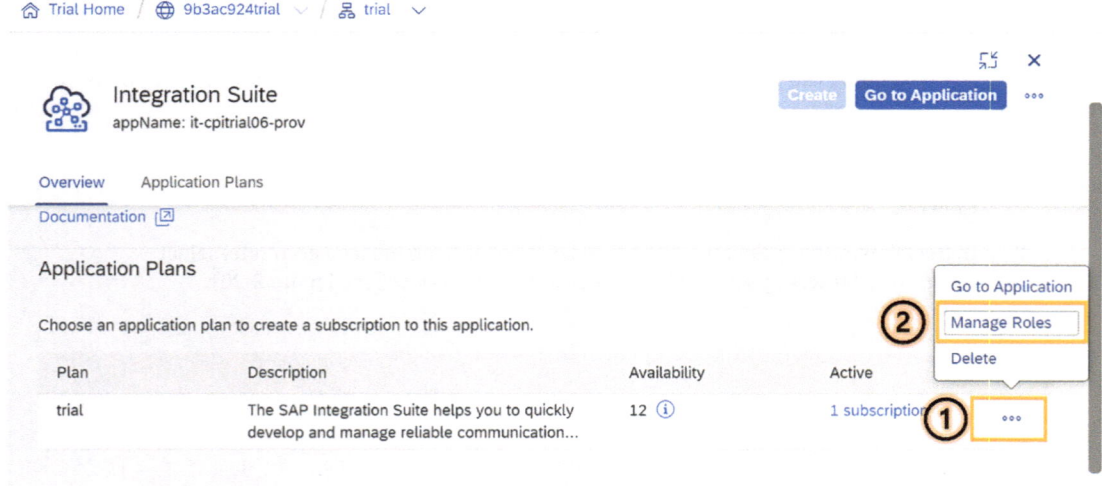

Figure 2-21. *Select Manage Roles*

4. Choose (+) to add a new custom role (see Figure 2-22).

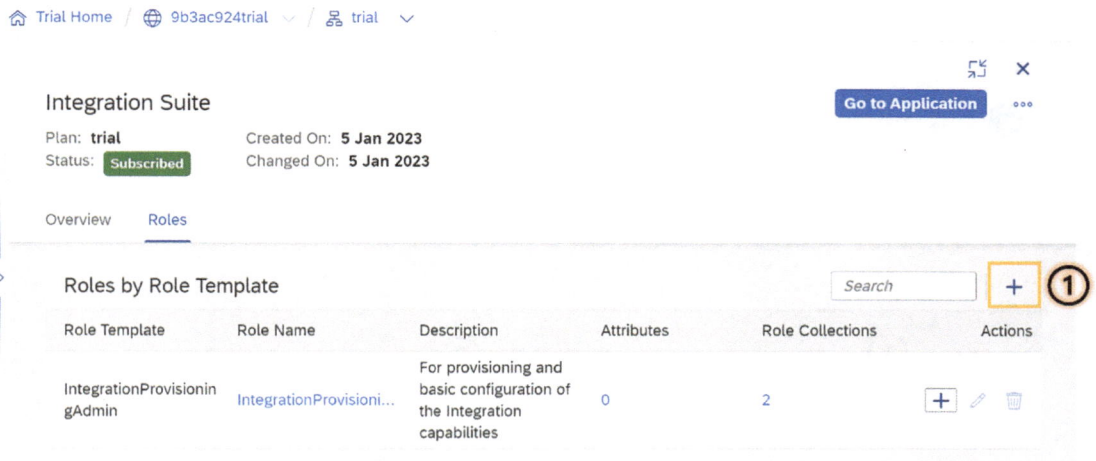

Figure 2-22. *Click (+) to add new role*

5. As shown in Figure 2-23, enter the following information in the Create Role dialog box.

Configure Role

- Role Name
- Description
- Role Template

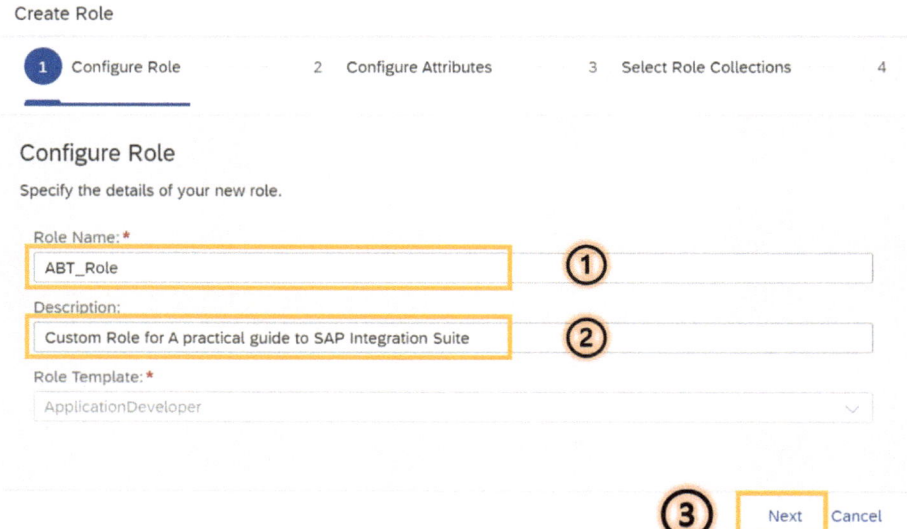

Figure 2-23. *Configure Role*

6. **Configure Attributes**—Keep the source value Static for the Custom Role Attributes. Provide the values of the attributes on the Values tab and click Enter (see Figure 2-24).

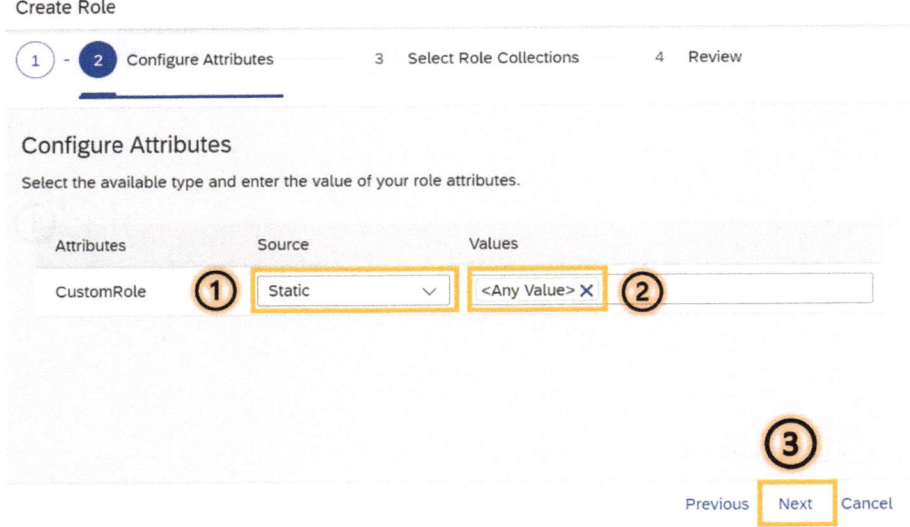

Figure 2-24. *Configure Attributes*

7. **Select Role Collections**—Search the desired Role Collection to which you want to assign the role. Select the Role Collection and click Next. Check the Review tab to see all the details and then click Create (see Figure 2-25). Your role will be added up to your Role Collection.

Figure 2-25. Create a role (select Role Collections)

8. You can see that the Custom Role has been created in the list, as shown in Figure 2-26.

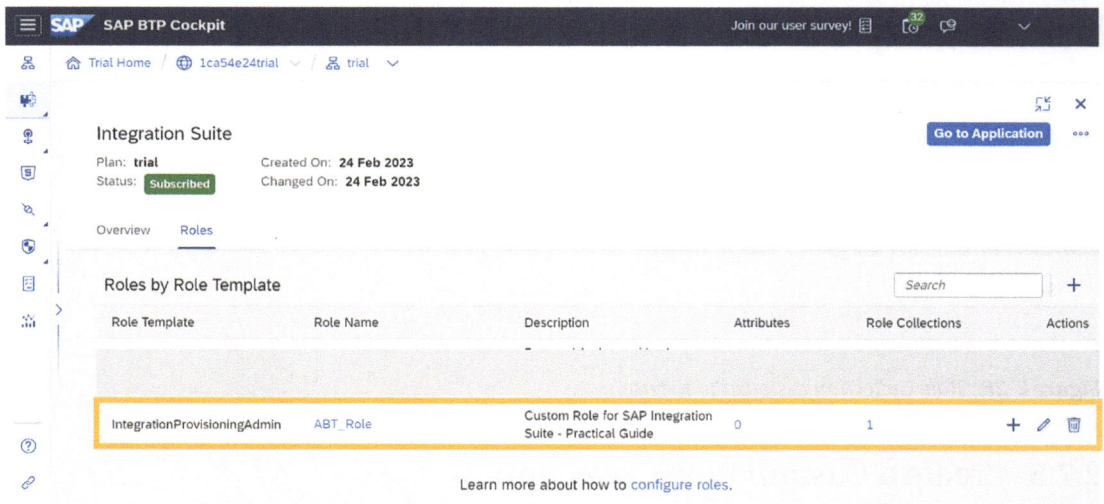

Figure 2-26. The custom role has been created

9. Add the created role to the Role Collection, as shown in Figure 2-27.

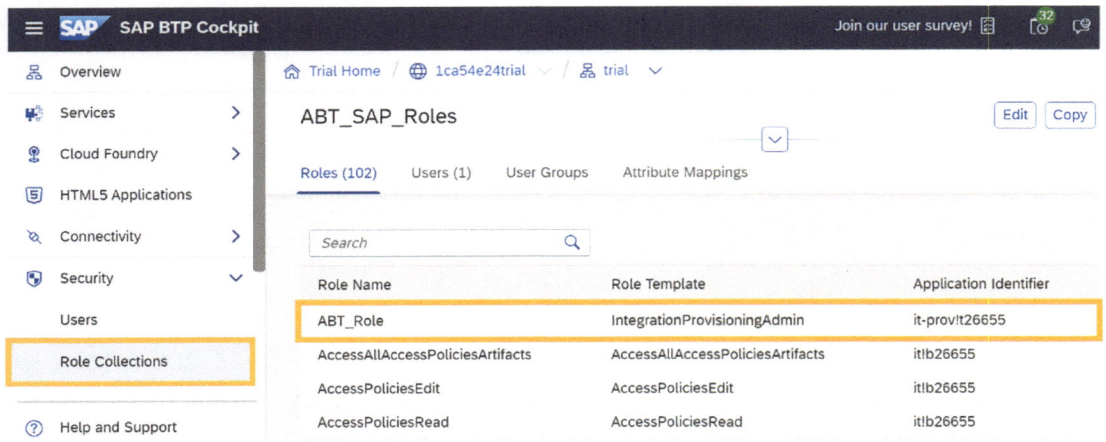

Figure 2-27. *Role has been added to the Role Collection*

10. Assign a Role Collection to the user, as shown in Figure 2-28.

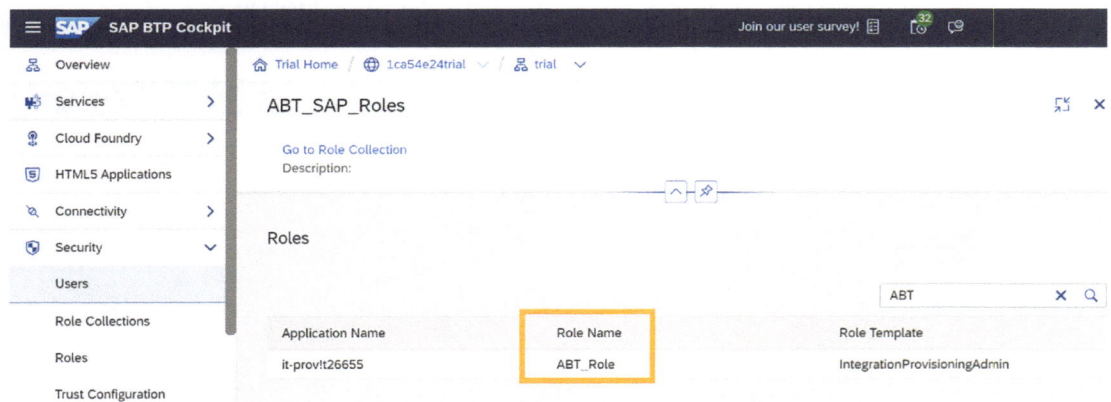

Figure 2-28. *Role Collection assigned to the user*

2.7.5 Create a Custom Role Collection

1. Every capability has a different Role Collection that must be assigned; you can create your Role Collection and assign it to the user after providing the correct user and roles.

2. Choose Security ➤ Roles Collection and click the + on the right panel (see Figure 2-29).

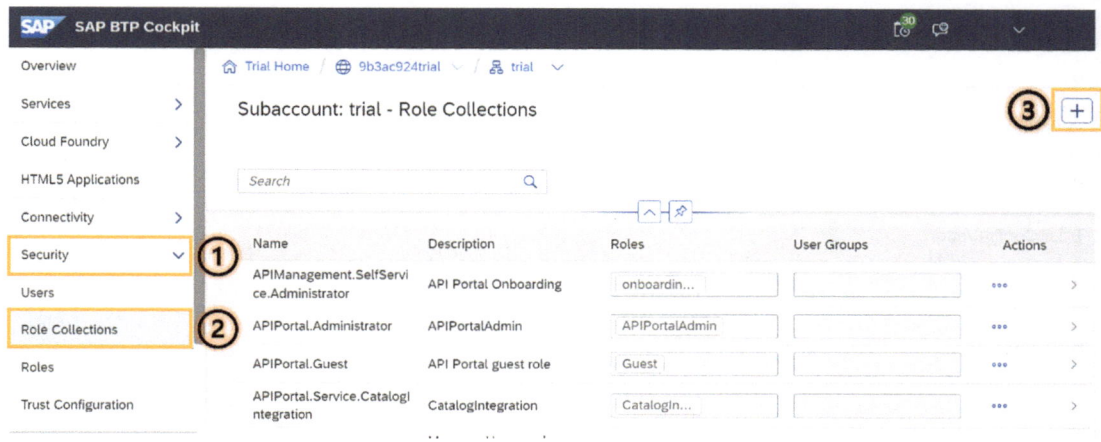

Figure 2-29. *Creating a custom Role Collection*

3. To create a new Role Collection, type a name and description (optional) and then click Create (see Figure 2-30).

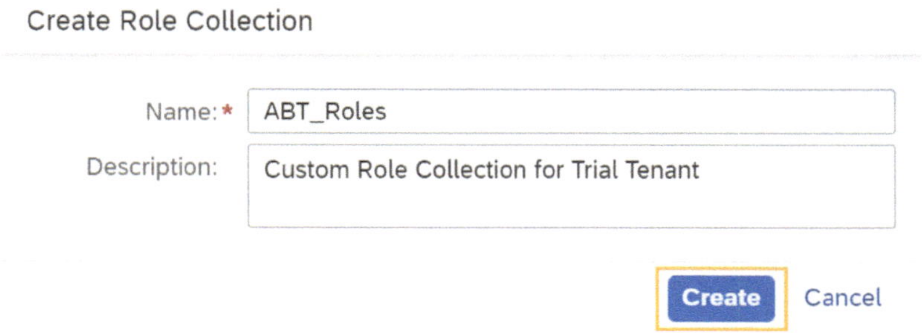

Create Role Collection

Name: *	ABT_Roles
Description:	Custom Role Collection for Trial Tenant

Create Cancel

Figure 2-30. *Basic info for a new Role Collection*

4. You can see that the custom Role Collection you created has been added to the list of Role Collections (see Figure 2-31).

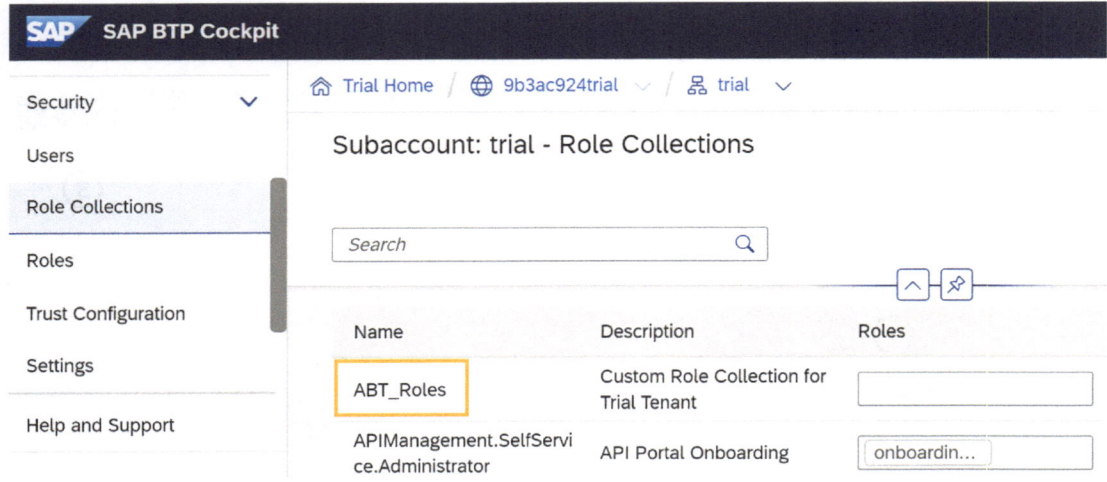

Figure 2-31. *ABT_Roles has been added to the list of Role Collections*

5. To add the roles to your custom Role Collection, click the Role Collection that you created. Open it in Edit mode. Search for your username and assign it to the Role Collection. Then search for the roles that you want to assign (see Figure 2-32). Click Save after you assign the roles.

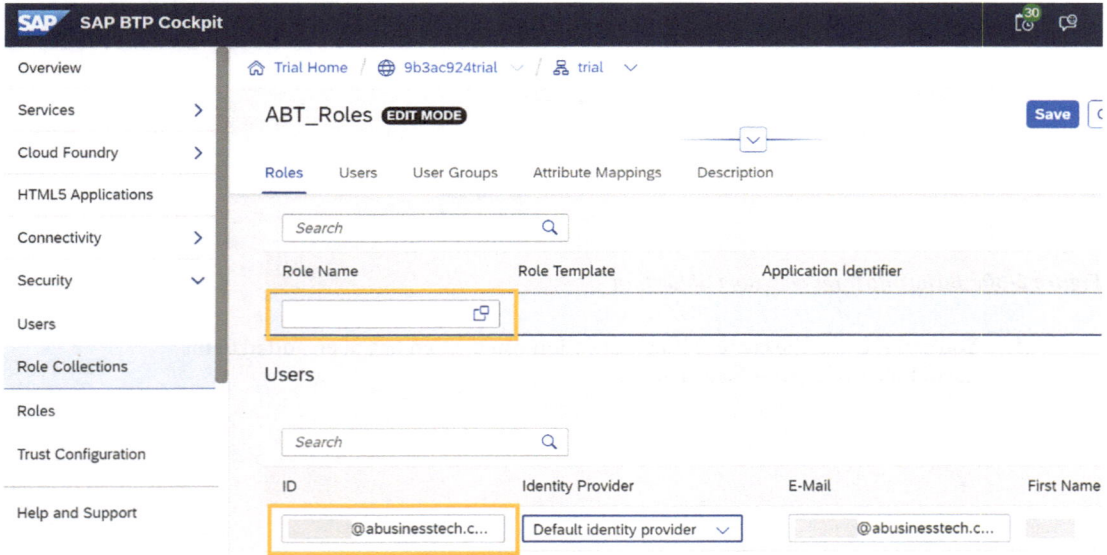

Figure 2-32. *Assign the Role Collections to the user*

6. Assign the different Role Collections according to your requirements.

2.7.6 Assigning Role Collections to Users

In order to access the Integration Suite, you must have the appropriate roles and Role Collections assigned.
You can organize the roles you create into collections using Role Collections. Users logged into the SAP ID service can be assigned any Role Collections you defined.

Procedure

1. Open the SAP BTP Cockpit in your web browser and select the appropriate subaccount.

2. Choose Security ➤ Role Collections from the left window.

3. Select the Role Collection you want to add users to.

4. Select Edit from the Users section.

5. The user ID of the user you want to add to the Role Collection should be entered. If the user is present in only one of the associated identity providers, you must select the provider and enter the user's email address.

6. Select (+) to add additional users.

7. Save your changes. See Figure 2-33.

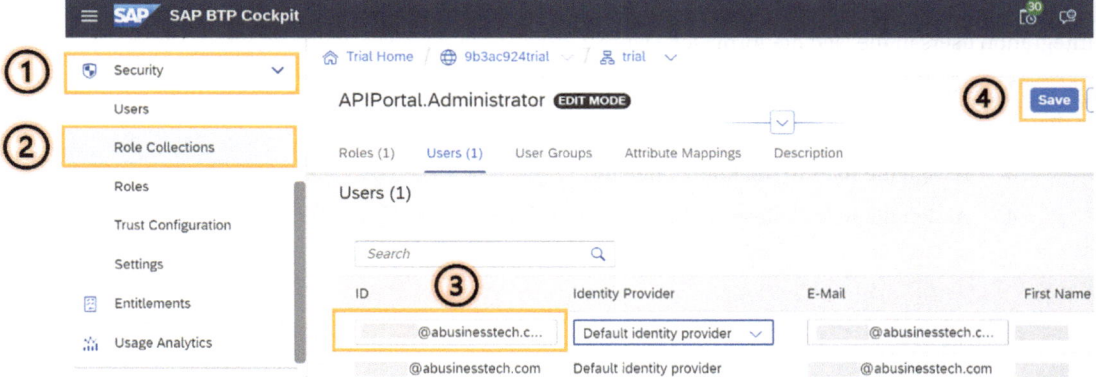

Figure 2-33. *Assigning a Role Collection to the user*

2.7.7 Access Management for Cloud Integration

There are several predefined roles that you can assign to account users when managing users using the SAP BTP Cockpit, as shown in Figure 2-34. These roles are connected to certain personas that are useful for integration projects based on the primary duties connected with integration projects.

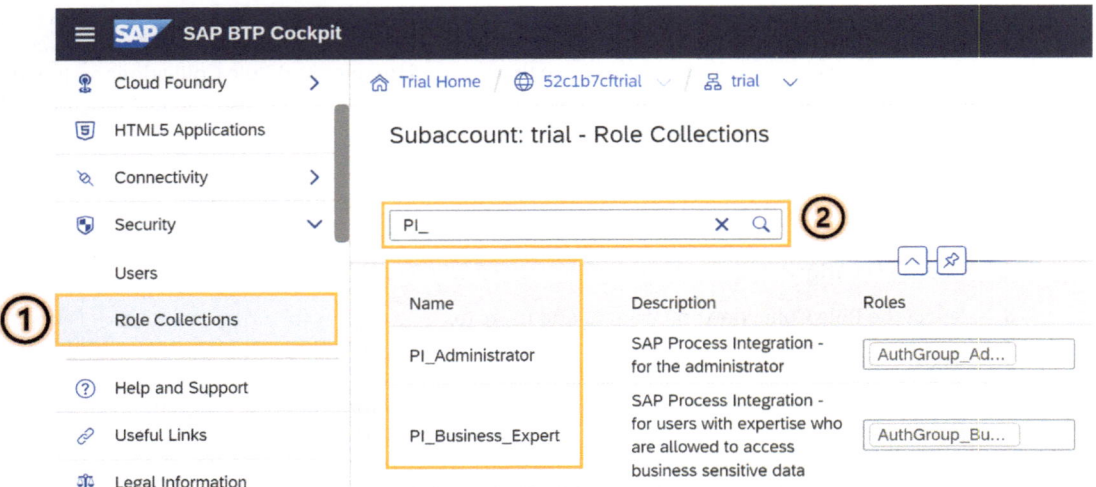

Figure 2-34. Role Collections for cloud integration in Cloud Foundry Environment

As with Cloud Foundry, in the Neo platform, you also have to assign the appropriate roles based on the user persona (Administrator, Developer, etc.). Use the following steps to assign roles to SAP Cloud Integration users in the Neo platform (see Figure 2-35).

1. Navigate to the Authorization tab in the SAP BTP Neo Cockpit.

2. Click New Group.

3. Assign the roles to the Role Collection.

4. Assign the Role Collection the appropriate user.

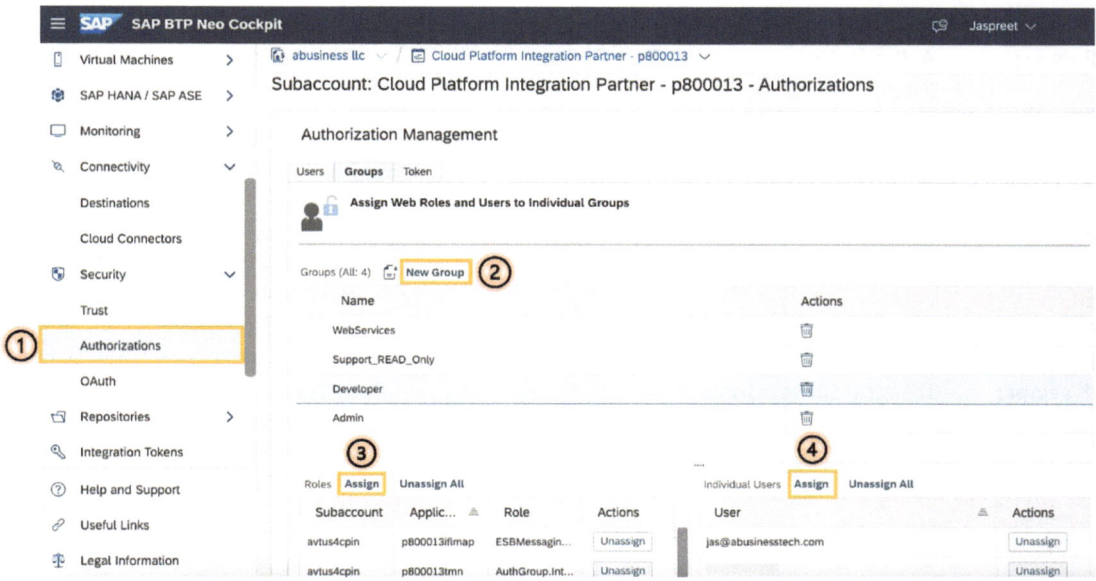

Figure 2-35. *Roles for cloud integration in the NEO environment*

In the Neo platform, you can create user personas (Administrator, Integration Developer, etc.) based on the best practices and assign them the roles, as shown in Figure 2-36.

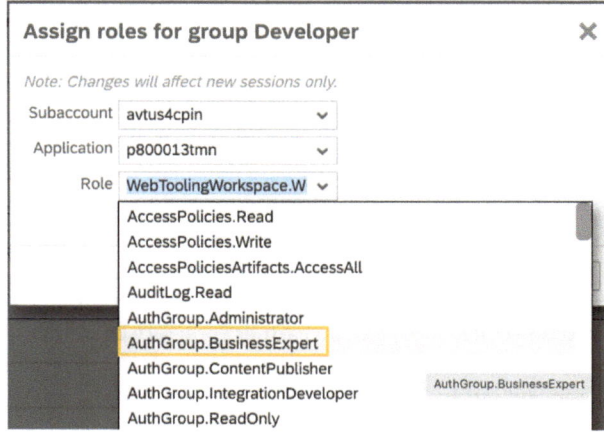

Figure 2-36. *Assigning roles to the group developer*

Persona

Cloud Integration comes with the predefined roles listed below. All users that need access to Cloud Integration based on their job responsibilities must be assigned one or more Role/Role Collections.

Authorization groups and Role Collections for each persona are listed in Table 2-3.

Table 2-3. *Authorization Groups/Role Collections*

Persona	Authorization Group (Neo)	Role Collection (Cloud Foundry)	Description
Business expert	AuthGroup. BusinessExpert	PI_Business_Expert	Allows a business expert to do business operations, such as looking at the payload.
Administrator	AuthGroup. Administrator	PI_Administrator	Allows the tenant administrator, also known as the administrator of the tenant cluster, to connect to a cluster and manage its operations.
Integration developer	AuthGroup. IntegrationDeveloper	PI_Integration_Developer	Enables an integration developer to use the Integration Designer to connect to a cluster to view, download, and deploy artifacts (for example, integration flows). To access Cloud Integration, you must belong to this authorization group.
Read-only persona	AuthGroup.ReadOnly	PI_Read_Only	Allows you to connect to a tenant and check messages.
			With the help of this permission group, you can download WSDL files and have read-only access to integration flow artifacts in the Monitoring area.
System developer	AuthGroup. SystemDeveloper	N/A	It gives a system developer the ability to carry out activities necessary for system support.
			You can use the Data Store viewer (read-only access) thanks to this authorization group.
Partner Directory configurator	AuthGroup. TenantPartner DirectoryConfigurator	For the Partner Directory configurator, there is no predefined role collection. The service instance for the associated API client must have the role template AuthGroup TenantPartnerDirectory.	Allows the administrator of the Partner Directory to read and write Partner Directory material.

Check out the SAP Help Portal for more details on access management for Cloud Integration.

Task and Permission

The responsibilities necessary to complete the various cloud integration-related tasks are summarized in the Table 2-4. It also lists to what extent the roles and duties relate to the key persona for cloud integration.

Table 2-4. *Summary of Cloud Integration Responsibilities*

Area	Task	Role Template (Cloud Foundry)	Persona
Discover	View packages	`CatalogPackagesRead`	Integration Developer Business Expert Read-Only Persona System Developer Tenant Administrator
	View package artifacts	`CatalogPackageArtifactsRead`	Integration Developer Business Expert Read-Only Persona/ System Developer Tenant Administrator
	Copy package to workspace	`CatalogPackagesCopy`	Integration Developer
Design	View packages and package artifacts	`WorkspacePackagesRead`	Integration Developer Business Expert Read-Only Persona/ System Developer Tenant Administrator
	Create, edit, export, and remove a package and all its artifacts	`WorkspacePackagesEdit`	Integration Developer
	Update packages	`WorkspacePackagesEdit`	Integration Developer
	Configure artifacts	`WorkspacePackagesConfigure`	Integration Developer
	Deploy/undeploy artifacts	`WorkspaceArtifactsDeploy`	Integration Developer Tenant Administrator
	Export Package for transport	`WorkspacePackagesTransport`	N/A
	Import package from transport	`WorkspacePackagesTransport`	N/A

2.8 Trial Account Setup: The SAP Integration Suite

If you have assigned the right role collection for Enterprise BTP accounts and have the subaccount viewer and administrator roles, you should be able to set up the Integration Suite subaccount. For the trial account, you can set up the BTP trial account first. SAP BTP offers the free trial account for 90-days by which you can play around with the SAP BTP and SAP Integration Suite, which we will be going to study in this book.

The SAP Integration Suite trial account is created inside the SAP BTP Cockpit. The SAP BTP Cockpit is a web-based tool that allows users to manage and monitor applications running on the SAP Business Technology Platform (BTP), formerly known as the SAP Cloud Platform.

SAP offered a trial system to let users learn more about their SaaS service, and as the first move in that direction. To access the SAP CI, the SAP Integration Advisor, the SAP API Management, the SAP Open Connectors, and the other capabilities of SAP Integration Suite, you first need to set up a trial account in the SAP Business Technology Platform (BTP).

2.8.1 Setting Up the BTP Trial Account

1. Navigate to SAP Business Technology Platform (BTP) through `https://account.hanatrial.ondemand.com/`. Sign in if you have an SAP account.

2. If you do not have an account, register for one using your email. Provide the necessary information and create the account.

3. When you log in, you will be asked to select your nearest region to create the account.

4. Select the nearest region and click Create Account (see Figure 2-37).

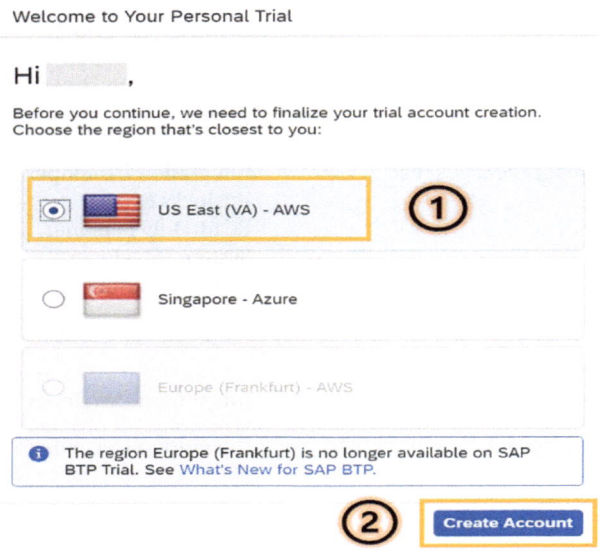

Figure 2-37. *Select the closest data center based on your location*

5. When you create an account, the popup shown in Figure 2-38 will be displayed to continue ongoing processing, which takes about 1-2 minutes. Click Continue to proceed.

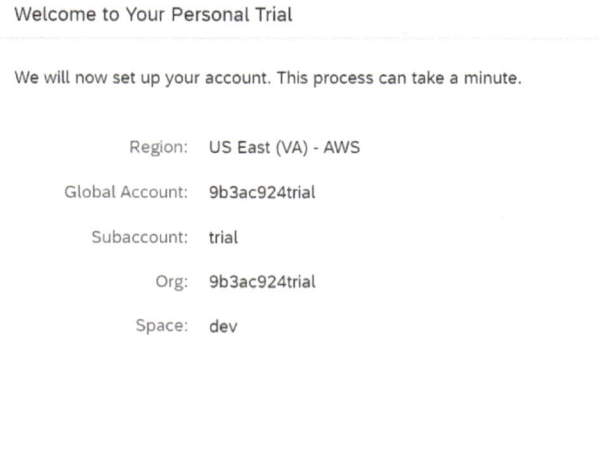

Welcome to Your Personal Trial

We will now set up your account. This process can take a minute.

Region:	US East (VA) - AWS
Global Account:	9b3ac924trial
Subaccount:	trial
Org:	9b3ac924trial
Space:	dev

Figure 2-38. The primary information of the trial account

6. On the next screen (shown in Figure 2-39), click Go to Your Trial Account. The trial account will be created along with the subaccount.

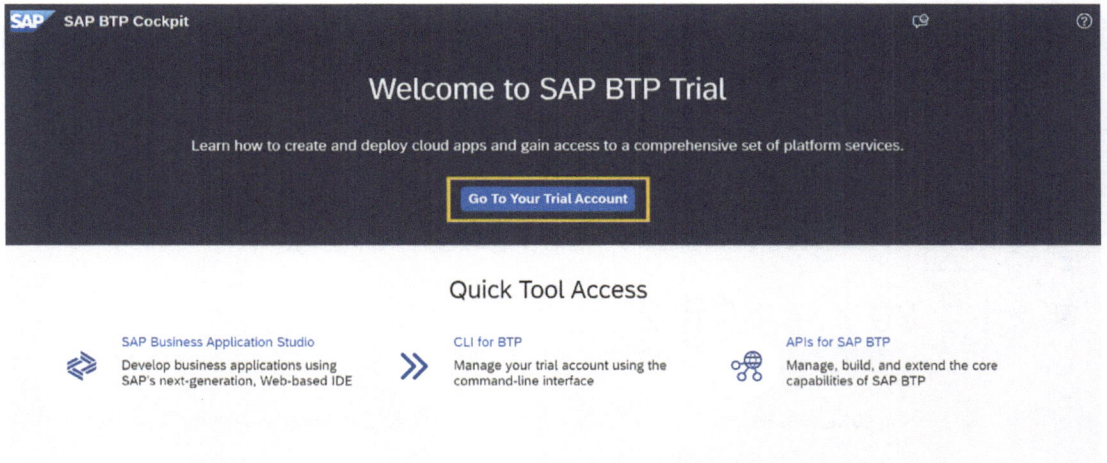

Figure 2-39. Dashboard of the SAP BTP Cockpit

7. You will be taken to the trial home page, where the account and subaccount have been created (see Figure 2-40). You can use this subaccount for your Integration Suite or you can create another subaccount dedicated to your needs.

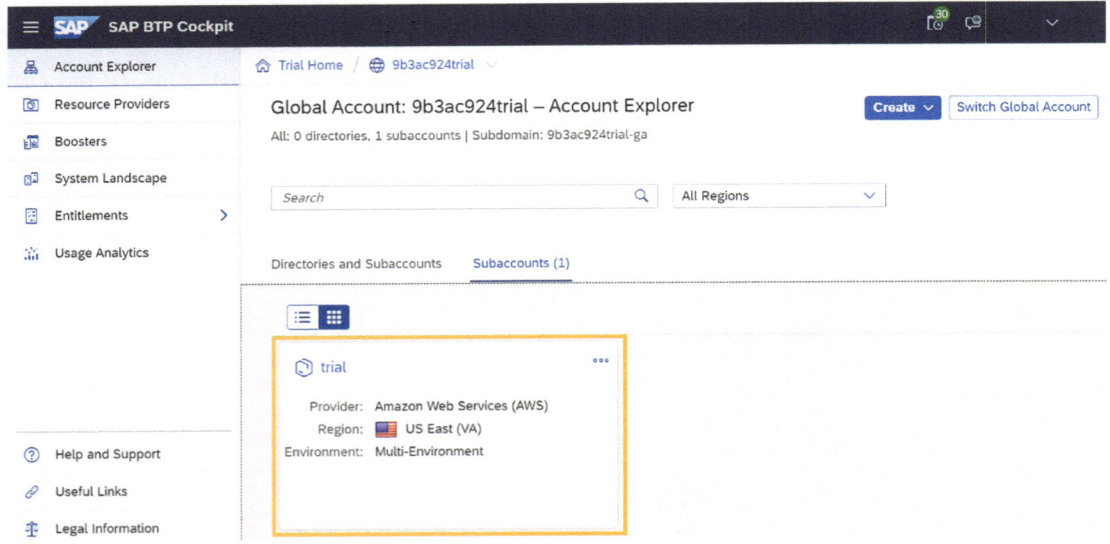

Figure 2-40. *The trial subaccount has been created*

8. After clicking a subaccount, you will be directed to the home page of the SAP BTP Cockpit, where you can see the 80 Entitlements and 2 Instances and Subscriptions (see Figure 2-41).

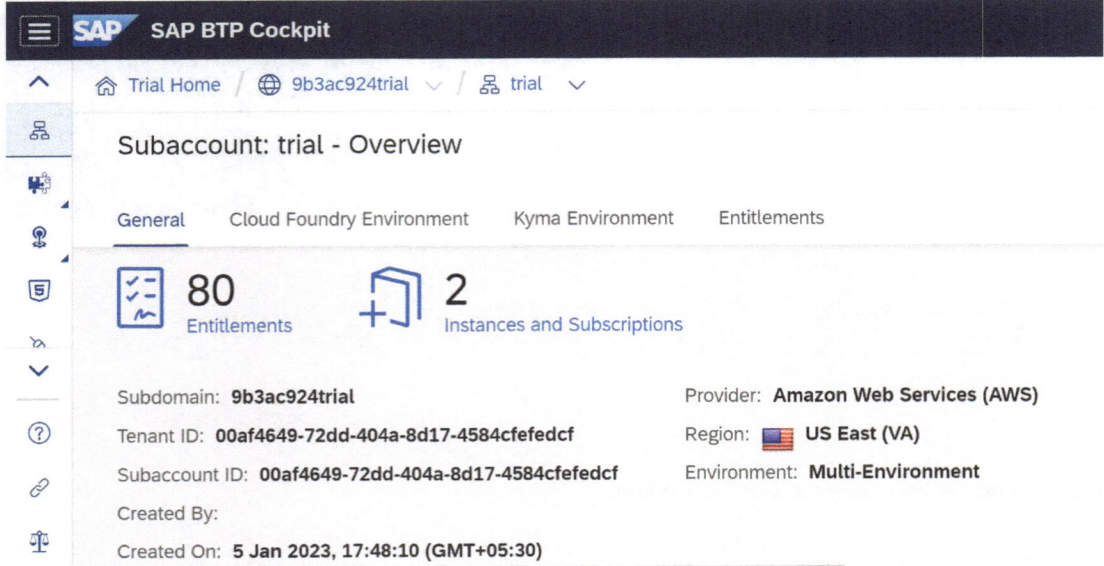

Figure 2-41. *Home page of the SAP BTP Cockpit*

2.8.2 Create a Space with Cloud Foundry

For Enterprise, you can create your own space in the Cloud Foundry environment or continue with the default Dev space in the trial account. To create a new space, follow these steps:

1. Navigate to Cloud Foundry ➤ Spaces ➤ Create Space in the left panel, as shown in Figure 2-42.

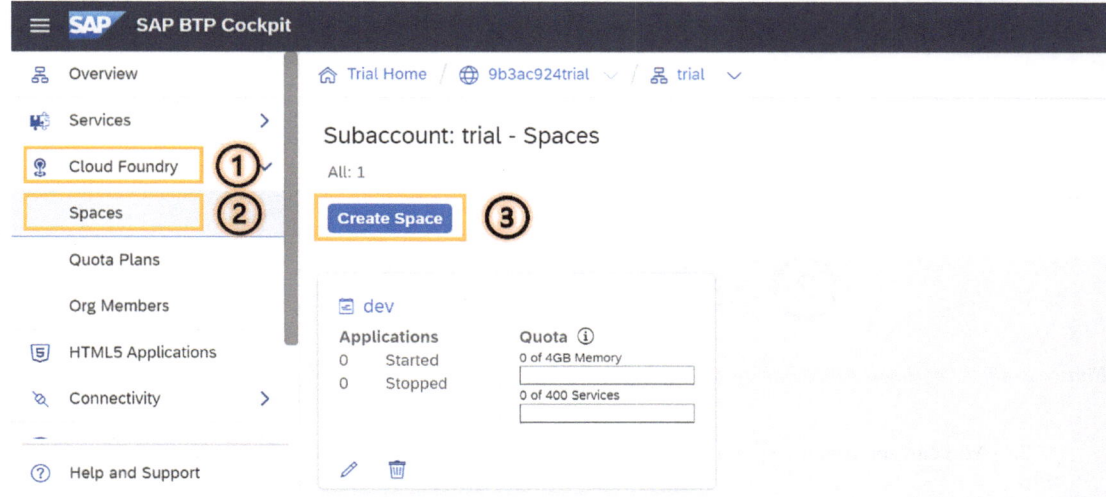

Figure 2-42. *Navigate to Cloud Foundry to create a space*

2. Provide a space name and check all the boxes. Click Create (see Figure 2-43).

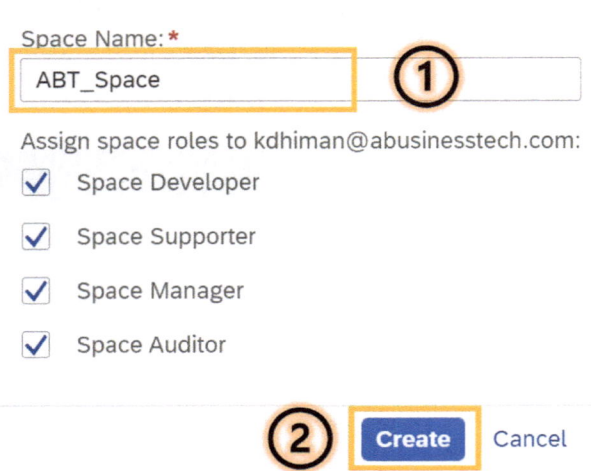

Figure 2-43. *Creating a space in Cloud Foundry*

3. You can see that the space has been created, as in Figure 2-44.

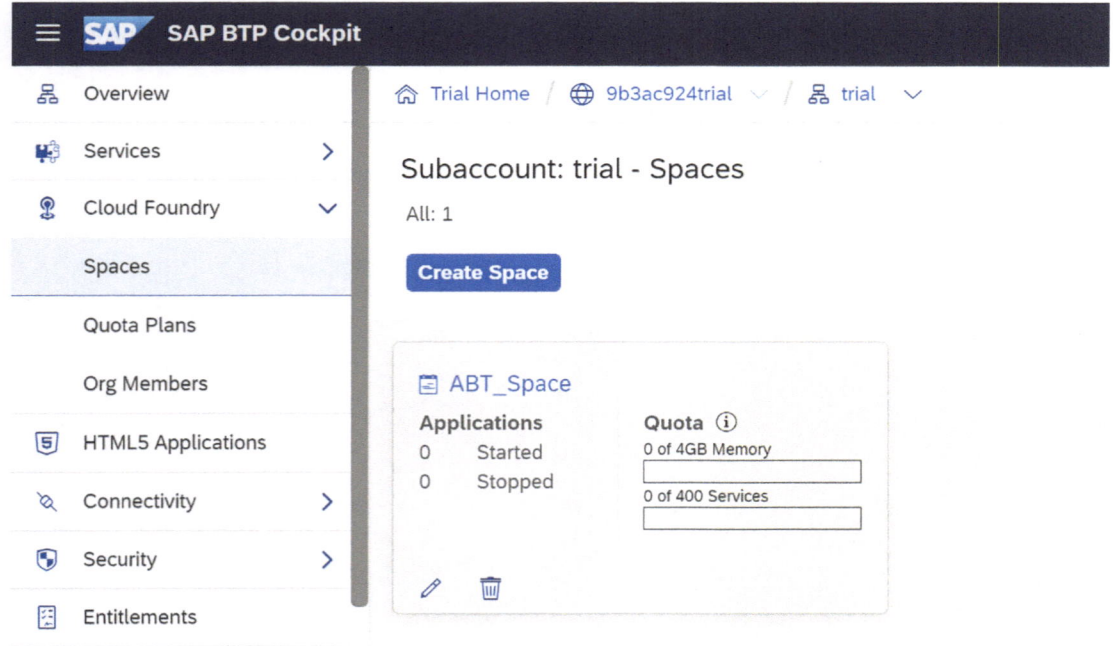

Figure 2-44. *A custom space has been created*

2.8.3 Managing Entitlements

To avoid confusion, you should know the differences between entitlements and quotas before you begin.

> **Entitlements**—Your right to provision and consume a resource is your entitlement.

> **Quota**—The quantifiable amount that specifies the resource's maximum allowable consumption. Or how much of a service package you are allowed to utilize.

At the global account level, entitlements and quotas are handled, distributed to directories and subaccounts, and used by subaccounts.

Quotas and entitlements can be assigned to other subaccounts if you remove them from a subaccount and make them available again at the global account level.

You automatically receive a subaccount called *trial* when you sign up for a trial account, and all trial entitlements are by default assigned to it. You must manually transfer the entitlements from the default subaccount to the new subaccount if you decide to create a second subaccount and experiment there. You can:

- Move all trial rights over

- Move just some of them

- Divide a service plan's quota between the two subaccounts

One or more service plans are offered for each service. A service plan is a breakdown of the advantages and costs of a specific service type. An example is the configuration of a database with numerous "T-shirt sizes," each of which represents a distinct service package.

Some service plans include numeric quotas, allowing you to alter the quantity of units accessible in a subaccount. These units represent various things depending on the service and can limit the number of service instances, apps, or routes you can have in a subaccount.

Entitlements and quotas for subaccounts can be viewed and modified in a number of locations in the Cockpit:

- In the global account

- In the Subaccount

Add Missing Entitlements

In the Enterprise when you buy the license for the SAP Integration Suite, it becomes necessary to configure the entitlements and add the service/quota to the tenant.

Sometimes, not all the service plans are added to the enterprise/trial account automatically. You can add any missing entitlements by using these steps:

1. Navigate to Entitlements ➤ Configure Entitlements to enter the edit mode of the subaccount, as shown in Figure 2-45.

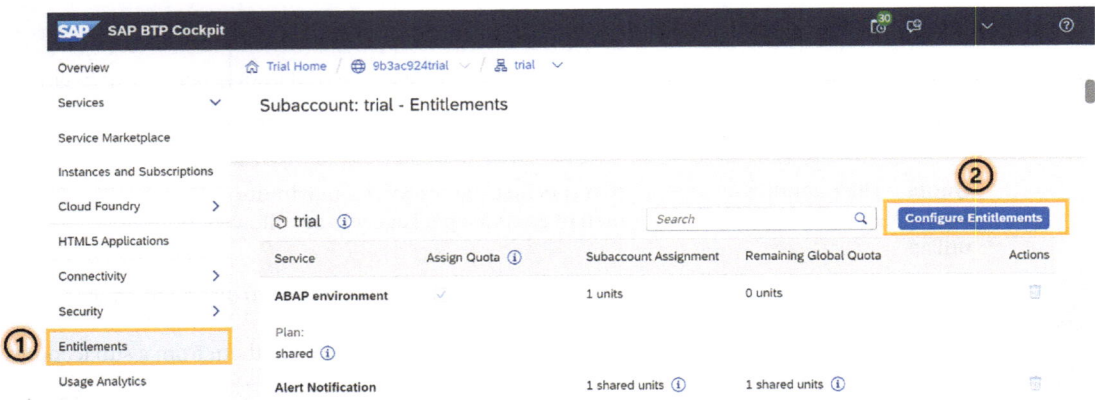

Figure 2-45. *Configuring entitlements in the SAP Integration Suite*

2. Click Add Service Plans.

3. Search for the desired service plan, select it, and then click Add Service Plan. Since mine is a Trial account, all the instances are already assigned in Figure 2-46.

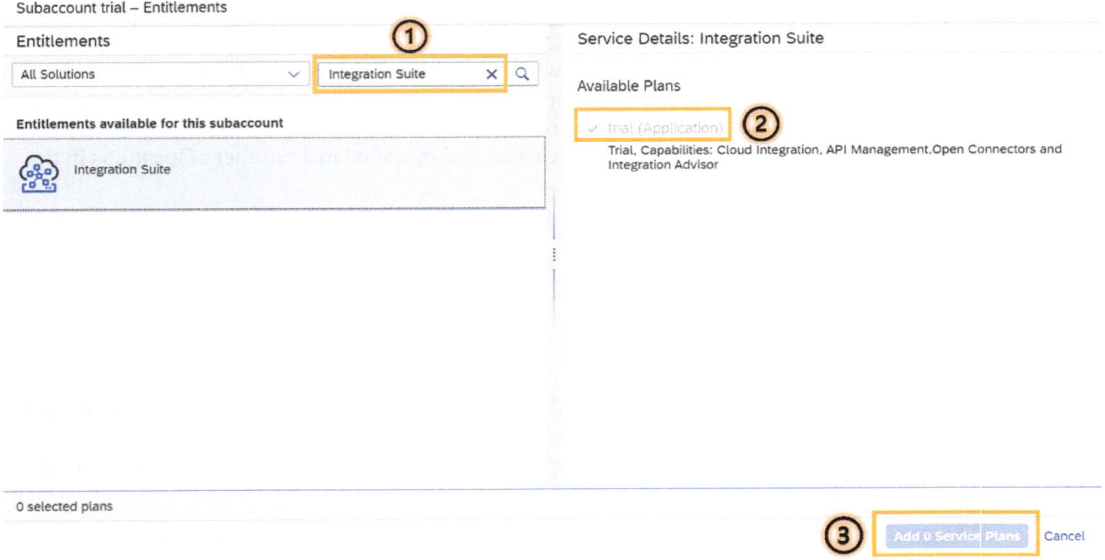

Figure 2-46. *Select the desired entitlement and it to your service plan*

2.8.4 Subscribing to the Service

Multiple global accounts and subaccounts can be created. This example utilizes the trial subaccount that was created by default, as you'll only be using one for the time being. Clicking this trial account will take you to the SAP Business Technology Platform (BTP) Cockpit, as shown in Figure 2-47.

The Cloud Foundry environment is activated by default when you navigate to this subaccount. You want to subscribe to the SAP Integration Suite, but you currently have some entitlements, instances, and subscriptions.

1. Select Services ➤ Instances and Subscriptions and then choose Create. See Figure 2-47.

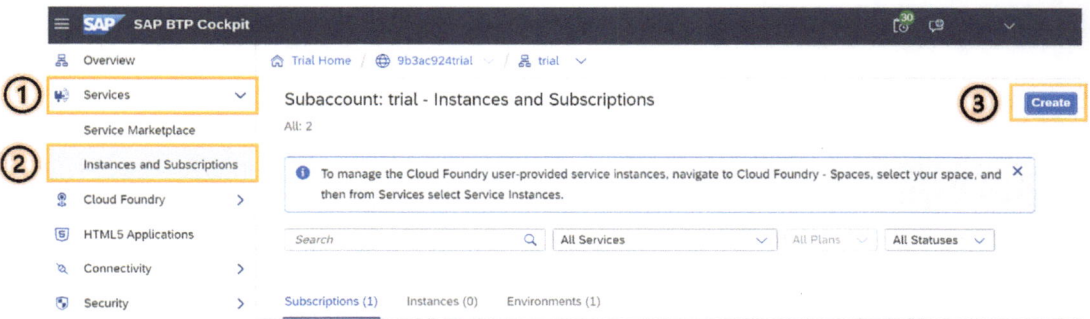

Figure 2-47. *Create a new service*

2. By default, one subscription (the SAP Business Application Studio) is already active. The SAP Integration Suite service will now be created. Choose the Integration Suite service and the trial package. Select Create. See Figure 2-48.

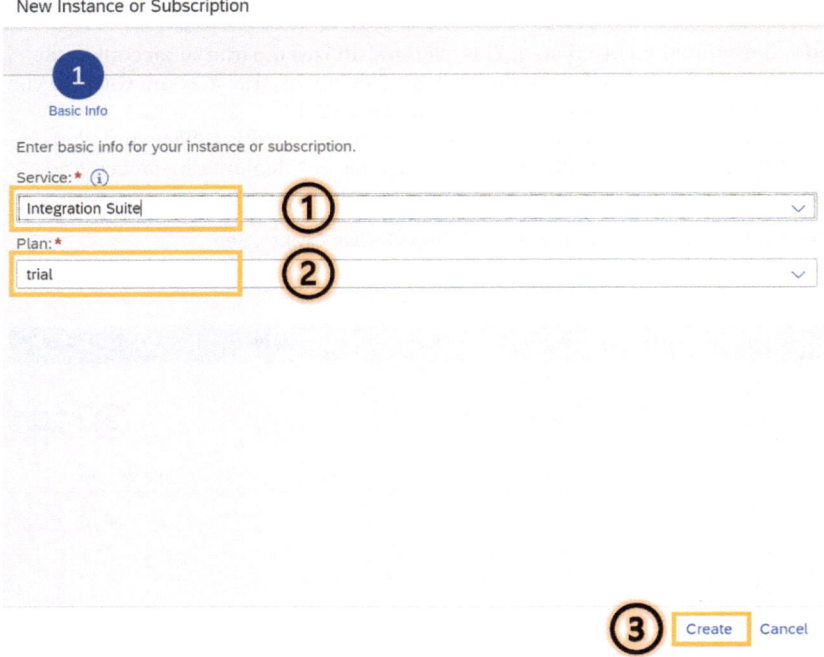

Figure 2-48. *Basic info for creating new service*

2.8.5 Assigning a Role Collection

The new service Integration Suite has been created, so it's now time to assign some roles. Follow these steps to do so:

1. Select Security ➤ Users. Choose the entry against your name. In the Role Collections section, choose Assign Role Collection, as shown in Figure 2-49.

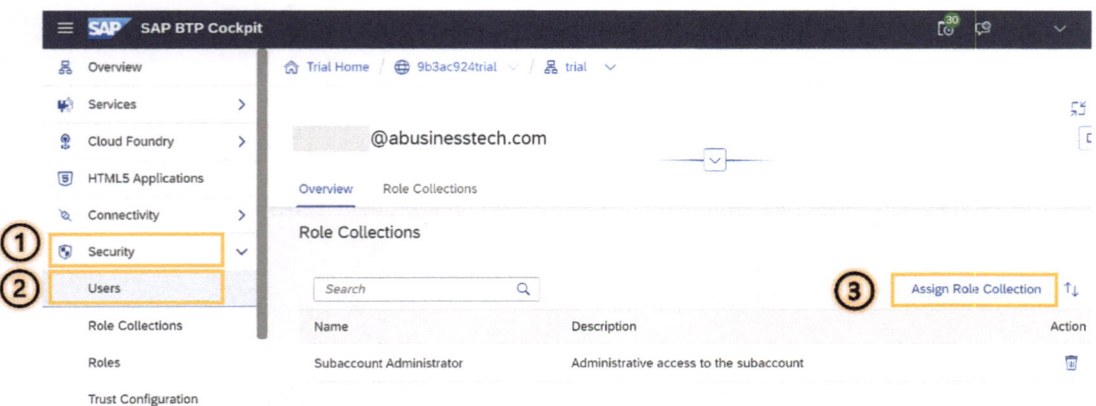

Figure 2-49. *Assigning a Role Collection to the user*

2. Select the Integration Provisioner as the Role Collection and assign it to the user (see Figure 2-50). Once the capabilities have been provisioned, You have to assign the necessary role collections as shown in Table 2-3.

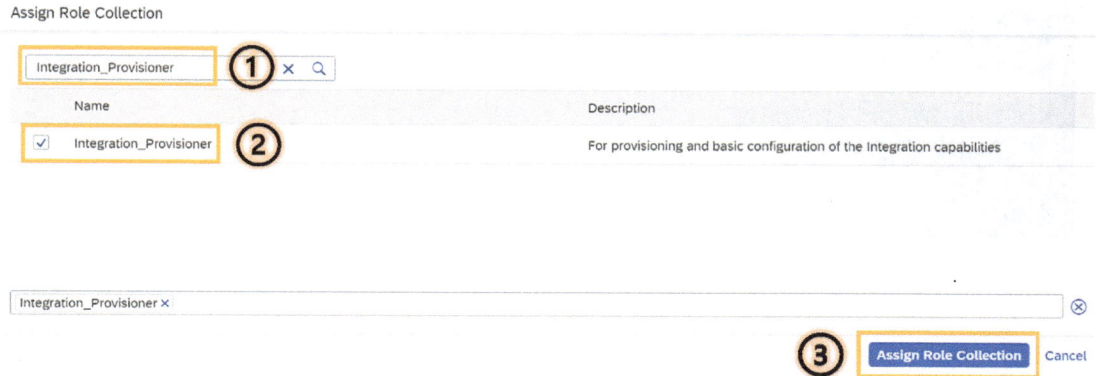

Figure 2-50. *Selecting Integration Provisioner as the Role Collection*

2.8.6 Provisioning Capabilities

The SAP Integration Suite works with the Cloud Foundry environment. While the Neo environment is still supported for cloud integration, any new capabilities will only be found in the Cloud Foundry environment.

1. In the SAP BTP Cockpit dashboard, navigate to the service (Integration Suite) to open the SAP Integration Suite UI. To use the services of the Integration Suite, you must add the capabilities of the SAP Integration Suite, which you will learn about while reading the chapter (see Figure 2-51).

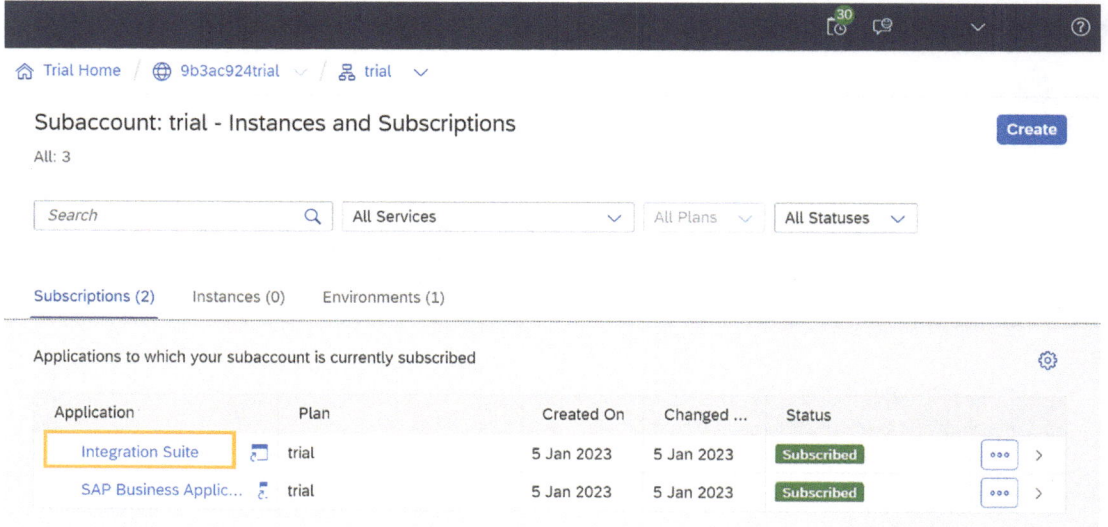

Figure 2-51. *Open the SAP Integration Suite UI*

2. In the Integration Suite Launchpad, choose Add Capabilities (see Figure 2-52) to activate the capabilities offered by the Integration Suite.

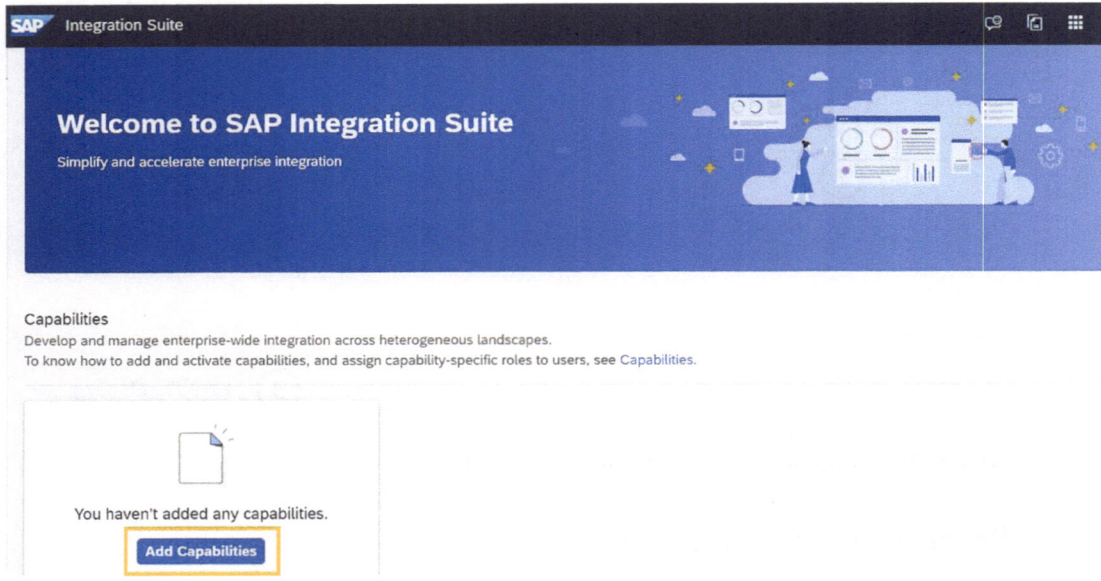

Figure 2-52. *Adding capabilities to the Integration Suite*

3. Select the capabilities according to the needs and requirements of the organization. Select the capabilities of Integration Suite to proceed further. After selecting the capabilities, click Next (see Figure 2-53).

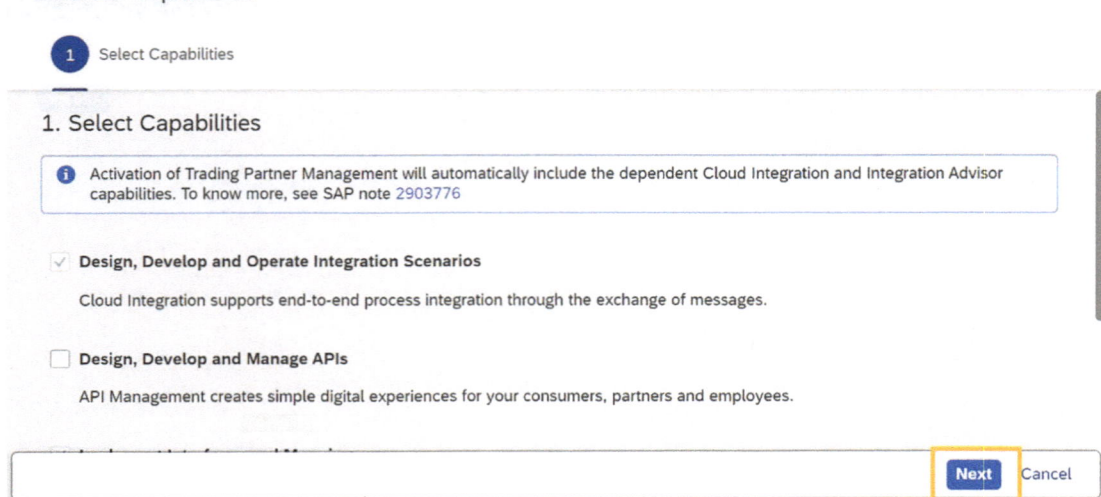

Figure 2-53. *Select the capabilities of the Integration Suite*

4. A popup will appear when you click Next, asking you to confirm these capabilities. Confirm the capabilities and click Next, as shown in Figure 2-54.

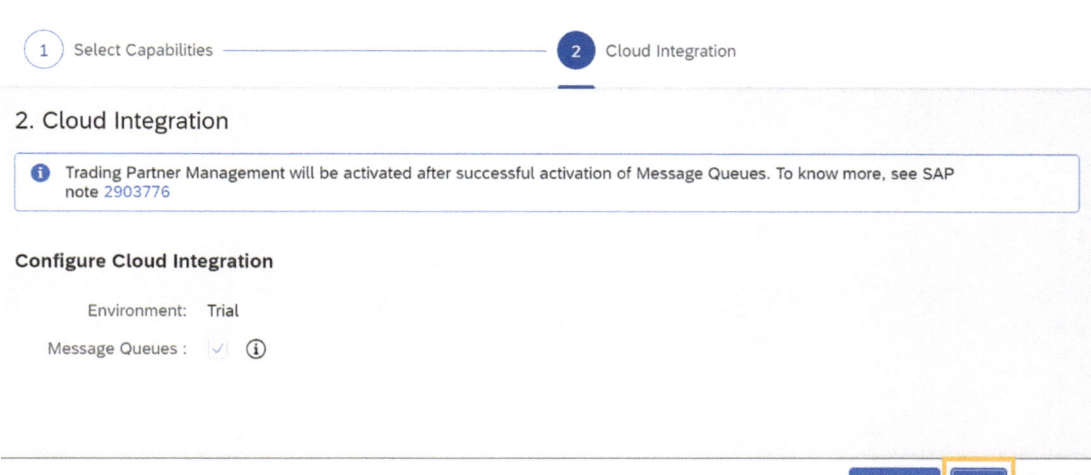

Figure 2-54. Confirm all the selected capabilities

5. After confirming all the capabilities, click Activate to provision the selected capabilities (see Figure 2-55).

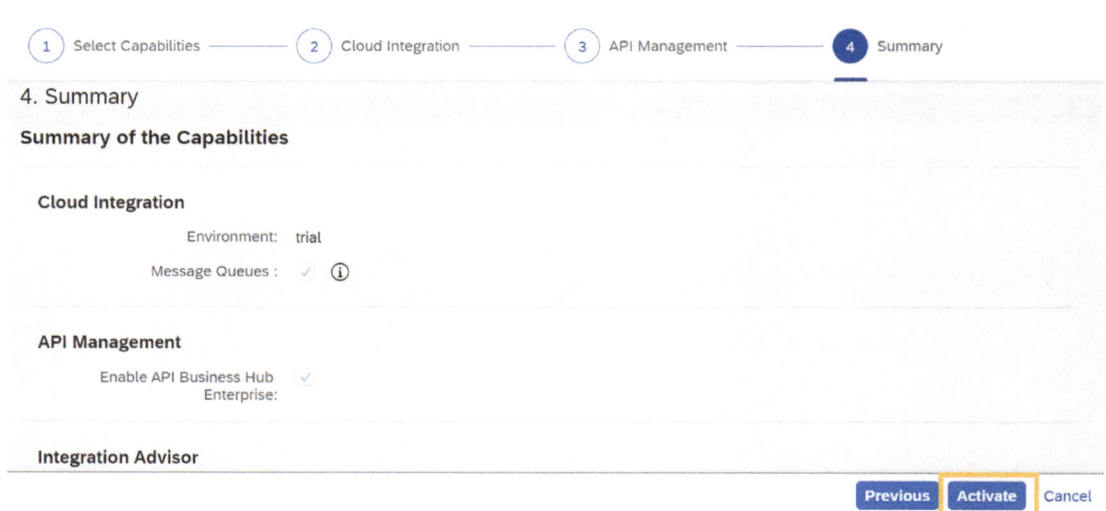

Figure 2-55. Activate all the capabilities

6.   Adding these capabilities takes around 60 minutes. See Figure 2-56.

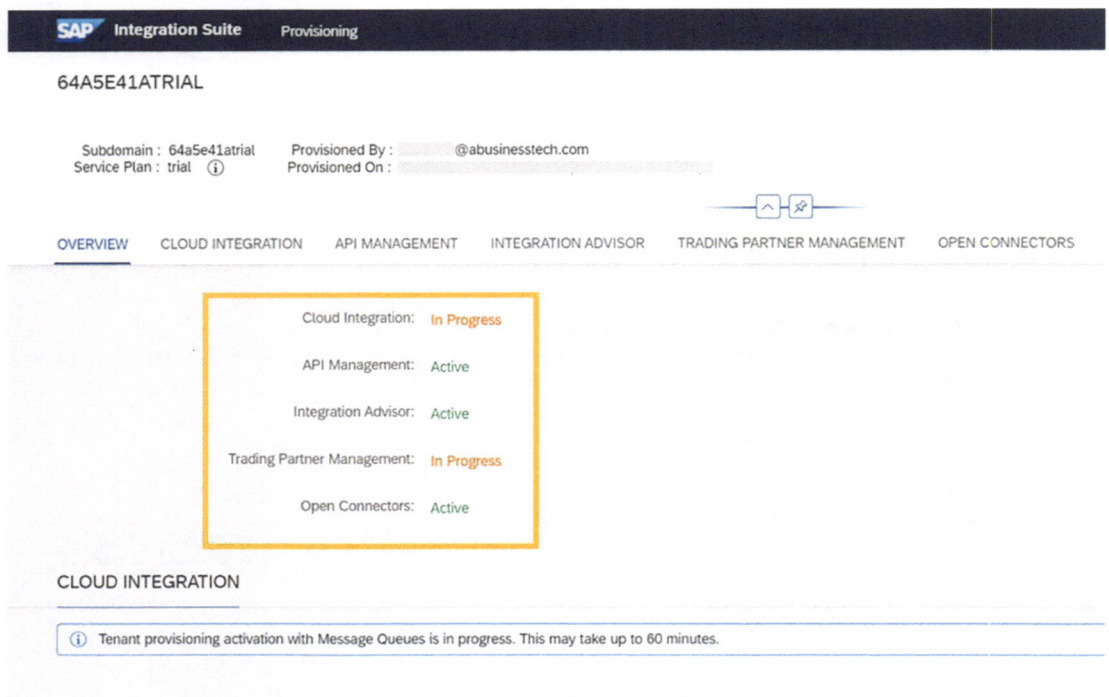

Figure 2-56. *Provisioning capabilities*

7.   After provisioning all the selected capabilities, you can see the tiles of the capabilities you added in the SAP Integration Suite dashboard, as shown in Figure 2-57.

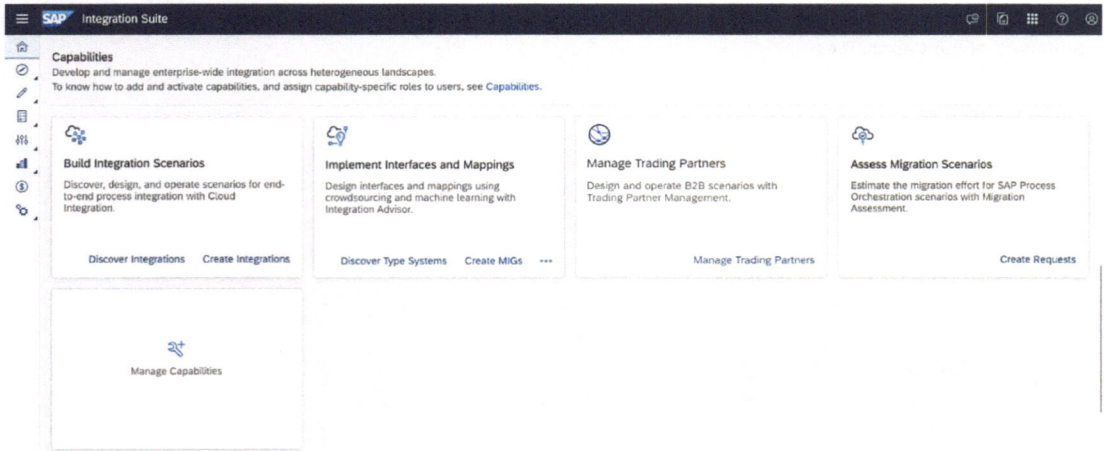

Figure 2-57. *Dashboard of the SAP Integration Suite with all the added capabilities*

2.8.7 Booster: Automatically Build Service Instances and Assign Roles

A *booster* is a series of guided and interactive actions that let you choose, set up, and use services on the SAP BTP to accomplish a particular technical objective. In this situation, the Integration Suite booster will assist you in creating service instances and assigning roles.

A service instance specifies how a remote component can call the Process Integration runtime service. The SAP BTP client is defined as a service instance in the context of cloud integration. The credentials and other data needed to invoke the integration flow are contained in the service key that's generated from the service instance.

1. Navigate to the overview page of your SAP BTP global account.

2. Choose Boosters on the left navigation pane.

3. Look for the Enable Integration Suite tile on the boosters list. Start the booster execution by clicking Start on the tile (see Figure 2-58).

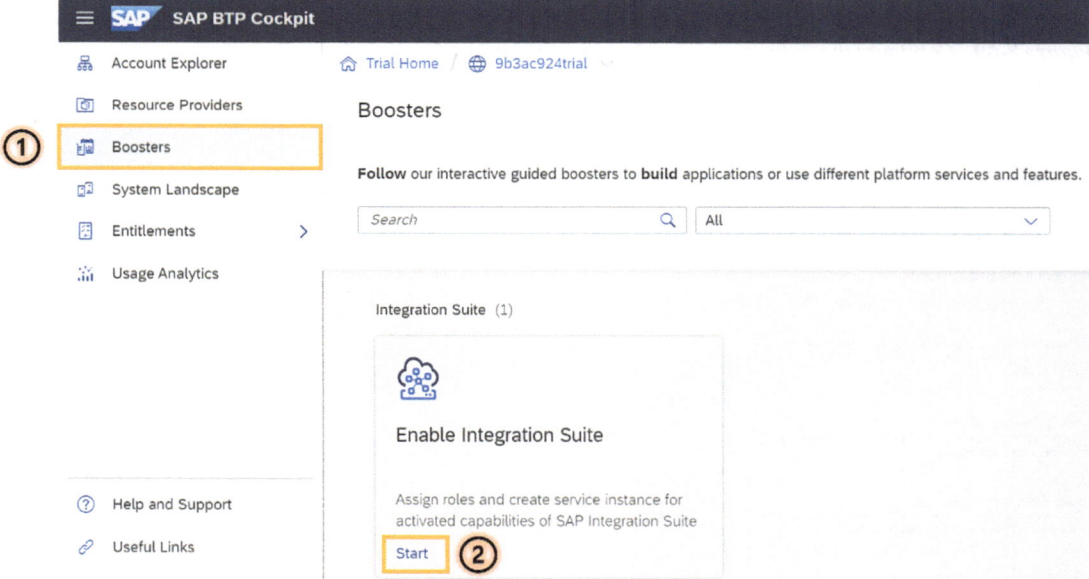

Figure 2-58. *Start boosters to assign roles and create service instances*

4. To enable the Integration Suite, you must first configure the subaccount. Provide the following information to proceed.

   - Subaccount—Trial

   - Org—Default

   - Space—Dev (For different environments, you can name your space accordingly—dev, test, prod)

   After providing all the details, click Next (see Figure 2-59) and move to the next screen.

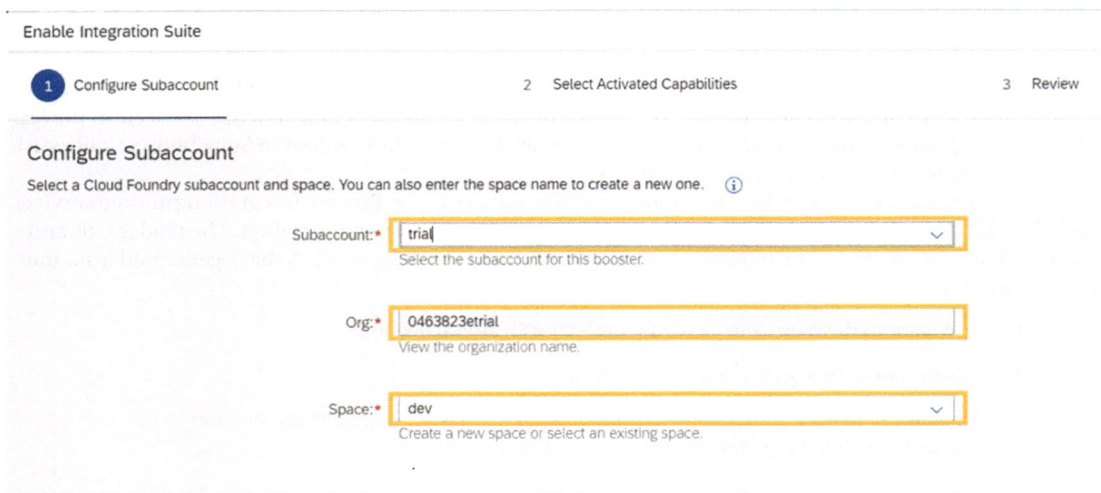

Figure 2-59. *Configure the Subaccount by providing the basic info*

5. To provision the capabilities, select the capabilities and click Next (see Figure 2-60).

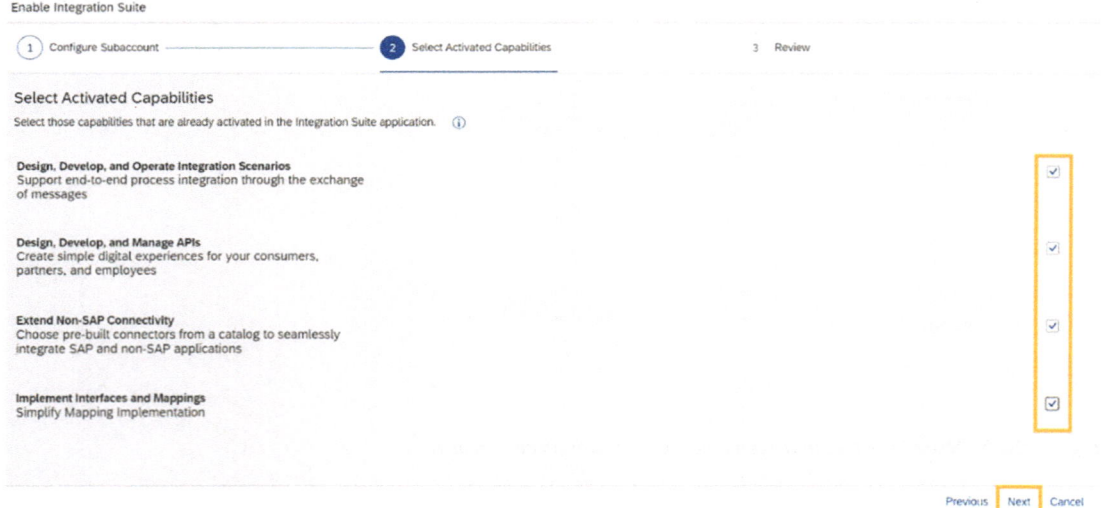

Figure 2-60. *Select the capabilities to provision*

6. We will see the progress status with some ongoing processing.

7. Within a few minutes, you will see the success status of the booster, in which the roles and services instances are available for activated capabilities of the subaccount.

8. You can navigate to the Subaccount after successfully assigning the roles and instances; see Figure 2-61.

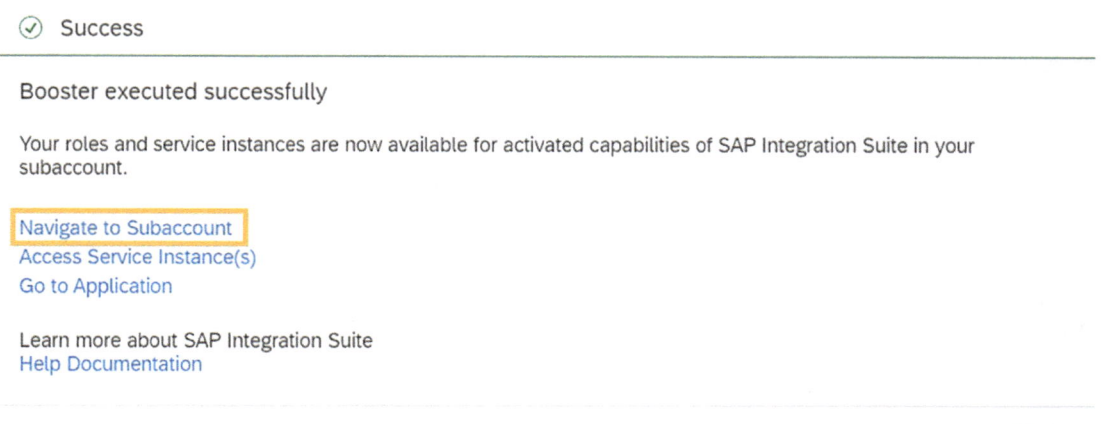

⊘ Success

Booster executed successfully

Your roles and service instances are now available for activated capabilities of SAP Integration Suite in your subaccount.

Navigate to Subaccount
Access Service Instance(s)
Go to Application

Learn more about SAP Integration Suite
Help Documentation

Close

Figure 2-61. *Navigate to Subaccount*

2.9 Setting Up the Process Integration Runtime (Optional)

This step is optional but recommended because this runtime is necessary in the Integration Suite when you're developing cloud integration scenarios that consume HTTPs endpoints as the sender and you want to expose web services.

1. To enable this service, choose Services ➤ Service Marketplace. Search for the Process Integration Runtime and click the package, as shown in Figures 2-62 and 2-63.

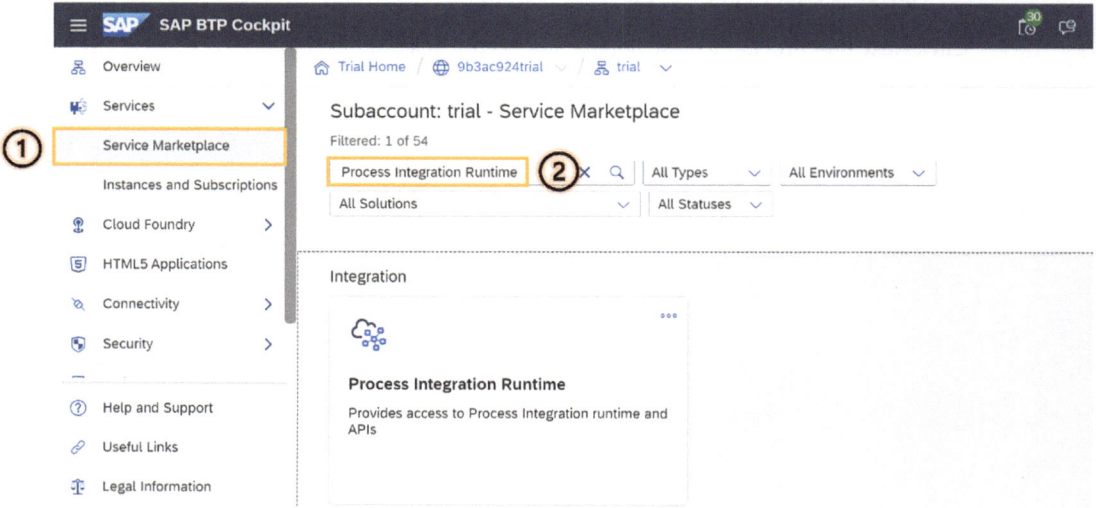

Figure 2-62. *Provisioning the Process Integration Runtime*

2. Go to the Process Integration Runtime tile and click the Create button to subscribe to the package (see Figure 2-63).

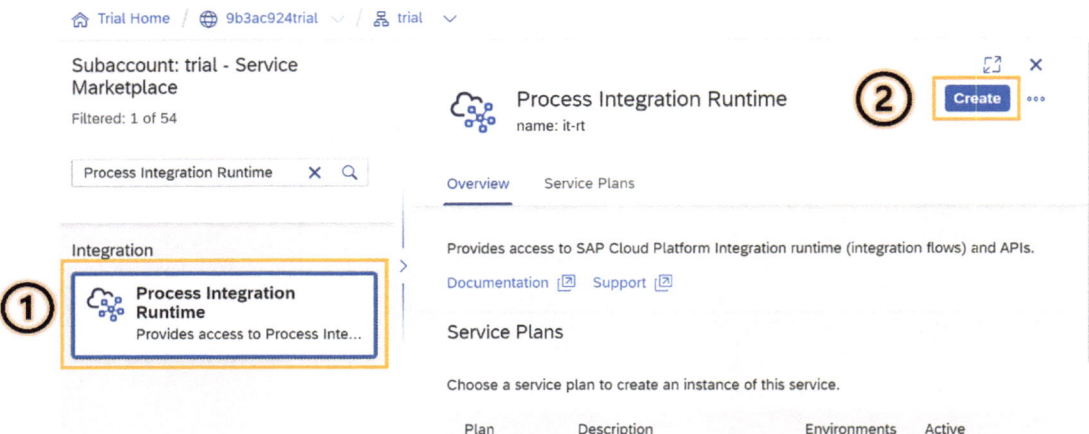

Figure 2-63. *Subscribe to the Process Integration Runtime*

3. For a new instance or subscription of the Process Integration Runtime package, provide the following details.

- Service—Process Integration Runtime

- Plan—Integration-flow

- Runtime Environment—Cloud Foundry

- Space—Dev or a custom created space

- Instance Name—ProcessIntegrationRuntime

Click Next after providing these details (see Figure 2-64).

New Instance or Subscription

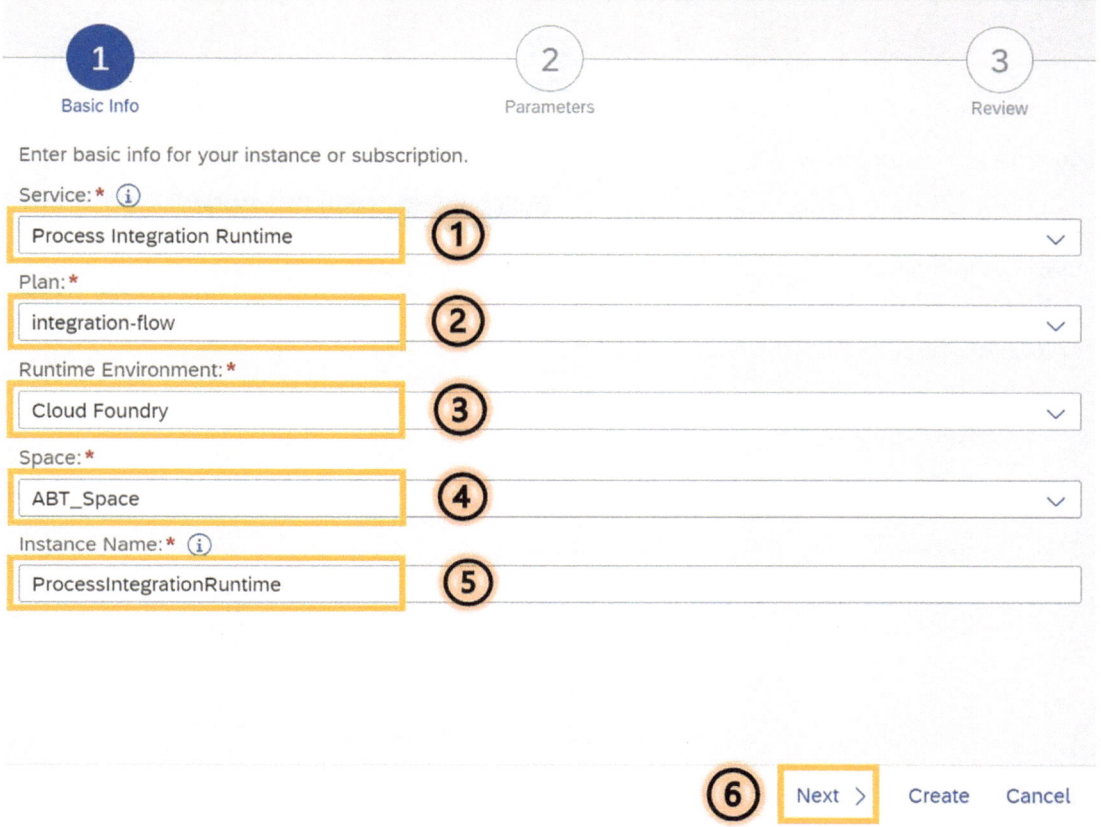

Figure 2-64. *Basic info to create a new instance for Process Integration*

4. This next step is very important. You have to add the JSON format code. Specify the parameters value as follows:

```
{
  "roles":[
    "ESBMessaging.send"
  ]
}
```

Enter this code on the Parameters editing area by copying and pasting it. After selecting Next, choose Create (see Figure 2-65).

New Instance or Subscription

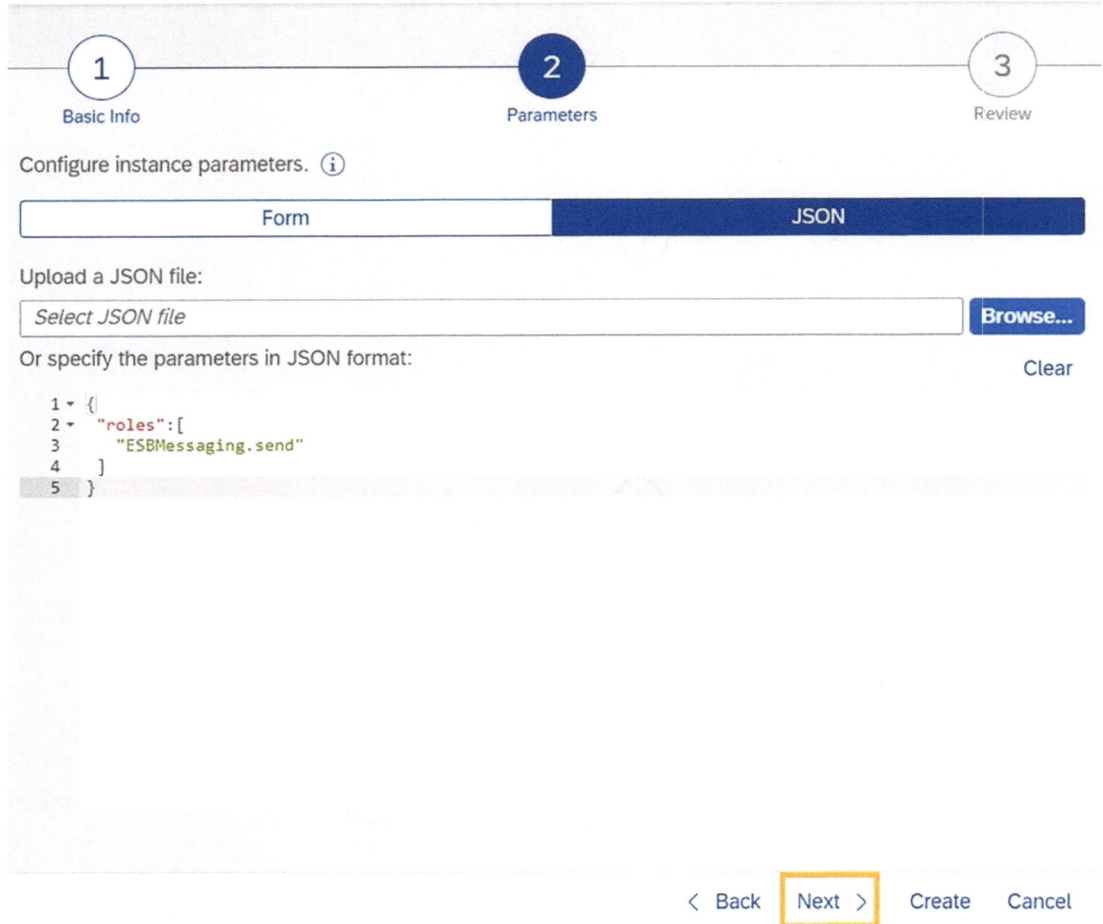

Figure 2-65. *Specify the parameters in JSON format*

5. You can see on the home page that the instance has been created (see Figure 2-66).

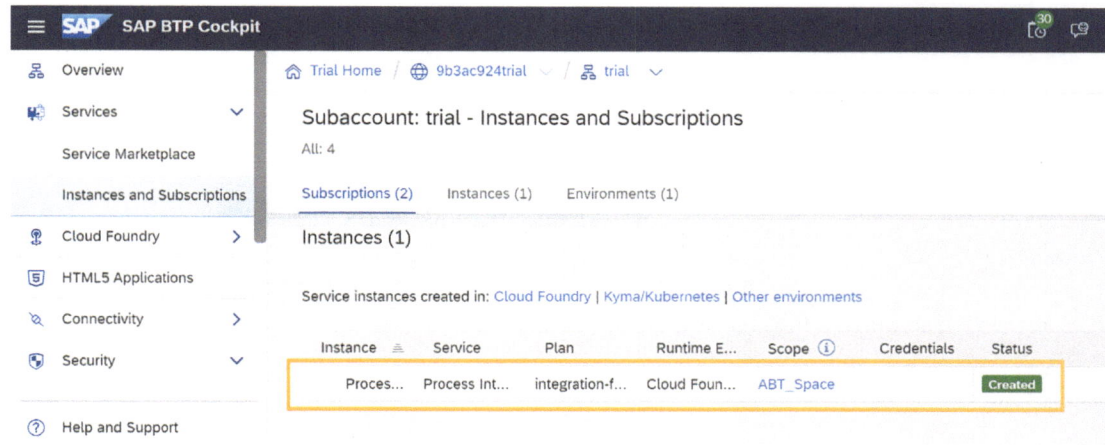

Figure 2-66. *The ProcessIntegrationRuntime instance has been created*

2.9.1 Creating a Service Key

1. In the `ProcessIntegrationRuntime` instance, click Create in the Service Key tab (see Figure 2-67).

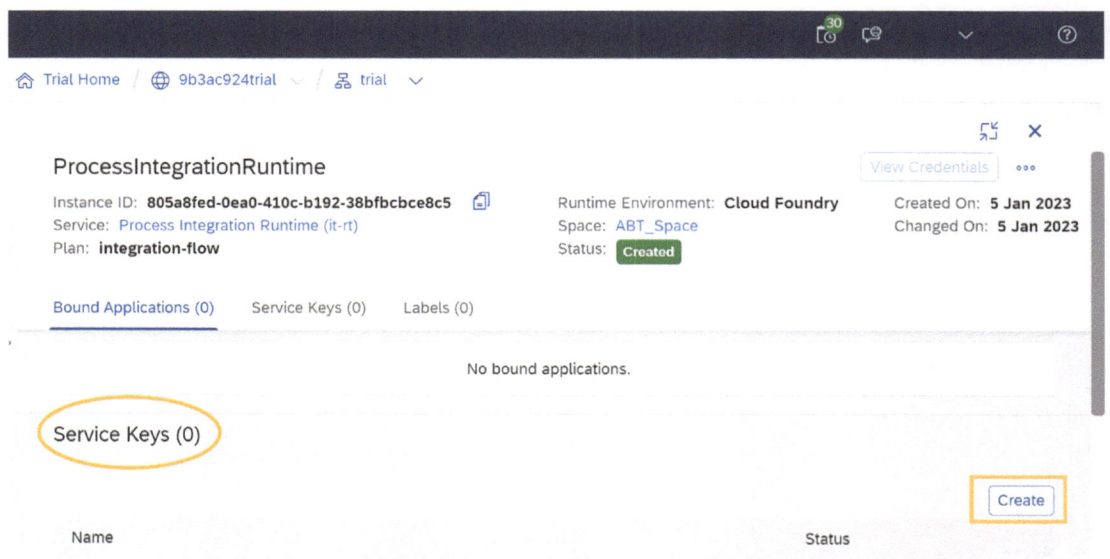

Figure 2-67. *Create a service key*

2. You will see the popup where you must name the service key without adding any JSON code. Click Create (see Figure 2-68).

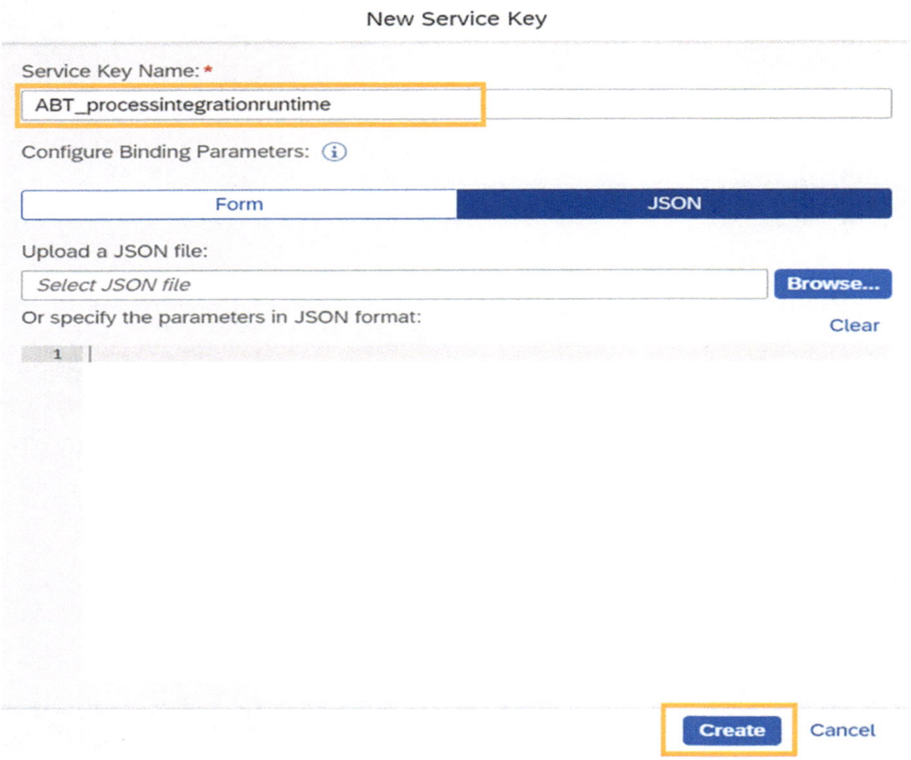

New Service Key

Service Key Name: *

ABT_processintegrationruntime

Configure Binding Parameters: ⓘ

Form	JSON

Upload a JSON file:

Select JSON file	Browse...

Or specify the parameters in JSON format: Clear

1 |

Create   Cancel

Figure 2-68. *Name the service key*

3. To see the value of the key, click the service key. The JSON code, as shown in Figure 2-69, will appear in the screen.

Credentials

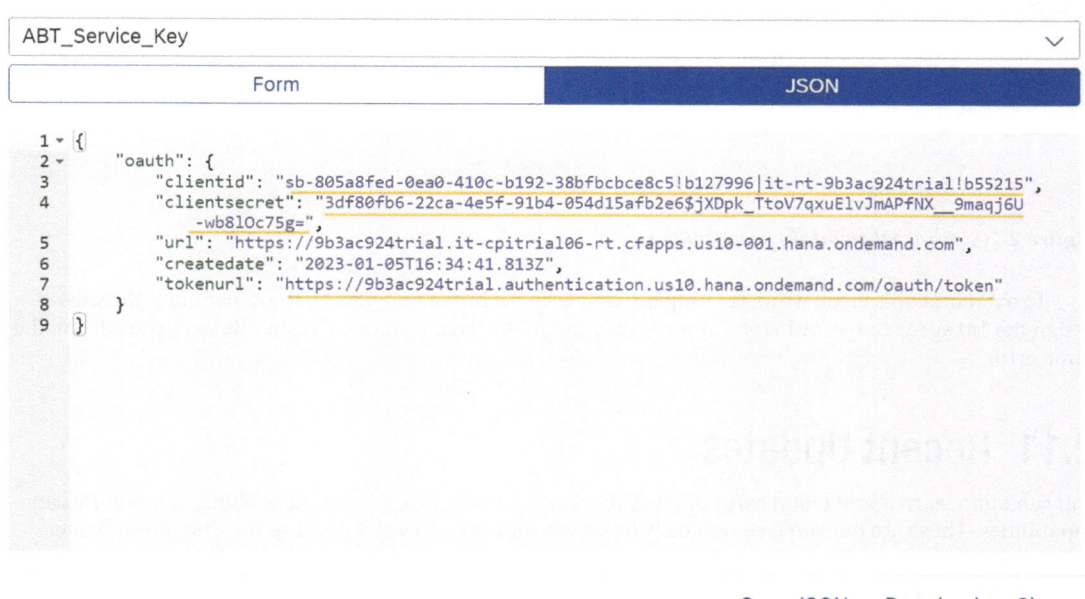

Figure 2-69. *The ClientID and Client secret of the trial tenant*

4. This ClientID is the OAuth 2.0 username, and the Client secret is the password to fetch the bearer token. These credentials are used when you test your integration flows in Postman, SOAP UI, and so on.

For example, say you are integrating an HR application or a Sales Order application that needs to send orders to cloud integration via HTTPs. You will share the client ID and the secret generated for OAuth 2.0 authentications of your webservices/APIs.

2.10 Common Errors (Installation)

If you do not have access to the requested page, as shown in Figure 2-70, this might mean you are supposed to go to the Integration Suite home page to add the capabilities.

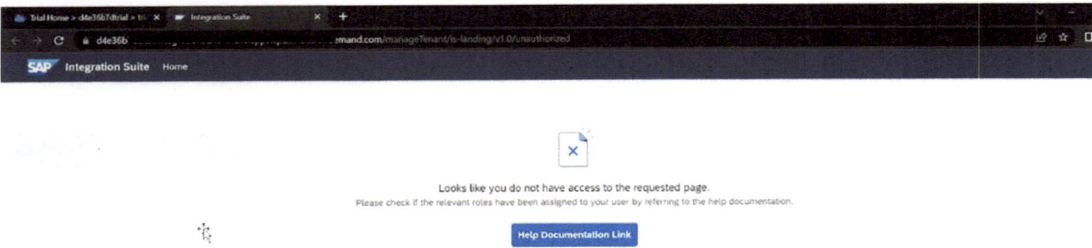

Figure 2-70. *Error when adding capabilities*

To overcome this error, while creating the trial account but before adding the capabilities directly, assign the `Integration_Provisioner` Role Collection to the user. If you try to skip this step, you will get the same error.

2.11 Recent Updates

You can often learn about changes/updates to the SAP Integration Suite, as well as about new features and capabilities. These are delivered periodically (every six months or a year) because the Integration Suite solution is SAP managed True iPass (cloud-based). You can stay informed about the platform updates that SAP is delivering periodically.

This following screen popup is presented to users each time they log in to Integration Suite, and it can be disabled by clicking the checkbox, as shown in Figure 2-71.

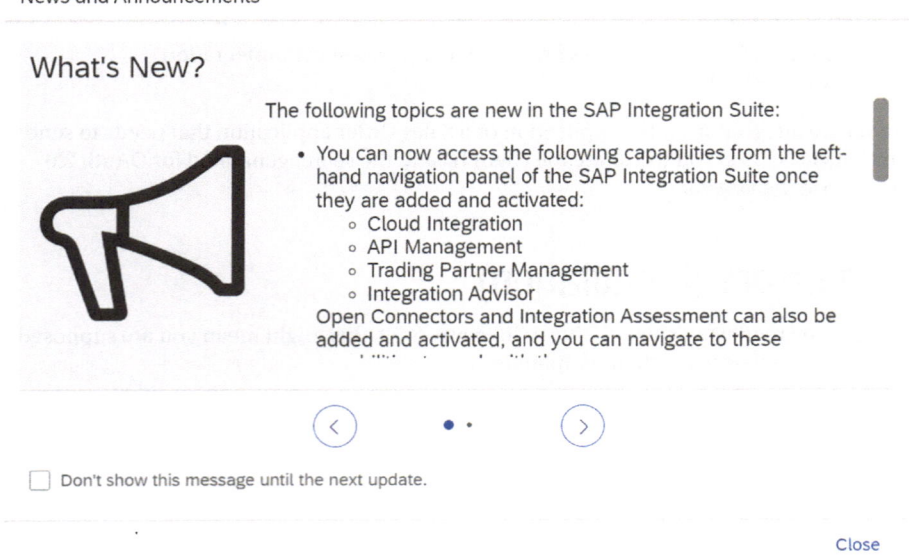

Figure 2-71. *What's new in the SAP Integration Suite*

2.12 Accessing the Integration Suite: Bookmark URLs

If you followed all the steps for the trial account setup correctly and assigned the right role collection, your SAP BTP integration suite is fully provisioned and ready to use for further learning.

After successfully installing the SAP BTP Integration Suite, you can access all the capabilities over the browser.

Depending on your roles, personas, and job description, you can access one or many of the capabilities of the Integration Suite. If you have access to the Integration Suite, you can navigate to the capabilities from the home page. If your role is defined to one capability, like cloud integration or API management, you can navigate to the capabilities separately by sharing the link to the application with your team individually. Every organization has different URLs depending on the organization, space, and service instance created.

For example, your trial account has generated URLs for the following Integration Suite capabilities.

1. Cloud Integration

    • Discover Integrations—https://4d5b5bf7trial.integrationsuite-trial01.
      cfapps.us10-001.hana.ondemand.com/shell/integration.

    • Create Integrations—https://4d5b5bf7trial.integrationsuite-trial01.
      cfapps.us10-001.hana.ondemand.com/shell/design

2. Trading Partner Management—https://4d5b5bf7trial.integrationsuite-
    trial01.cfapps.us10-001.hana.ondemand.com/shell/tpm

3. Integration Advisor

    • Discover Type Systems—https://4d5b5bf7trial.integrationsuite-trial01.
      cfapps.us10-001.hana.ondemand.com/shell/typesystems

    • Create MIGs—https://4d5b5bf7trial.integrationsuite-trial01.cfapps.
      us10-001.hana.ondemand.com/shell/migs

    • Create MAGs—https://4d5b5bf7trial.integrationsuite-trial01.cfapps.
      us10-001.hana.ondemand.com/shell/mags

2.13 Summary

In this chapter, you learned about the main characteristics and advantages of the SAP Integration Suite, including its support for hybrid integration situations, management and security of APIs, and end-to-end visibility and monitoring of integrations. There are many capabilities of the SAP Integration Suite, such as the SAP Cloud Platform Integration, the SAP API Management, the SAP Integration Advisor, Trading Partner Management, Migration Assessment, Open Connectors, Migration Assessment, and Integration Assessment.

This chapter also covered the functionalities and the Role Collections that are necessary to assign to the users. The chapter also dug deep into the roles and Role Collections in the SAP BTP Cockpit. You learned how to set up the trial account for SAP Integration Suite.

In upcoming chapters, you will explore the one of the most important capability of SAP Integration Suite i.e., SAP Cloud Integration. We will go through in detail about the Cloud Integration in the upcoming chapters.

CHAPTER 3

■ ■ ■

SAP Cloud Integration: Features and Connectivity

This chapter covers connectivity and introduces SAP Cloud Integration. The chapter examines the salient aspects of SAP Cloud Integration and explains how it works with SAP Integration Suite. It also examines SAP Cloud Integration's features in more detail; you get a glimpse of its web user interface as well.

You learn about all of SAP Cloud Integration's features, including its compatibility with Process Orchestration and its integration capabilities. Then, the chapter gets into the specifics of SAP Cloud Integration, providing you with a rundown of the Web UI and a real-world illustration of how to build an interface utilizing tools like the Content Modifier, Outbound OData Channel, and Groovy Script.

Next, the chapter covers the essential aspects of adapters—senders and receivers—and explains how to configure them using practical examples. You also learn about connectivity options and communication security, including connecting inbound and outbound systems, supporting protocols, and setting up inbound and outbound communication channels using various authentication methods, such as OAuth, basic authentication, and client certificate authentication.

To provide you with comprehensive knowledge of how to set up and configure SAP Cloud Integration for your organization, the chapter highlights the important aspects. This chapter is an essential resource for anyone looking to gain knowledge and understanding of SAP Cloud Integration and its capabilities.

Having read the first two chapters, you should now understand system integration concepts and what the SAP BTP offers. The Integration Suite is a platform as a service (iPaaS) that offers many capabilities, as discussed in the previous two chapters.

3.1 What Is Cloud Integration?

A cloud-based integration platform called SAP Cloud Integration, also known as SAP Cloud Platform Integration (CPI), enables businesses to link and integrate their SAP systems and applications with other cloud-based services and systems. This enables companies to take advantage of the benefits of SAP with cloud computing, such as increased scalability, flexibility, and cost savings.

SAP Cloud Integration provides various tools and features for connecting, managing, and automating data flow between systems and services. For integration with well-known cloud services such as Salesforce, ServiceNow, and Google Cloud, it offers built-in connections. Additionally, it enables the development of unique connectors to integrate with systems not supported by built-in connectors.

Neo and Cloud Foundry (CF) environments both support the operation of this service. Integration of content artifacts created in the Neo environment are compatible with the CF environment, but there are some restrictions.

Cloud Integration enables businesses to link their SAP and applications to other cloud-based services. The next section discusses the features of SAP Cloud Integration.

© Jaspreet Bagga 2023
J. Bagga, *A Practical Guide to SAP Integration Suite*, https://doi.org/10.1007/978-1-4842-9337-9_3

3.1.1 Key Features of SAP Cloud Integration

SAP Cloud Integration is an enterprise-level, cloud-based solution that allows organizations to integrate their business processes, data, and applications across various systems and platforms.

Some of the key features of SAP Cloud Integration include:

- **Prebuilt connectors**—Provides a range of prebuilt connectors for integration with popular cloud services, such as Salesforce, ServiceNow, and Google Cloud. These connectors can easily connect and integrate with these services without extensive IT resources or expertise.

- **Custom connectors**—Enables the development of unique connectors for syncing with other systems not supported by the built-in connectors.

- **Integration flows**—Provides a user-friendly, web-based interface for designing, deploying, and managing integration flows. It enables you to build, test, and execute integration flows without needing in-depth programming skills.

- **Mapping and transformation**—To convert, enrich, and transform the data between various systems, SAP Cloud Integration offers mapping and transformation capabilities.

- **Monitoring and management**—It has monitoring and administration features that can assist you in optimizing and troubleshooting your integration flows.

- **Security**—Provides a range of security features, such as authentication, authorization, and encryption, to ensure that integration flows are secure.

- **Data governance and quality**—To validate the data, guarantee data completeness, and enhance data quality for improved decision-making, it offers data governance and quality features.

- **Scalability**—Built on a highly scalable architecture, SAP Cloud Integration can support numerous concurrent users and massive data volumes.

- **Support for various integration patterns**—It offers several integration patterns, including real-time integration, data migration, and file transfer, which can be applied to several integration scenarios.

The following essential components make up SAP Cloud Integration:

- Core runtime for message processing, transformation, and routing between the customer systems involved. It ensures that information concerning to various clients connected to Cloud Integration is segregated. For example, leveraging Cloud Integration for business-to-business scenarios is crucial.

- Support for connectivity is included out of the box (for example, IDoc, SFTP, SOAP/HTTPS, SuccessFactors, OData, and HTTPS).

- Security elements include certificate-based communication and content encryption.

Without needing hardware or integration expertise on the part of the customer, SAP can make preconfigured, ready-to-use prepackaged integration content available upon purchase. This considerably reduces resource consumption and the lead times for integration projects.

By using the following methods, cloud integration provides complete freedom in how messages can be transferred between customer systems:

- Utilizing integration patterns that have already been set up: These integration patterns offer various configuration choices, such as routing rules, for setting up the data flow between participants.

- Utilizing a range of connecting methods: This refers to a group of adapters (or endpoint types) that enable users to connect to SAP Cloud Integration using various communication protocols.

Now that you are aware of the features of SAP Cloud Integration and some of its essential components, the next section explains how SAP is compatible with Process Orchestration.

3.1.2 Compatibility with Process Orchestration

The on-premises runtime of SAP Process Integration allows you to run integration content well. You can choose the SAP Process Integration version for the Cloud Integration Web UI so that the integration content designer's feature set is tailored to the destination runtime's capabilities.

You can learn more about Process Orchestration. in Chapter 1; see the section entitled "Integration Technologies."

You have seen how SAP Cloud Integration and Process Orchestration are connected. There are some integration capabilities available in SAP Cloud Integration. The next section focuses on these integration capabilities.

3.1.3 Integration Capabilities

Numerous integration capabilities exist that specify various methods for processing messages on the integration platform and exchanging them between the sender and receiver systems.

SAP Cloud Integration offers many integration patterns that apps can use to be integrated with one another. By adding a specific integration flow step or a mix of several integration flow steps to an integration flow when using SAP Cloud Integration, you can describe the desired integration pattern.

The available integration capabilities are listed in Tables 3-1 through Table 3-5, organized according to the many types of associated integration flow steps.

Table 3-1 defines the first integration capability provided by SAP Cloud Integration, i.e., message transformations. Message transformations allow users to manipulate and transform messages as they flow through integration scenarios.

Table 3-1. *Message Transformation*

Feature	Description
Mapping	Converts the sender's data structure and format into a structure and format that the recipient can understand.
ID Mapping	Connects a target message ID to the source message ID. This feature allows you to create message-processing scenarios with precisely one message processing.
Content Modifier	Alters the message's body or the header to change the content of an incoming message. The message exchange's message body, the message header, and properties section can all be read from and written to by the content modifier. In this manner, a message's content can be adaptably changed and readied for a receiver or additional processing processes.
XML Modifier	Removes external DTDs and XML declarations to alter the content of an incoming message.
Converter	Changes the format of an input message. Here are the following converters: **XML to JSON**: Converts XML messages to JSON messages. **JSON to XML**: Converts JSON messages to XML messages. **XML to CSV**: Converts XML messages to CSV messages. **CSV to XML**: Converts CSV messages to XML messages. **XML to EDI**: Converts an XML message to an Electronic Data Interchange (EDI) message. **EDI to XML**: Converts an EDI message (in EDIFACT or ASC-X12 format) to an XML message.
Decoder	To recover the original data, it decodes the incoming message (for example, if a base64-encoded message has been received). **Base64 Decode**: Decodes the content of base64-encoded messages. **GZIP Decompress**: GNU ZIP is used to decompress the message's content (GZIP). **ZIP Decompress**: Uses ZIP to decompress the message's content (only ZIP archives with a single entry are supported). **MIME Multipart Decode**: Converts a MIME multipart message into an attachment-containing message.
Encoder	Secures any sensitive content during transit over the network by encoding the message using an encoding method. **Base64 Encode**: Base64 is used to encode the message's content. **GZIP Compress**: GNU zip is used to compress the message's content (GZIP). **ZIP Compress**: Uses ZIP to compress the message's content (only ZIP archives with a single entry are supported). **MIME Multipart Encode**: Converts the message's content into a MIME multipart message.
Filter	Extracts a particular node from the incoming message using an XPath expression to filter information.
Message Digest	Creates a payload digest, or partial payload digest, and puts the result in the message header.
Script	Executes custom Groovy or JavaScript to process messages.

Table 3-2 defines the external systems and subprocesses provided as an integration capability by SAP Cloud Integration. These external systems allow users to establish communication between SAP Cloud Integration and other applications.

Table 3-2. *Calling External Systems and Subprocesses*

Feature	Description
Request-Reply	Receives an answer after making a synchronous call to an external receiver system.
Send	When a response is not anticipated, it calls an external receiver system.
Content Enricher	It calls a remote system, uses its resources, and combines the content it returns with the original message.
Poll Enrich Step	Collects information from an outside source and adds it to the original message.
Process Call	A local integration process is called. A container for a distinct subprocess that can be invoked from the main process is defined by a local integration process. A complex message-processing sequence can be divided into smaller pieces using local integration techniques.
Looping Process Call	Calls a looping local integration procedure.
Idempotent Process Call	Determines whether a message ID has already been successfully processed and records the results in the idempotent repository. The called subprocess can be skipped, or the message can be flagged as a duplicate if there is a duplicate execution with the same message ID (for instance, if the sender system retries). The duplicate in the subprocess can then be dealt with however you choose.

One of SAP Cloud Integration's most important features is routing integration, as listed in Table 3-3. It gives customers a flexible and effective means for transferring data across various applications and systems. Users can set up integration flows for data routing based on predetermined criteria, such as message content, sender or receiver systems, or other factors, using routing integration.

Table 3-3. Routing

Feature	Description
Router	Sends a message to a receiver or recipients.
	Additionally, SAP Cloud Integration provides routing based on the message's content (content-based routing). For example, the tenant can identify a message's specific field value and deliver the message to the receiver participant that responds to requests from the sender participant.
Multicast	Sends an identical message to multiple recipients. There are the following Mulciast:
	Parallel multicast: Starts simultaneous message delivery to all receiver nodes.
	Sequential multicast: Refers to the order in which message transmission to recipients begins.
Splitter	Sends a receiver a composite message that has been divided into a number of separate messages.
	There are the following types of Splitters:
	General splitter: Splits a composite message made up of n messages into n separate messages using a general splitter. The same components encircled the composite message also around each individual message.
	Iterating splitter: Splits a composite message into a number of smaller messages iteratively without replicating the composite message's wrapping components.
	PKCS#7/CMS splitter: Splits a message that is PKCS#7 Signed Data and has both a signature and content (and breaks down the signature and content into separate files).
	IDoc splitter: Splits a composite IDoc message into a number of separate IDoc messages using the composite IDoc message's enveloping elements.
	EDI splitter: The EDI splitter validates and acknowledges the incoming message while dividing a bulk EDI message into a number of separate messages.
	Zip splitter: Creates separate files from an incoming ZIP archive file.
Join	Creates a single message by combining messages from many routes.
	This function is combined with the Gather function. Join does not change the content of the messages; it merely brings the messages from several routes together.
Gather	Merges messages from many routes (into a single message), with the possibility to provide specific methods for doing so.

With the help of SAP Cloud Integration, businesses can establish seamless connectivity between their enterprise systems and receive real-time insights into their data. SAP Cloud Integration offers extensive capabilities for storing, processing, and integrating data from a variety of sources, as explained in Table 3-4.

Table 3-4. *Storing Data During Processing*

Feature	Description
Persist Message	A message's payload is kept so that you can retrieve it and examine it at a later time.

Data Store Operations	Momentarily saves messages for processing later.
	Operations that are supported include:
	
	`SELECT`
	`GET`
	`WRITE`
	`DELETE`

Write Variables	Values for variables that must be specified for message processing.

A key integration capability of SAP Cloud Integration is Protecting Messages, which are explained in Table 3-5. These are intended to protect the security and privacy of data when it is sent between various systems and apps. Users can utilize a variety of security tools, such as encryption, digital signatures, and authentication processes, with Safeguarding Messages to safeguard important information and ensure that it is not compromised during transmission.

Table 3-5. *Protecting Messages*

Feature	Description
Encryptor	Encrypts a message's content. Standards that are supported include: PGP PKCS#7/CMS Enveloped Data and Signed Data
Decryptor	Decrypts a message's content. Standards that are supported include: PGP PKCS#7/CMS Enveloped Data and Signed Data
Signer	Signs a message. Standards that are supported include: PKCS#7/CMS Enveloped Data and Signed Data Digital XML Signature
Verifier	Checks a message. Supported standards include: PKCS#7/CMS Enveloped Data and Signed Data Digital XML Signature

You should now be aware of the integration capabilities of SAP Cloud Integration that are used for message transformations, calling external systems, routing, storing data during processing, and protecting messages. To learn more about these capabilities, you must first learn about the Cloud Integration UI, which is covered in the next section.

3.2 Overview of the SAP Cloud Integration Web UI

The SAP Cloud Integration Web UI is a web-based user interface that enables users to manage and keep track of integration flows on the SAP Cloud Integration. It offers a variety of features and tools to assist users in designing, implementing, and managing integration flows between various systems.

You can develop integrations all at once using the SAP Cloud Integration Web UI. Keep in mind that the Web UI URL ends with /itspaces. The sections of the Web UI are as follows:

- **Discover**—The SAP Cloud Integration Web UI's Discover section gives users access to a variety of built-in integration material that can facilitate the quick and simple creation of integration processes between various systems. You can discover SAP-provided preconfigured integration content here that you can use right out of the box or customize to your needs.

- **Design**—Designing and creating integration processes across various systems takes place primarily in the Design part of the SAP Cloud Integration Web UI. To assist customers in designing, setting up, and testing their integration flows, it offers a variety of tools and functionalities. It includes the environment for graphical integration flow modeling.

- **Monitor**—The SAP Cloud Integration Web UI's Monitor feature offers a thorough overview of the integration flows that are active at any one time. It enables users to manage and debug integration flows by keeping track of their status, performance, and problems. You can keep an eye on your integration flow here. Additionally, you can manage other artifacts that must be installed on your tenant in addition to your integration flows.

Now that you have a good understanding of the SAP Cloud Integration Web UI, the next section takes a closer look at a practical example of interface development using the platform. In the next section, you explore a sample interface development scenario and see how to use the SAP Cloud Integration Web UI to create and configure integration flows.

By following this practical example, you will gain a deeper understanding of how the SAP Cloud Integration platform works and how it can be used to connect different systems and applications. The section walks through the steps involved in developing an interface and shows you how to leverage the various features and tools of the SAP Cloud Integration Web UI to build powerful and effective integration flows.

3.3 Sample Interface Development: Practical Example

This is a very simple test to verify that your SAP Cloud Integration is working as expected. You do not need a receiver system to perform this test.

In the following scenario, you will access an OData API and obtain information about a product. The result is written into the message processing log, which you can directly inspect with the message monitoring application.

In the example shown in Figure 3-1, you will develop the following integration flow. To make it easy, you will not configure a sender system. This will save the effort of setting up the sender system and connecting it to SAP Cloud Integration. Instead, message processing is triggered by the Timer event, as shown in Figure 3-1.

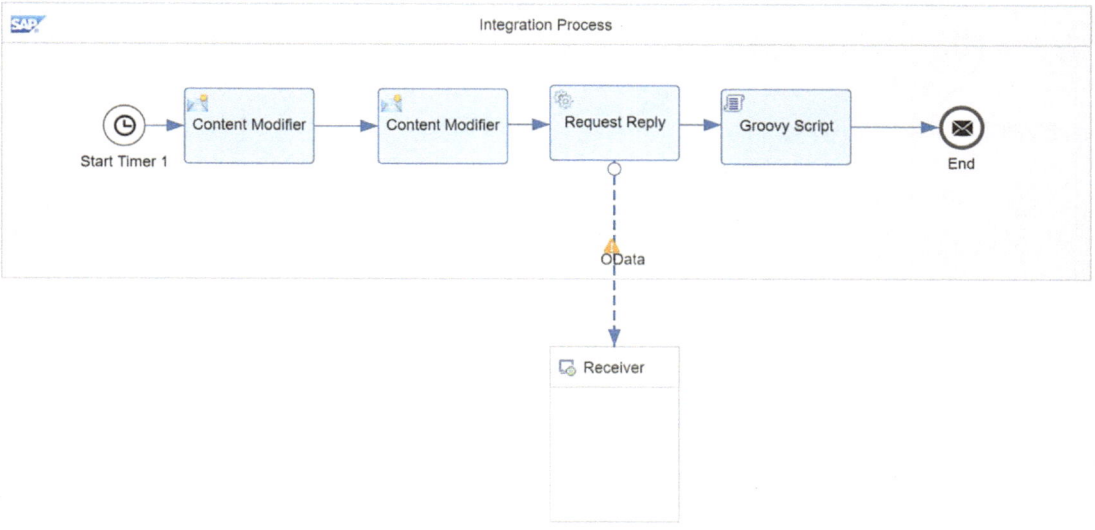

Figure 3-1. *Sample integration flow*

This is how the integration flow will process the message at runtime:

1. The Timer event triggers the processing of the message.

2. The first Content Modifier step creates a message with only one element, a `productIdentifier`.

3. The second Content Modifier creates a message header and writes the actual value of the `productIdentifier` element into it.

4. The Request-Reply step passes the message to an external data source and retrieves data (about products) from there.

5. The ODATA API provides the details of one specific product (according to the product identifier in the inbound message).

6. The Groovy Script step logs the message's payload (that is, it writes the message content into the message processing log).

The next sections explain which elements are used in this scenario.

3.3.1 Start Timer

The SAP Cloud Integration Start Timer functionality enables users to activate integration processes at predetermined intervals. With the help of this effective technology, businesses can plan the execution of integration flows and automate their integration operations.

Start Timer is the first step in the interface development process. Follow these steps:

1. Open the integration flow in Edit mode. Delete the Sender by clicking the Delete icon; similarly, delete the Start Message event.

2. From the palette, select the Events entry and scroll down to bottom. You will see the Timer event, as shown in Figure 3-2.

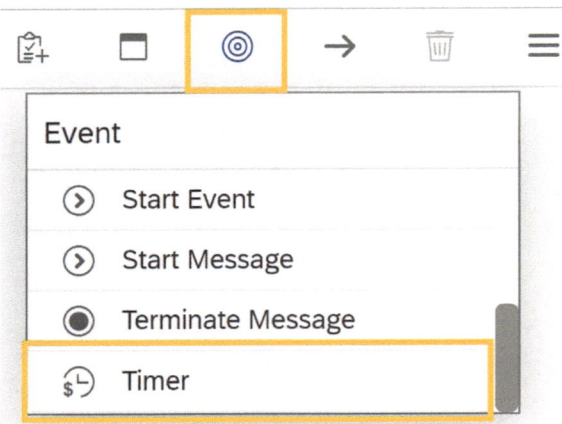

Figure 3-2. *Choose the Timer event*

3. Select the Timer event in place of the Start message. Connect the Timer and the End Message by clicking the arrow icon and dragging and dropping it over the End Message.

4. The results are shown in Figure 3-3.

Figure 3-3. *Start Timer 1*

3.3.2 Content Modifier

A component in SAP Cloud Integration called the Content Modifier enables users to alter, adapt, and pass data as it moves through integration channels. Modifying the message content, headers, and properties in real time is a configurable step that can be introduced to an integration flow.

Add a Content Modifier step to define the Message Body. Follow these steps to do so:

1. Place the Content Modifier in the flow after the Timer Start event. To do so, go back to the palette and select the Transformations; then click the Content Modifier. Place the Content Modifier between the Start Timer and the End Message.

2. Go to the Message Body tab in the Content Modifier Properties section and type the following string in the entry field, as shown in Figure 3-4:

```
<root>
<productIdentifier>HT-1080</productIdentifier>
</root>
```

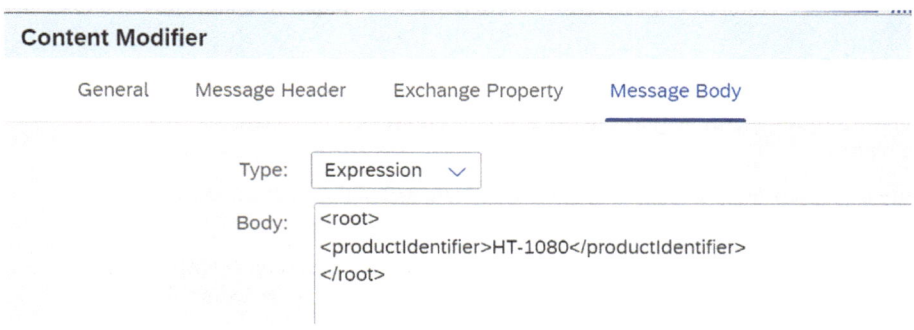

Figure 3-4. *Message body of the Content Modifier*

3. Add another Content Modifier to your model to define a header, which will be used in a later step to filter data from the external source. Similar to the first Content Modifier, drag and drop another near it.

4. In the Properties section of the second Content Modifier, go to the Message Header tab and choose Add, as shown in Figure 3-5.

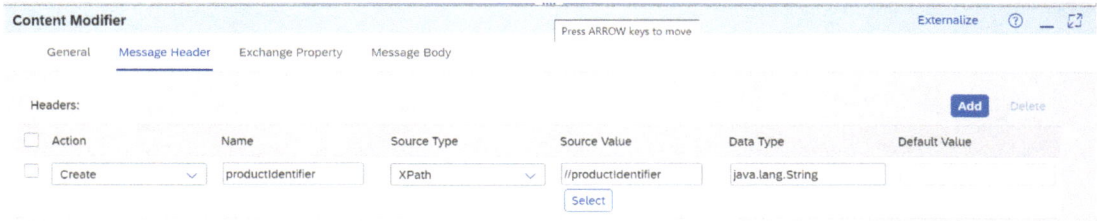

Figure 3-5. *Message header of the Content Modifier*

5. Specify the following parameters in the message header of the Content Modifier, as shown in Figure 3-5.

- Action—Create

- Name—productIdentifier

- Source Type—XPath

- Source Value–//productIdentifier

- Data Type—java.lang. String

Now that you have set up the Content Modifier to define the message header, you can establish an outbound communication to fetch the data from the external source (Northwind).

3.3.3 Outbound OData Channel

To call the external data source, add a Request-Reply step to the integration flow model and connect this step to the external system using an OData channel.

To send a request message to the external ODATA API (to retrieve the required data), you need to do the following:

- Create a Request-Reply step.

- Connect the Request-Reply step to a Receiver shape and select the OData adapter type.

- Configure the OData adapter to specify how the OData API of the external service should be called (to define query options, for example).

    1. Place the Request-Reply Step in the flow and connect it to the receiver with the OData adapter type. To do this, from the palette, choose Call ➤ External Call ➤ Request Reply and place it near the second Content Modifier.

    2. Connect the Request Reply to the receiver using the OData adapter. Click Request Reply and you will see the arrow icon pointing right. Click that arrow and drag it to the receiver.

    3. You will be asked to select the adapter type from the list. Select OData ➤ OData V2. You will find that the Request-Reply will be connected to the receiver using the OData adapter, as shown in Figure 3-6.

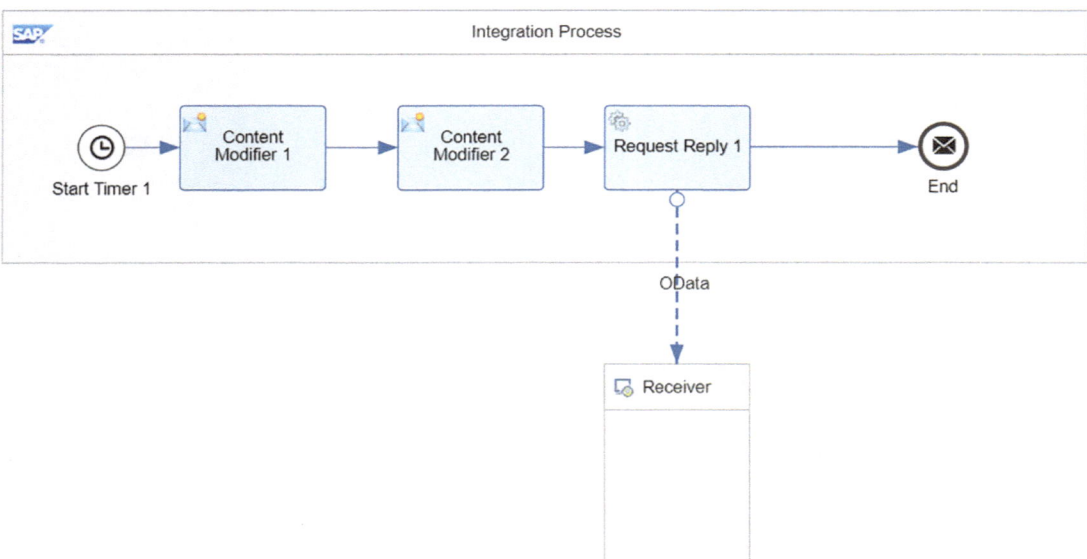

Figure 3-6. *Integration flow*

    4. In the Connection tab of the OData adapter, enter the following address: refapp-espm-ui-cf.cfapps.eu10.hana.ondemand.com/espm-cloud-web/espm.svc, as shown in Figure 3-7.

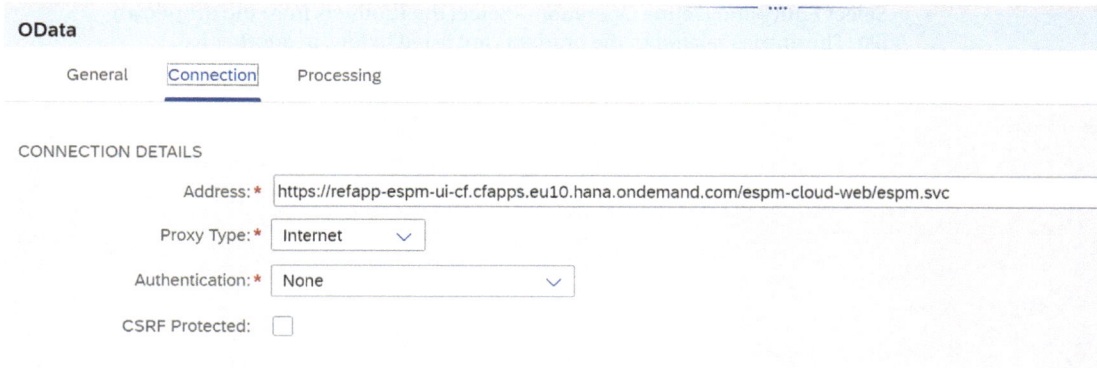

Figure 3-7. *OData Connection tab*

5. In the Processing tab, configure the resource path by making the ProductId the unique Identifier.

6. In the Resource path of the Processing tab, click Select.

7. A wizard will open and ask for three steps. The three steps have the following additional steps:

   • Connect to the System—This is Step 1, in which you have to connect to the system. Check all the entries, as shown in Figure 3-8, and then click Step 2.

Model Operation

① Connect to System ──────────── ② Select Entity & Define Operation

1. Connect to System

Connection Source : *	Remote
Address : *	https://refapp-espm-ui-cf.cfapps.eu10.hana.ondemand.com/es...
Proxy Type : *	Internet
Authentication : *	None

Finish   Cancel

Figure 3-8. *Connect to the system*

- Select Entity and Define Operation—Select the Products from the dropdown list. The entries related to the products are listed below, in another list. Choose Select All Entries, as shown in Figure 3-9, and then click Step 3.

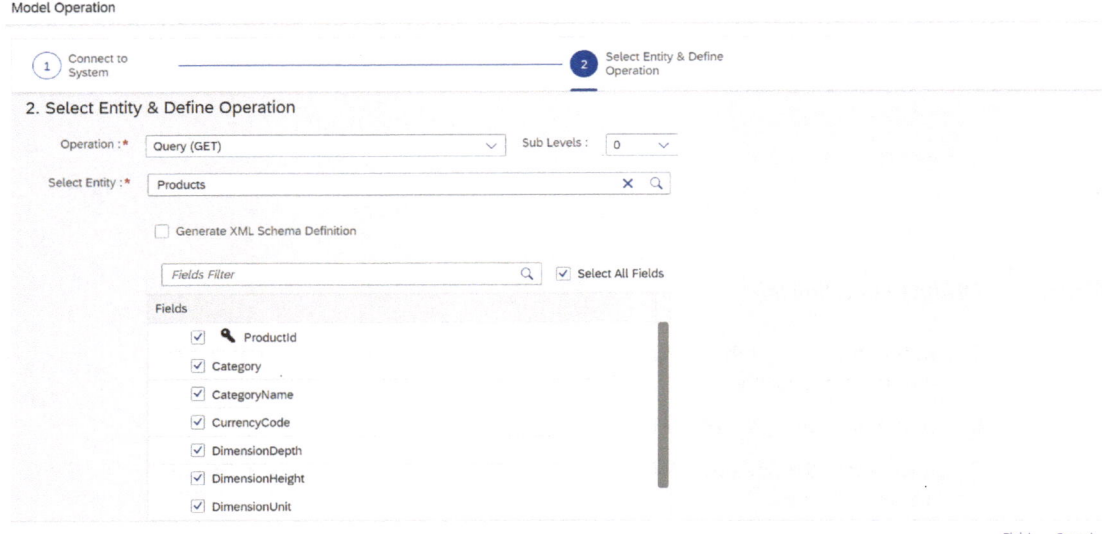

Figure 3-9. *Select the entities*

- Configure Filtering and Sorting—This is Step 3, in which you have to configure your filtering and sorting parameters. In the Filter By tab, select the unique identifier as ProductID and mark it as equal to the ${header. ProductId}. This is the ID that you give in the header. Click Finish, as shown in Figure 3-10.

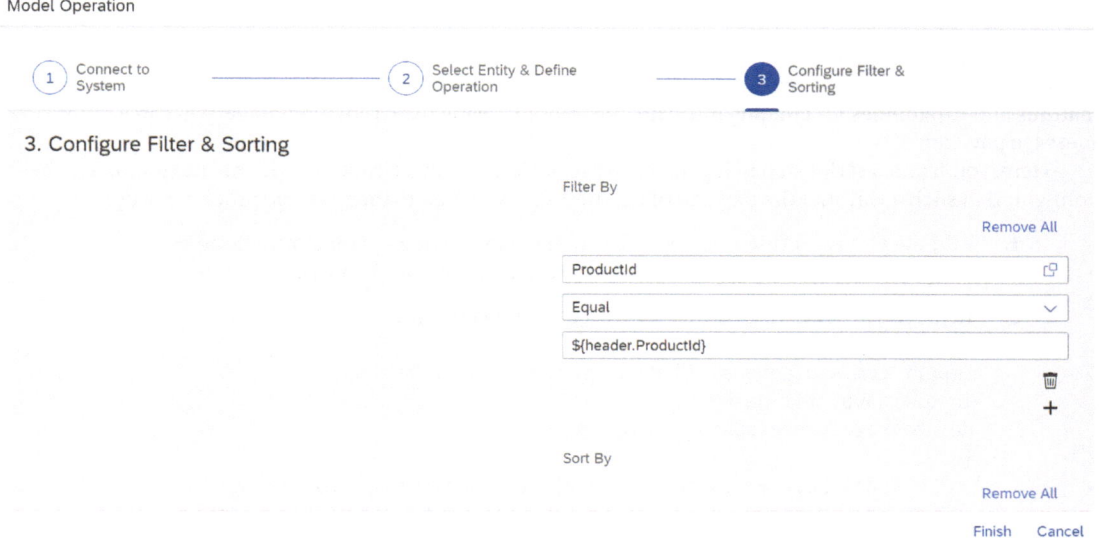

Figure 3-10. *Configuring filter and sorting*

8. You will find that the EDMX file has been saved in the backend, resulting in the query option. You can find the entire query written automatically in the Query Options box, as shown in Figure 3-11.

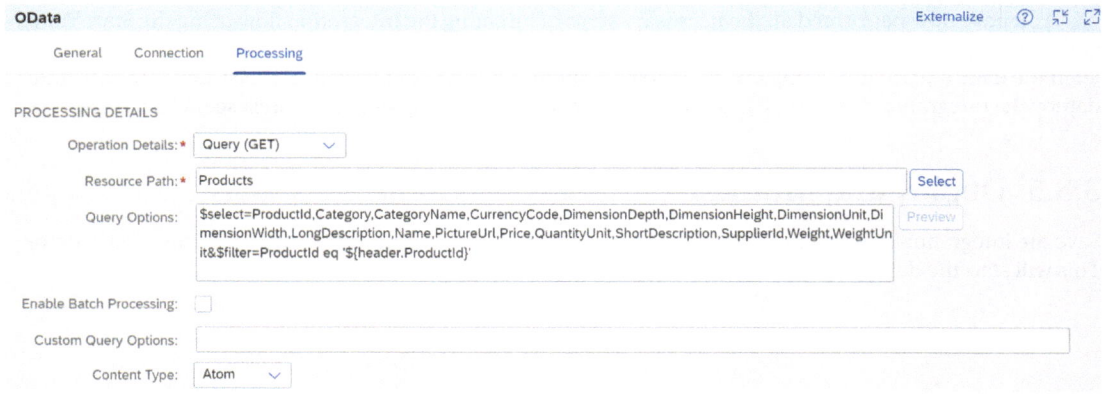

Figure 3-11. *Processing OData*

Once the connection is successfully established using the outbound OData channel, you can get the result that you fetched—but how? You will use a Groovy script to print the payload, which you will learn about in the next section.

3.3.4 Groovy Script

To create unique integrations and extensions, SAP Cloud Integration uses the Groovy programming language. Based on the Java Virtual Machine (JVM), Groovy is a scripting language that offers a wealth of features and capabilities for creating intricate integrations. Using the Groovy language, you can log the message payload.

Here, you'll add a script step to log the message payload. With a Groovy script, the integration can be configured in such a way that the payload of the message is written to the message processing log.

1. Add the Groovy script to the flow. From the palette, choose Transformations ➤ Script ➤ Groovy Script. Add the Groovy script before the End message.

2. Replace the contents of the script with the following code:

```
import com.sap.gateway.ip.core.customdev.util.Message;
import java.util.HashMap;
def Message processData(Message message)
{
        def body = message.getBody(java.lang.String) as String;
        def messageLog = messageLogFactory.getMessageLog(message);
        if(messageLog != null)
        {
        messageLog.addAttachmentAsString("Log current Payload:", body,
        "text/plain");
    }
        return message;
}
```

You have now performed all the necessary steps of initiating the integration flow using the Start Timer, adding the Content Modifier to modify the content payload, setting up the outbound communication to fetch the data, and finally setting up the Groovy script to show the log payload. The last step is to save and deploy the integration flow. You will learn about this save and deploy step in the next section.

3.3.5 Deploy and Monitor

Save the integration flow by clicking Save on the right side of the screen. After saving the flow, click Deploy. This will start the deployment process, as shown in Figure 3-12.

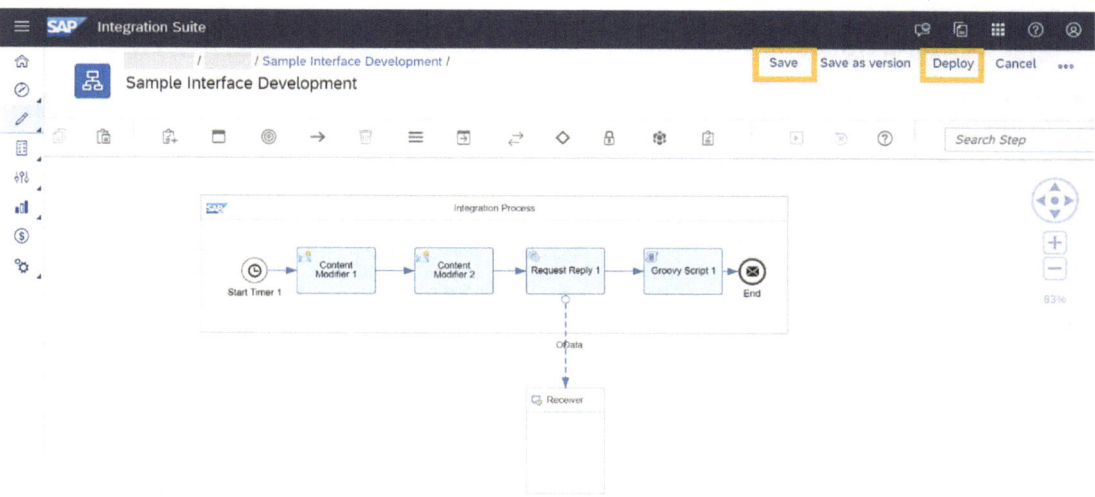

Figure 3-12. *Save and deploy*

After successfully saving and deploying the integration flow, follow these steps:

1. Go to the Monitor section of the Web UI and select the Manage Integration Content tile, as shown in Figure 3-13.

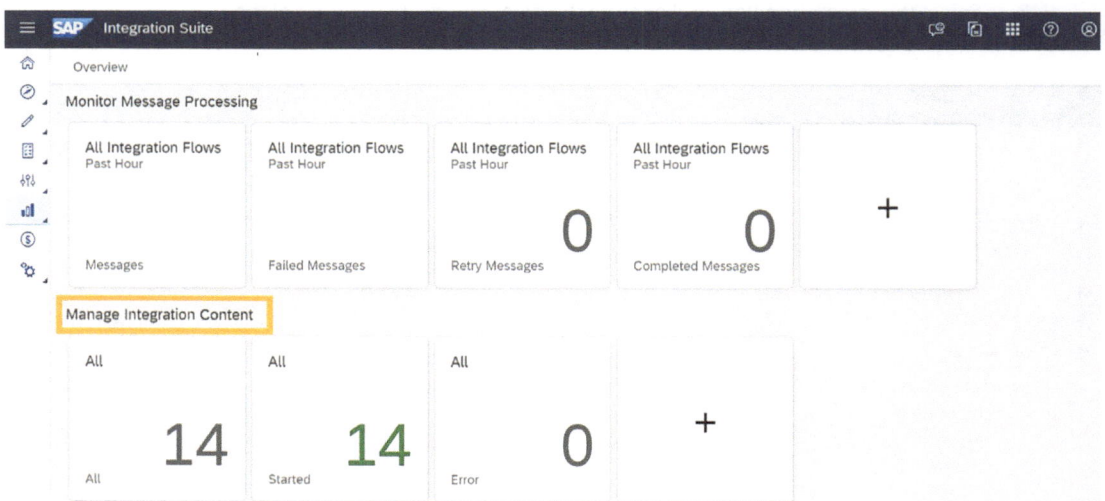

Figure 3-13. *Manage Integration Content tile*

2. You will see that the integration flow shows the status of Starting or Started, and you will learn about these statuses in upcoming chapters.

If the status is Started, then in the Artifacts Details tab, click Monitor Message Processing, as shown in Figure 3-14.

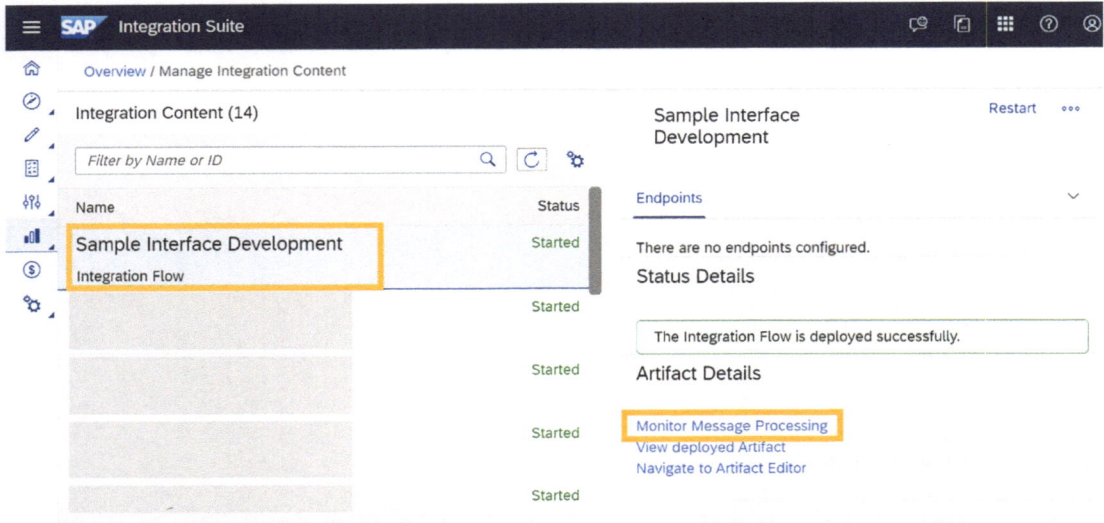

Figure 3-14. *Monitoring message processing*

3. Select the integration flow and analyze the detailed area to the right of the integration flow list, as shown in Figure 3-15.

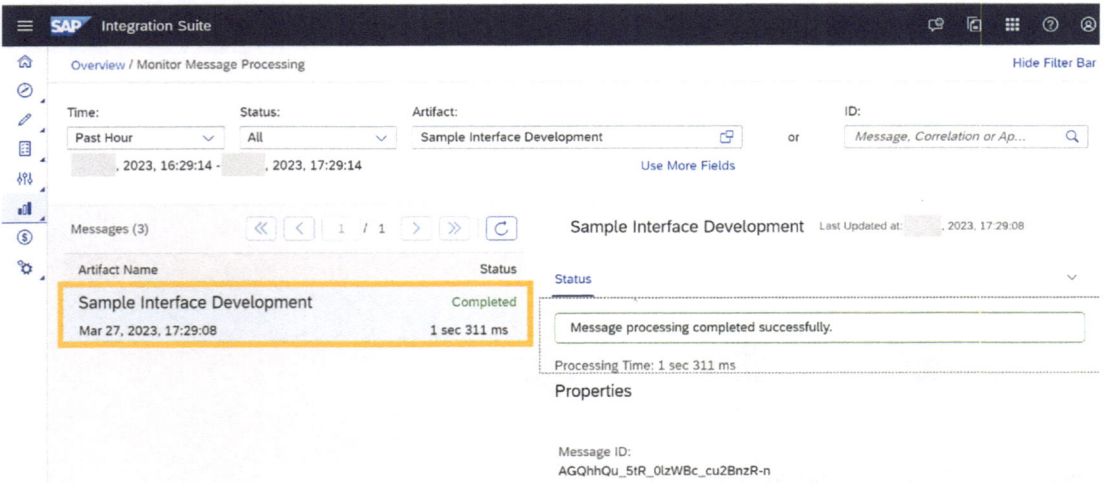

Figure 3-15. *Completed integration flow*

4. Under Attachments, click Log Current Payload. You will get the log current payload, as shown in Figure 3-16.

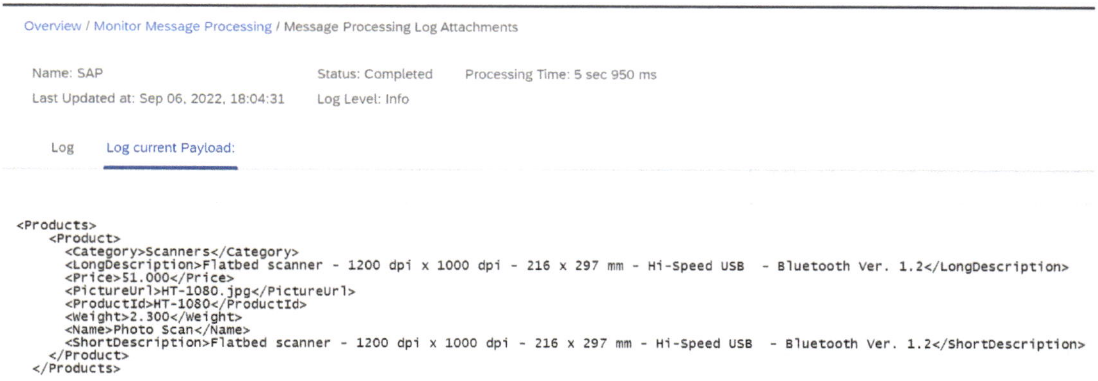

Overview / Monitor Message Processing / Message Processing Log Attachments

Name: SAP	Status: Completed	Processing Time: 5 sec 950 ms
Last Updated at: Sep 06, 2022, 18:04:31	Log Level: Info	

Log   Log current Payload:

```
<Products>
    <Product>
        <Category>Scanners</Category>
        <LongDescription>Flatbed scanner - 1200 dpi x 1000 dpi - 216 x 297 mm - Hi-Speed USB  - Bluetooth Ver. 1.2</LongDescription>
        <Price>51.000</Price>
        <PictureUrl>HT-1080.jpg</PictureUrl>
        <ProductId>HT-1080</ProductId>
        <Weight>2.300</Weight>
        <Name>Photo Scan</Name>
        <ShortDescription>Flatbed scanner - 1200 dpi x 1000 dpi - 216 x 297 mm - Hi-Speed USB  - Bluetooth Ver. 1.2</ShortDescription>
    </Product>
</Products>
```

Figure 3-16. *The log current payload of the sample interface*

You can now see the message content, which consists of the details of the product associated with the value of productIdentifier that was entered in the first Content Modifier. This indicates that the message has been processed successfully.

This was a practical example of interface development in the SAP Integration Suite. It is now time to delve deeper into the various components of SAP Cloud Integration. One of the most essential components of SAP Cloud Integration is the adapter—both sender and receiver. It facilitates the exchange of data between different systems and applications. The next section explores inbound and outbound communication using the adapters of SAP Cloud Integration.

3.4 Adapters—Sender and Receiver

Adapters in SAP Cloud Integration are connectors that enable communication between different systems and applications. They provide a standard interface for connecting different systems and allow for the exchange of data in a structured and secure manner. Adapters can be used to connect to various types of systems, such as databases, enterprise resource planning (ERP) systems, and cloud services. They can also be used to accomplish tasks, such as routing and filtering, as well as data transformations from one format to another, such as from JSON to XML.

Table 3-6 shows the various adapters available in SAP Cloud Integration.

Table 3-6. *Adapters (Sender and Receiver)*

Adapter	Description	
	Sender Adapter	Receiver System
Amazon Web-Services	Connects Amazon Web Services and SAP Cloud Integration.	Connects Amazon Web Services and SAP Cloud Integration.
	The following protocols are supported by the adapter:	The following protocols are supported by the adapter:
	Simple Cloud Storage (S3)	Simple Cloud Storage (S3)
	Simple Queue Service (SQS)	Simple Queue Service (SQS)
		Simple Notification Service (SNS)
		Simple Workflow Service (SWF)
AMQP	Using SAP Cloud Integration, it is possible to consume messages from a queue or a topic subscription in an exterior messaging system.	Using SAP Cloud Integration, messages can be sent to a queue or a topic in an exterior messaging system.
AMQP—Event Mesh	Enables SAP Cloud Integration to ingest messages from queues or topics subscribers on SAP Event Mesh.	Enables sending messages to a SAP Event Mesh queue or topic using SAP Cloud Integration.
AMQP for Microsoft Azure Service Bus	Enables SAP Cloud Integration to ingest messages from a Microsoft Azure Service Bus queue or topics subscriptions.	Enables Microsoft Azure Service Bus queues or topics to receive messages sent by SAP Cloud Integration.
AMQP for Solace PubSub+	Allows SAP Cloud Integration to use Solace PubSub+ to ingest messages from a subscription or queue.	Allows the use of SAP Cloud Integration to send messages to a Solace PubSub+ queue or topic.
AMQP—Apache Qpid Broker-J	Enables the intake of messages from topic subscriptions or Apache Qpid Broker-J queues via SAP Cloud Integration.	Allows messages to be sent to Apache Qpid Broker-J topics or queues.
AMQP for Apache ActiveMQ 5/Apache ActiveMQ Artemis	Allows SAP Cloud Integration to ingest messages from topic subscriptions or Apache ActiveMQ 5/Apache ActiveMQ Artemis queues.	Allows sending messages via SAP Cloud Integration to topics or queues in Apache ActiveMQ 5 or Apache ActiveMQ Artemis.
AMQP for IBM MQ	Enables IBM MQ queues or topic subscriptions to receive messages for SAP Cloud Integration.	Enables IBM MQ queues or topics to receive messages sent by SAP Cloud Integration.

(continued)

Table 3-6. (*continued*)

Adapter	Description	
	Sender Adapter	**Receiver System**
Ariba	Establishes a link between SAP Cloud Integration and the Ariba Network. The Ariba network can communicate business-specific documents in the Commerce Extensible Markup Language (cXML) format to cloud apps from SAP and other vendors. With the help of the sender adapter, you can determine a schedule for collecting data from Ariba.	Connects SAP Cloud Integration and the Ariba network. With the help of this adapter, SAP and non-SAP cloud apps can send business-specific documents in commerce eXtensible Markup Language (cXML) format to the Ariba network. receiver adapter
AS2	With the help of the Applicability Statement 2 (AS2) protocol, business-specific documents can be shared with a partner using SAP Cloud Integration. The sender adapter is capable of sending back to the AS2 message's sender an electronic receipt in the form of an MDN (Message Disposition Notification).	With the help of the Applicability Statement 2 (AS2) protocol, business-specific documents can be shared with a partner using SAP Cloud Integration.
AS4	Enables the secure processing of incoming AS4 messages by SAP Cloud Integration using web services. The AS4 sender adapter is an ebMS 3.0 specification-based product that supports the ebMS handler conformance profile.	Allows any two message service handlers (MSHs) to connect to each other in order to exchange business documents via SAP Cloud Integration. The AS4 receiver adapter uses the Lite Client compliance policy, which only permits selective message pulling to receive MSH and message pushing to transmit MSH.
Data Store	Enables messages from a data storage to be consumed by SAP Cloud Integration.	N/A
ELSTER	N/A	Enables sending a tax document to the ELSTER server via SAP Cloud Integration.
Facebook	N/A	Enables SAP Cloud Integration to access Facebook and retrieve data based on search terms or user information. With OAuth, a Facebook user can grant the SAP BTP tenant access to Facebook resources.

(*continued*)

Table 3-6. (*continued*)

Adapter	Description	
	Sender Adapter	Receiver System
FTP	Establishes a TCP (Transmission Control Protocol) connection to a distant system to download files from the system, allowing SAP Cloud Integration to do so. With the sender adapter, you can schedule data from the associated system to be polled.	Enables uploading files to a remote system via a TCP (Transmission Control Protocol) connection established via SAP Cloud Integration.
HTTPS	Creates an HTTPS connection between a sender system and SAP Cloud Integration.	N/A
HTTP	N/A	Creates an HTTP connection between a receiver system and SAP Cloud Integration. Only works with HTTP 1.1. (The HTTP Content-Length header must not be relied upon by the target system, which must support squished transfer encoding.) The GET, HEAD, POST, PUT, TRACE, DELETE, and GET methods are supported.
IDoc	Exchange of Intermediate Document (IDoc) messages is possible with SAP Cloud Integration when the sender system supports communication via SOAP Web services.	Enables communication between SAP Cloud Integration and a recipient system that accepts Intermediate Document (IDoc) messages using SOAP Web services.
JDBC	N/A	Connects to a JDBC (Java Database Connectivity) database and enables SAP Cloud Integration to run SQL commands on the database.
JDBC for DB2 (On-Premises)	N/A	Enables the use of JDBC (Java Database Connectivity) by SAP Cloud Integration to connect to DB2 (on-premise) and execute SQL commands on the database.
JDBC for Microsoft SQL Server (Cloud)	N/A	Allows SAP Cloud Integration to connect to a cloud-based instance of Microsoft SQL Server and execute SQL commands on the database using JDBC (Java Database Connectivity).

(*continued*)

Table 3-6. (*continued*)

Adapter	Description	
	Sender Adapter	Receiver System
JDBC for Microsoft SQL Server (On-Premises)	N/A	Enables JDBC-based SQL commands to be executed on the database by SAP Cloud Integration when it connects to Microsoft SQL Server (on-premise) (Java Database Connectivity).
JDBC for Oracle (Cloud)	N/A	Enables SAP Cloud Integration to connect to Oracle in the cloud and execute SQL queries against the database using JDBC (Java Database Connectivity).
JDBC for Oracle (On-Premises)	N/A	Enables the use of JDBC (Java Database Connectivity) to connect to Oracle (on-premise) and execute SQL queries against the database through SAP Cloud Integration.
JDBC for PostgreSQL (Cloud)	N/A	Enables SAP Cloud Integration to establish a JDBC connection to PostgreSQL on the cloud and execute SQL queries against the database.
JDBC for SAP ASE Service (Neo)	N/A	Enables the execution of SQL queries on the database using JDBC (Java Database Connectivity) to connect to the SAP ASE Service (Neo).
JDBC for SAP HANA Cloud	N/A	Enables the execution of SQL queries against the database using JDBC (Java Database Connectivity) to connect to the SAP HANA Cloud.
JDBC for SAP HANA Platform (On-Premises)	N/A	Allows for the execution of SQL commands on the database utilizing a JDBC (Java Database Connectivity) connection to the SAP HANA Platform (on-premises).
JMS	Message queues are used to enable asynchronous messaging.	Uses message queues to provide asynchronous messaging.
	A queue of messages is used by the sender adapter. The messages are handled simultaneously.	The receiver adapter queues up messages for processing and saves them. The messages are handled simultaneously.
	To prevent situations when the JMS adapter tries again to process a failed (large) message, you can place messages (where processing suddenly halted) in a dead-letter queue after two retries.	

(continued)

113

Table 3-6. (*continued*)

Adapter	Description	
	Sender Adapter	**Receiver System**
Kafka	Enables SAP Cloud Integration to connect to a third-party Kafka broker and obtain Kafka records via the Kafka protocol (messages).	Enables SAP Cloud Integration to establish a connection to an outside Kafka broker and send records via the Kafka protocol (messages).
Mail Sender for IMAP	Enables SAP Cloud Integration to read email via the Internet Message Access Protocol (IMAP) protocol from an email server.	N/A
	While logging in to the email server, you can choose to send the username and password in plain text or encrypted.	
	With IMAPS and STARTTLS, you can secure inbound emails at the transport layer.	
	You can set a polling schedule for data from the linked system using the sender adapter.	
Mail Sender for POP3	Enables the Post Office Protocol (POP3) protocol-enabled email servers used by SAP Cloud Integration to read emails.	
	While authenticating against the email server, you can send the username and password in plain text or encrypted.	
	With POP3S and STARTTLS, you can secure inbound emails at the transport layer.	
	You can set a polling schedule for data from the linked system using the sender adapter.	
Mail	N/A	Allows emails to be sent to a mail server using SAP Cloud Integration.
		While authenticating against the email server, you can send the username and password in plain text or encrypted.
Microsoft Dynamics CRM	N/A	A connection between Microsoft Dynamics CRM and SAP Cloud Integration (CRM).

(*continued*)

Table 3-6. (*continued*)

Adapter	Description	
	Sender Adapter	Receiver System
OData	Connects systems in ATOM or JSON format to SAP Cloud Integration via the Open Data (OData) interface .	Uses the Open Data (OData) standard to connect systems to SAP Cloud Integration.
ODC	N/A	Connects the SAP Gateway OData Channel to the SAP Cloud Integration (through transport protocol HTTPS). POST (Create), DELETE (Delete), GET (Read/Query), and PUT are a few of the protocols that are available (update).
Open-Connectors	N/A	Connects more than 150 non-SAP cloud applications to SAP Cloud Integration that are enabled by SAP Open Connectors.
		Data from certain programs developed by third parties is fetched through APIs.
		Possesses the capacity to manage enormous inflows of data.
		Both request and answer calls can accept messages in the JSON and XML formats.
		Enables you to declare specified values for variables.
ProcessDirect	Ties together two integration flows that have been installed on the same tenancy.	Ties together two integration flows that have been installed on the same tenancy.
	Data from one integration flow is consumed by another integration flow that has a ProcessDirect sender adapter (as a consumer).	Via the use of a ProcessDirect receiver adapter, data is transmitted from one integration flow to another integration flow (as producer).
RFC	N/A	SAP Cloud Integration is linked to a remote receiving system using a remote function call (RFC).
		On-premise ABAP systems are linked to cloud-hosted systems via the RFC standard interface provided by the SAP Cloud Connector.
Salesforce	Integrates Salesforce with SAP Cloud.	Integrates Salesforce with SAP Cloud.
ServiceNow	N/A	Enables the integration of SAP Cloud with ServiceNow. Supports OAuth and simple authentication.

(*continued*)

Table 3-6. (*continued*)

Adapter	Description	
	Sender Adapter	Receiver System
SFTP	By establishing an SSH File Transfer connection to a distant system, SAP Cloud Integration can read files from that system. SSH File Transfer Protocol is also known as Secure File Transfer Protocol (SFTP) (or SFTP).	SSH File Transfer protocol is used to establish a connection between SAP Cloud Integration and a remote system and upload files to the system. SSH File Transfer Protocol is also known as Secure File Transfer Protocol (SFT) (or SFTP).
SOAP SOAP 1.x	Sends and receives messages using a sender system that can use SOAP 1.1 or 1.2. The sender adapter supports one-way messaging and request-reply as message exchange patterns.	Exchanges messages with a receiver system that is compatible with SOAP 1.1 or 1.2 (Simple Object Access Protocol).
SOAP SAP RM	Messages are exchanged with a sender system using the message protocol SAP Reliable Messaging (SAP RM) and the SOAP communication protocol. The adoption of Web Service Reliable Messaging standards is not necessary when using the streamlined communication protocol SAP RM for asynchronous Web service communication.	Uses SAP Reliable Messaging (SAP RM) as the message protocol and the SOAP communication protocol to exchange messages with a receiver system. The adoption of Web Service Reliable Messaging standards is not necessary when using the streamlined communication protocol SAP RM for asynchronous web service communication.
SuccessFactors REST	REST messaging protocol is used to link a SuccessFactors sender system to SAP Cloud Integration. The supported adapter is GET.	SAP Cloud Integration can be connected to a SuccessFactors receiver system via the REST communication protocol. The GET and POST operations are supported by the adapter.
SuccessFactors SOAP	Connects a SuccessFactors sender system's SOAP-based web services to SAP Cloud Integration. Query is the only supported protocol in this adapter.	Connects a SuccessFactors receiving system's SOAP-based web services to SAP Cloud Integration. INSERT, QUERY, UPDATE, and UPSERT are some of the supported protocols in this adapter.
SuccessFactors OData V2	N/A	Uses OData V2 to link a SuccessFactors system to SAP Cloud Integration.
SuccessFactors OData V4	N/A	Uses OData V4 to link a SuccessFactors system to SAP Cloud Integration.
SugarCRM	N/A	Connects SAP Cloud Integration and SugarCRM.

(*continued*)

Table 3-6. (*continued*)

Adapter	Description	
	Sender Adapter	**Receiver System**
Twitter	N/A	Enables reading and posting tweets from Twitter with SAP Cloud Integration.
Workday	N/A	Connects SAP Cloud Integration and Workday. Offers Workday SOAP API basic authentication functionality.
XI	Establishes a connection between SAP Cloud Integration and a remote sender system capable of handling the XI message protocol.	Connects SAP Cloud Integration to a remote receiver that can handle the XI message protocol.

Now you are familiar with the adapters supported by Cloud Integration. The next section dives into the world of adapters, starting with a practical example of the JDBC adapter configuration.

3.4.1 Configuration of the JDBC Adapter: Practical Example

The JDBC (Java Database Connectivity) adapter, which enables data transmission between cloud-based apps and on-premise databases, is a crucial part of SAP Cloud Integration. Setting up the necessary connections and providing the necessary parameters for flawless integration between these two environments are important steps when configuring the JDBC adapter.

Follow these steps to reach the Integration Flow editor:

1. Open the Cloud Integration Web UI and navigate to the Design ➤ Integrations tab.

2. You will be navigated to the integration packages. You will learn in depth about these integration packages and their functionalities in upcoming chapters. The initial step is to create an integration package. On the right side of the screen, click Create.

3. The Header tab will ask you to enter the Name, Short Name, and Short Description as mandatory fields. Provide the necessary details. Click Save, as shown in Figure 3-17.

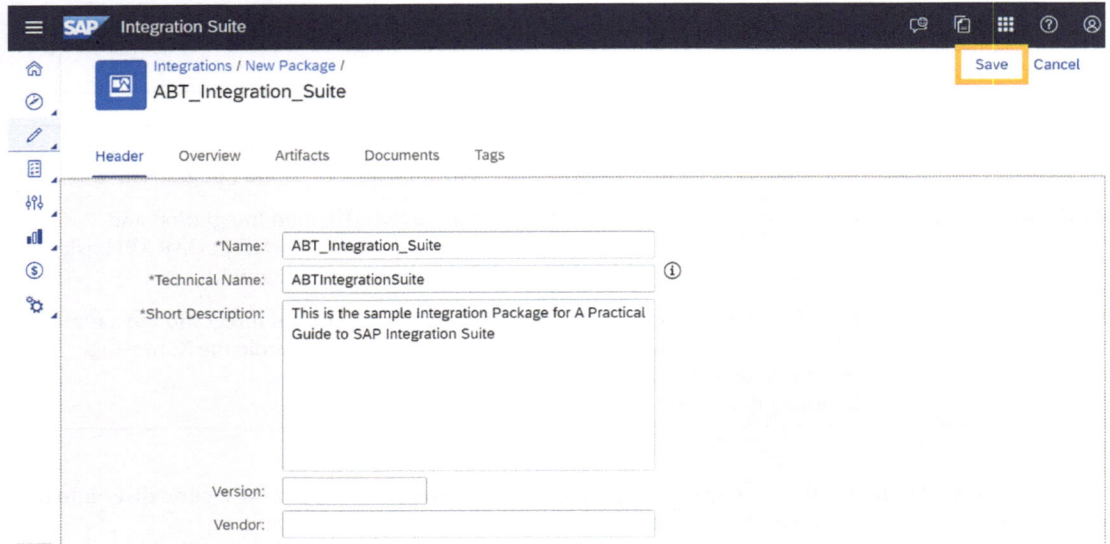

Figure 3-17. *Integration package*

4. Navigate to the Artifacts tab and click Add. From the dropdown list, select Integration Flow, as shown in Figure 3-18.

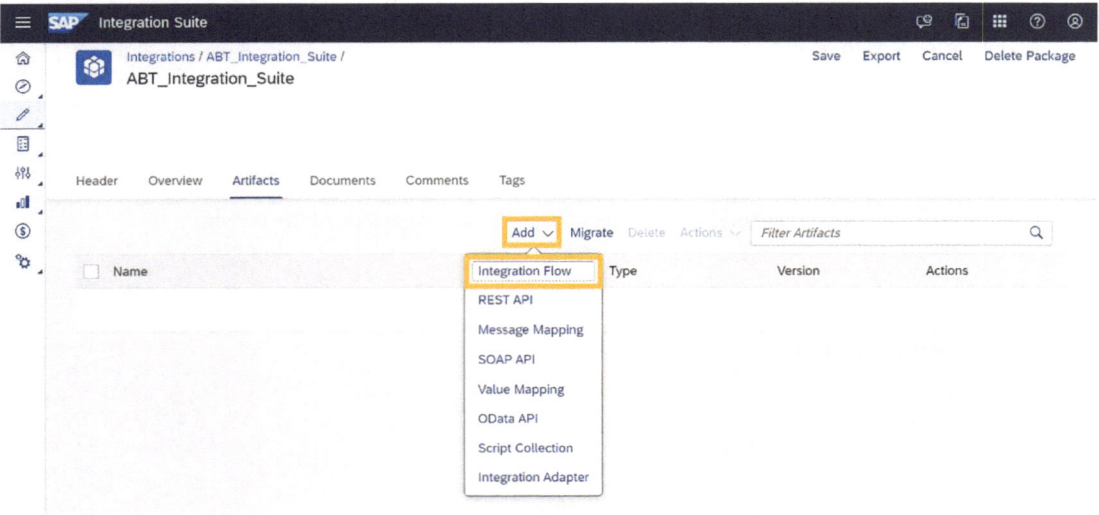

Figure 3-18. *Add an integration flow*

5. Provide the name of the integration flow and click Create. Your integration flow will be created and listed, as shown in Figure 3-19.

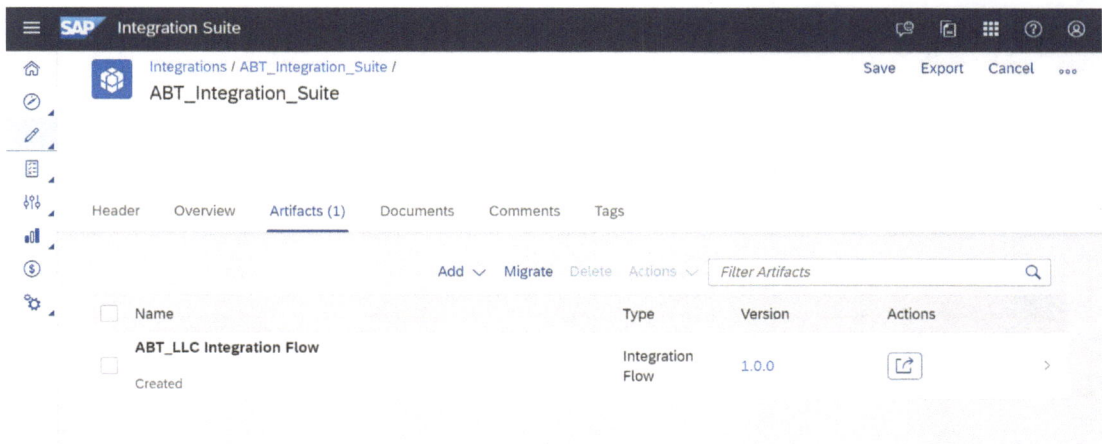

Figure 3-19. *Integration flow*

6. Click the Integration Flow to access the Integration Flow Editor. Open it in Edit mode by clicking the Edit button in the top-right corner, as shown in Figure 3-20.

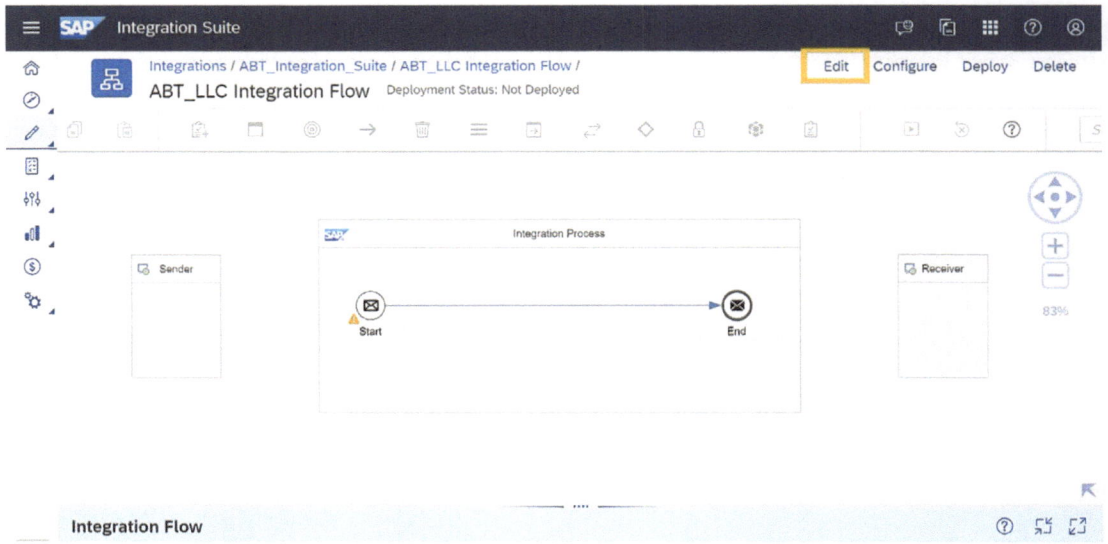

Figure 3-20. *The Integration Flow Editor*

You have now learned how to create an integration flow in the integration package. Let's get back on track. You'll now learn how to configure the JDBC adapter in Cloud Integration. Follow these steps:

1. Create an integration flow and open it in Edit mode.

2. Delete the Send event and replace it with a Timer event as done in section 3.3.1.

3. Add the Content Modifier to the flow and define its exchange property, as shown in Figure 3-21.

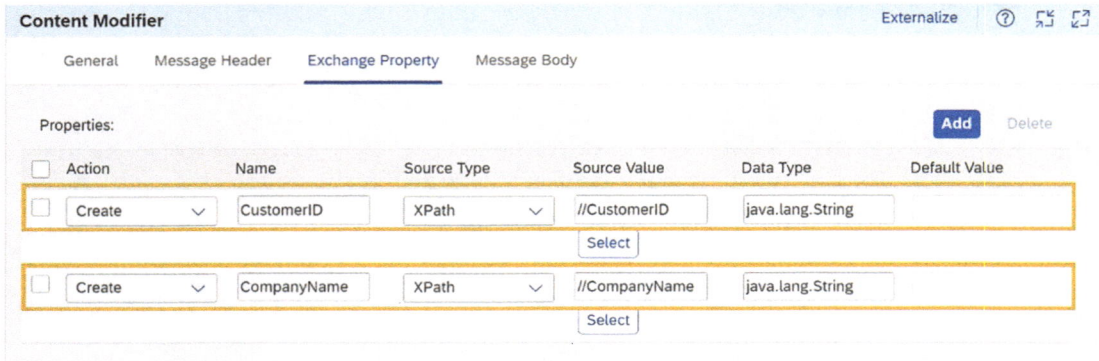

Figure 3-21. *The exchange property of the Content Modifier*

4. Add another Content Modifier to the flow to produce the query in the message body, as shown in Figure 3-22. The agenda here is to POST the data to the database.

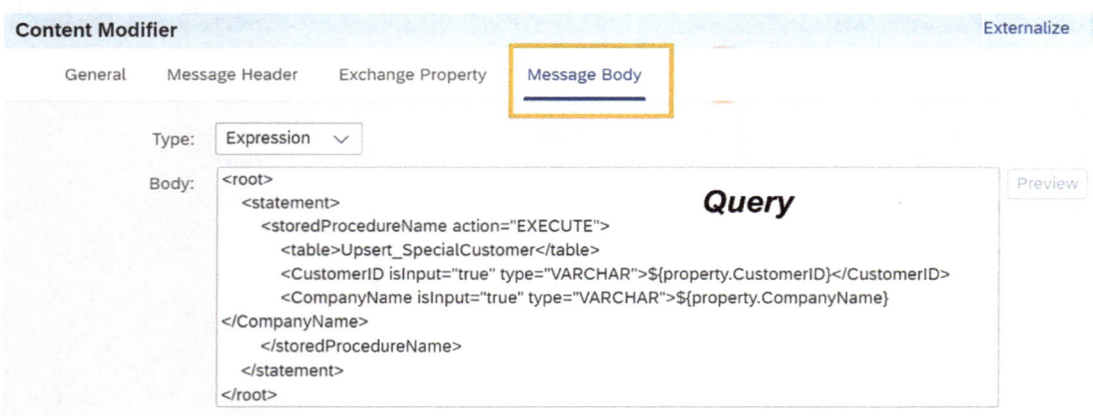

Figure 3-22. *Query to post the data to the existing database*

5. Connect the flow to the JDBC adapter using Request-Reply, as shown in Figure 3-23.

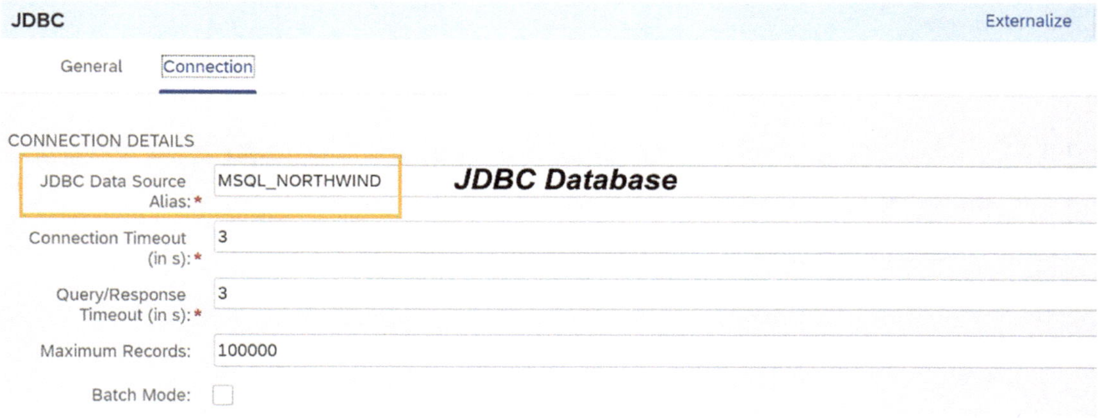

JDBC   Externalize

General   Connection

CONNECTION DETAILS

JDBC Data Source Alias: *	MSQL_NORTHWIND	**JDBC Database**
Connection Timeout (in s): *	3	
Query/Response Timeout (in s): *	3	
Maximum Records:	100000	
Batch Mode:	☐	

Figure 3-23. Configuration of the JDBC connection

6. Configure the JDBC adapter using the following details, as shown in Figure 3-23.

    - JDBC Data Source Alias—Give your JDBC Data Source. You will learn about configuring an alias in upcoming chapters.

    - Connection Timeout—3

    - Query Timeout—3

    - Maximum Records—100000

    - Uncheck the Batch Mode

7. If you have not created the JDBC alias, navigate to the Monitor tab of SAP Cloud Integration. Navigate to the Security Material ➤ JDBC Material, as shown in Figure 3-24.

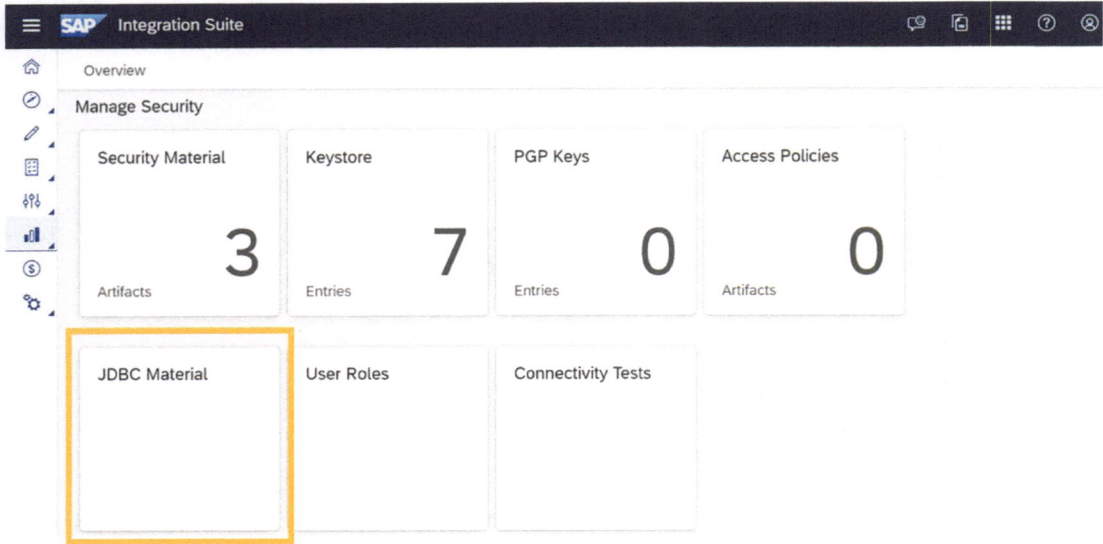

Figure 3-24. *JDBC material*

8.   Add the JDBC Driver by uploading the JDBC Driver File, as shown in Figure 3-25.

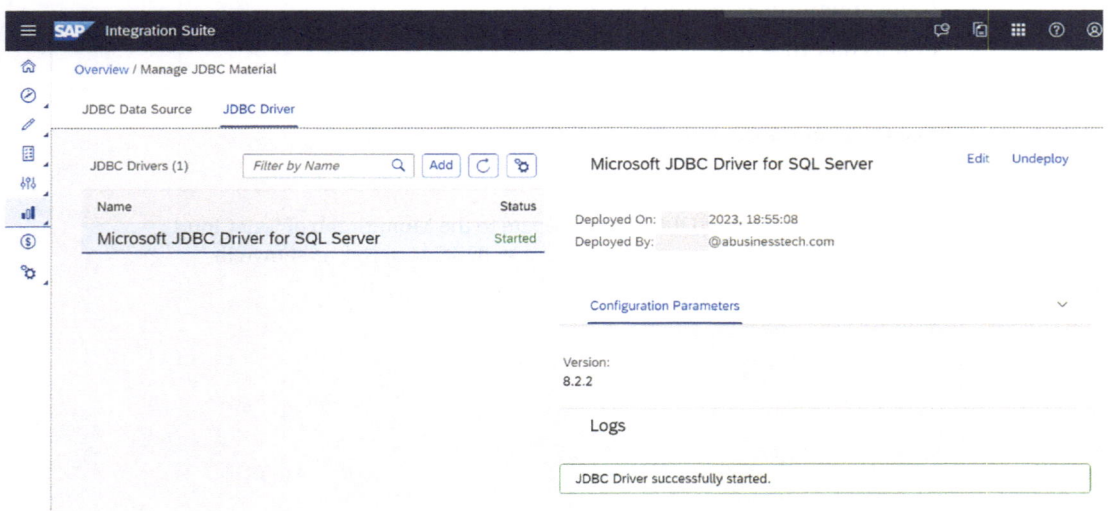

Figure 3-25. *JDBC driver*

9. Maintain the JDBC data source by providing the necessary information, as shown in Figure 3-26.

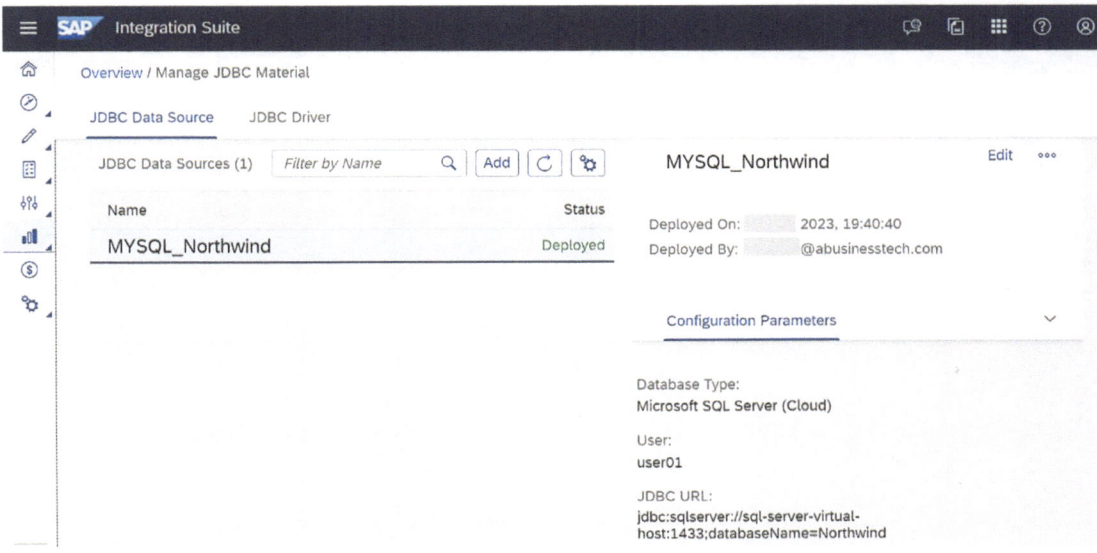

Figure 3-26. *JDBC data source*

10. Save and deploy the integration flow, as shown in Figure 3-27.

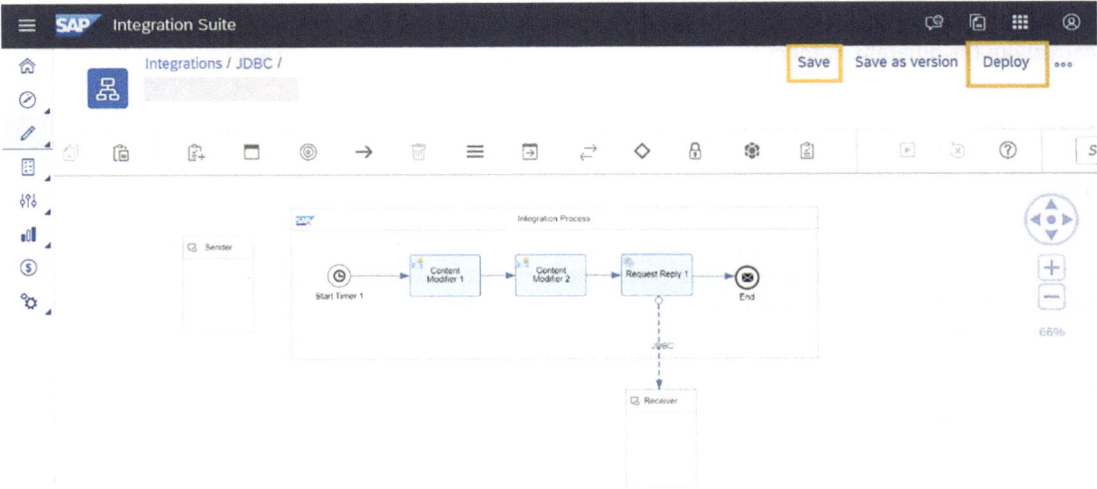

Figure 3-27. *Save and deploy*

11. Navigate to Monitor Integrations in the SAP Integration Suite. Open the Manage Integration Content tile. Find your integration flow, and on the right side under the endpoints, copy the endpoints of the integration flow, as shown in Figure 3-28.

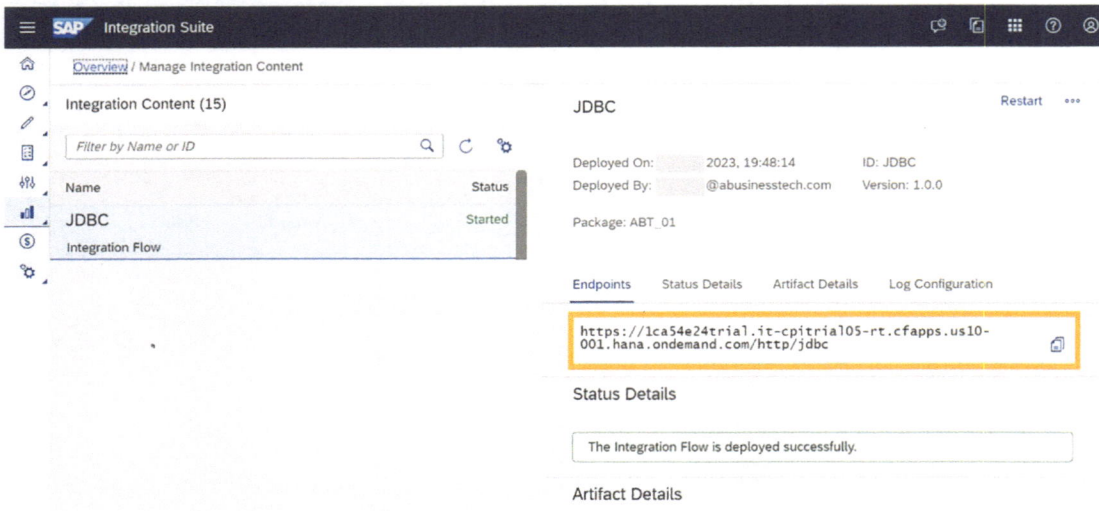

Figure 3-28. *Copy the endpoints*

12. Copy the endpoints and run it in the Postman. You'll see that the data has been posted to the JDBC database.

In this example, you established JDBC communication using the JDBC adapter. In the next section, you learn about the different connectivity options and the security of these communications.

3.4.2 Connectivity Options and Communication Security

The integration platform can be linked to remote systems utilizing a variety of technical communication protocols, thanks to adapters.

Based on the configured transport protocol, the linked remote systems communicate with one another during a scenario. To prevent unwanted access to the sent data, these protocols offer a number of options. Together with security at the transit level, messages can also have their contents safeguarded using digital signatures and encryption.

More about transport and message level security can be found in Chapter 2. You have seen the connectivity options and the message and the transport level security in SAP Cloud Integration. What are these communications called when connected from the sender side and when done from the receiver side? In the next section, you learn about inbound and outbound communications.

The chapter has briefly discussed the connectivity options and communication security of SAP Cloud Integration. They enable you to establish reliable and secure communication channels between your various enterprise systems and applications. However, once these channels are in place, it is important to ensure that inbound and outbound communication are properly connected and optimized to support business needs. Whether dealing with batch data transfers, real-time messaging, or event-driven workflows, effective communication is the key to successful enterprise integration. The next section explores the various options for connecting inbound and outbound communication within SAP Cloud Integration, providing guidance on how to optimize your channels for maximum performance, reliability, and scalability. Let's get started.

3.4.3 Connecting Inbound and Outbound Systems

You can establish a technological link between a tenant and various remote systems (in many cases, located in the customer landscape). One of the tenants given to the client is linked to a distant system (whose name is not mentioned). The remote system provides message sending and receiving capabilities. Different configurations and thorough configuration procedures are required, depending on whether a remote system is meant to send a message to the integration platform or the other way around. The terms inbound and outbound used in this book refer to the integration platform's viewpoint.

> **Inbound:** The term "inbound" describes the processing of messages coming from an external system to a cloud integration (often one that is situated in the customer landscape). Here, the server works as the integration platform.

> **Outbound:** When a message is transmitted from the integration platform to a remote system, this is known as outbound message processing (where the integration platform is the client).

Figure 3-29 shows an illustration of SAP Cloud integration between remote systems using inbound and outbound communications.

Figure 3-29. *Architecture of inbound and outbound communications*

You can link several kinds of external systems to the cloud-based integration platform using protocols, including HTTP/S, SSH, and SMTP/S. There are specific settings for message exchange protections in each communication protocol (security options).

The next section explains how to connect your system to SAP Cloud Integration.

3.4.4 Connecting a System to Cloud Integration

Here are some common instances (not a complete list) of the different systems that can be linked to the integration platform:

- On-premises systems, such as SAP NetWeaver-based SAP systems

- SFTP servers

- SAP SuccessFactors or SAP Cloud for Customer (two examples of cloud apps)

- Additional systems, including email servers and SOAP clients

A specific communication protocol must be considered, depending on the type of system being connected, as is discussed next.

3.4.5 Supported Protocols

To set up an integration scenario, a secure transport channel must first be established between the remote system and cloud integration. Subsequent protocols can be applied: the SSH File Transfer Protocol (SFTP), HTTPS, and Simple Mail Transfer Protocol (SMTP) are all protocols that are secured using transport layer security (SMTPS). Table 3-7 lists the different protocols associated with SAP Cloud Integration. You can see the different options associated with all these protocols.

Table 3-7. *Supported Protocols of SAP Cloud Integration*

Protocol	Call Direction	On-Premise (Mandatory)	On-Premise (Recommended)	Further Aspects to Consider
HTTP, HTTPS	Inbound	HTTP/S sender system (for example, SAP ERP Central Component)	HTTP/S proxy	Setting up and configuring a firewall
HTTP, HTTPS	Outbound	HTTP/S receiver system (for example, SAP ERP Central Component)	Web Dispatcher or SAP Cloud Connector	Setting up and configuring a firewall
SSH	Outbound	SFTP server (to store files)	Tools for managing SSH keys	Scanning the inbound directory for viruses
SMTP, SMTPS	Outbound	Mail server	Support for SMTPS (SMTP over SSL/TLS) by the mail server	Scanning incoming mailboxes with a virus scanner

Each protocol provides a variety of authentication methods, which are ways for connected systems to build confidence before connecting. Different connection setup processes are employed depending on whether outbound communication (when Cloud Integration calls a remote system that is judged to be the receiver) or inbound communication (when a distant system calls Cloud Integration as the sender) is configured. The particular process is also affected by the authentication protocol and mechanism that are used.

In today's digital landscape, seamless communication between different systems and applications is essential for success. Whether you're working with on-premises systems or cloud-based applications, ensuring that data flows smoothly between them is critical for driving business value and achieving your objectives. This is where inbound communication comes in. By enabling systems and applications to receive data and requests from external sources in a controlled and secure manner, enterprises can operate seamlessly and efficiently. The next section discusses inbound communications in detail.

3.4.6 Inbound Communications

Inbound communication configuration entails establishing a link between a remote sender system and the integration platform. The term "inbound communication" describes the transfer of messages from a remote system to Cloud Integration, which is frequently situated in the customer landscape. In this case, the server is the integration platform.

For inbound communications, you can set up HTTP, SFTP, and Mail adapter. The next section explains how to set up an HTTP inbound connection.

Setting Up an Inbound HTTP Connection

With SAP Cloud Integration, preparing the platform to accept data from external systems is known as setting up inbound communication. You can integrate data from external systems into SAP Cloud Integration through inbound communication, where you can process, transform, and route the data to the right destination systems.

To download a certificate, follow these steps:

1. Download the certificate from the TLS connectivity.

2. Navigate to the Test Connectivity tile in the Monitor section of Cloud Integration.

3. From the TLS tab, enter your `<tenant>.cfapps.<data center>.hana.ondemand.com`, as shown in Figure 3-30.

4. You should see a successful test result.

5. Download the certificate by clicking the Download button, as shown in Figure 3-30.

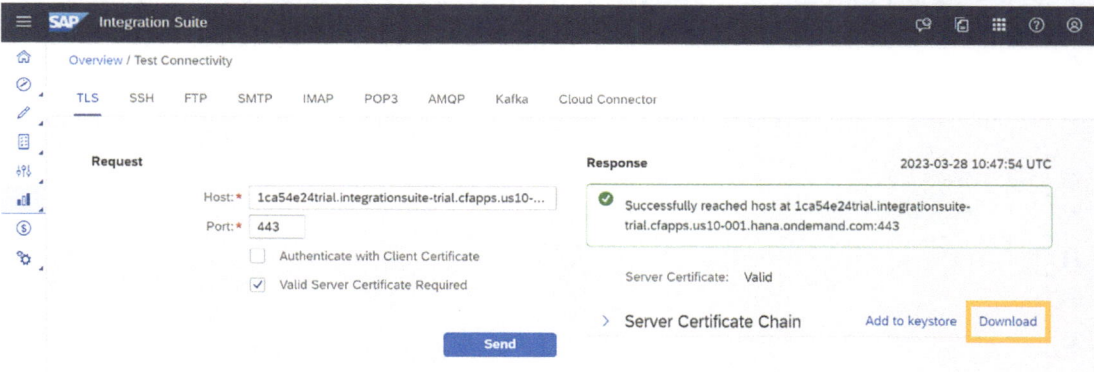

Figure 3-30. *Downloading the certificate*

The following sections briefly discuss the different options available in SAP Cloud Integration to set up the inbound communication for the HTTP connection. The next section starts with the client certificate authentication, which can be used in the integration flow process.

The Client Certificate Authentication for Integration Flow Process

When a sender uses a client certificate to call an integration flow installed on a worker node, the sender authenticates itself. The system (where the SAP Cloud Integration service is running) verifies whether a service key containing the sender-supplied client certificate is available at runtime. After determining whether a service key is available, the system determines if the corresponding service instance has a role that authorizes calling the integration flow endpoint. Figure 3-31 shows an illustration of authentication using a client certificate between the sender and SAP Cloud Integration.

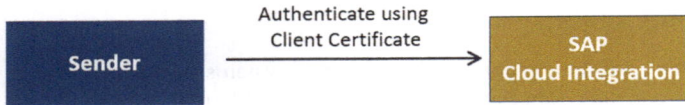

Figure 3-31. *Authentication using a client certificate*

Procedure:

1. Find the role that will give the sender permission to call the integration flow endpoint.

   • For the associated sender adapter of the integration flow to be invoked, this role must be supplied as the User Role argument.

   • This could be a custom role or the default role called `ESBMessaging.send`. Select the User Roles tile under Manage Security in the SAP Cloud Integration Monitor section to view the roles established for your tenant, as shown in Figure 3-32.

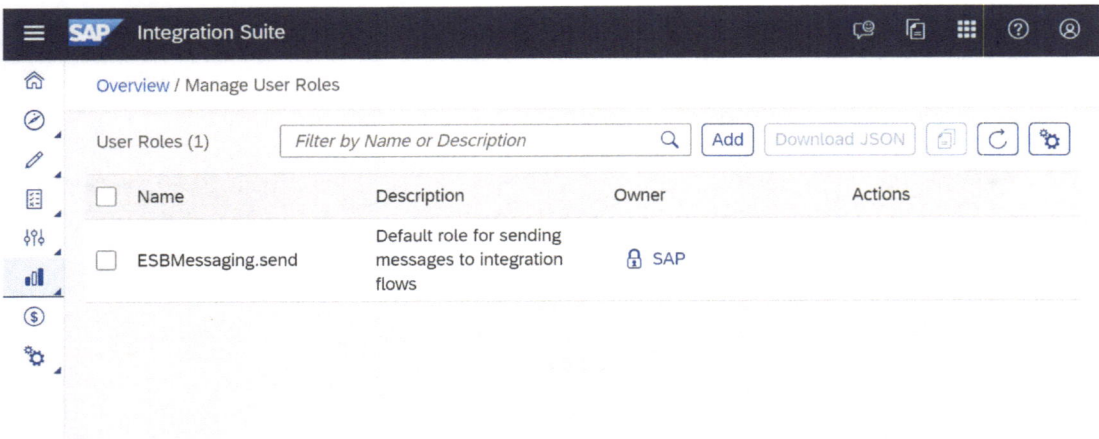

Figure 3-32. *User role*

2. Ask the sender system administrator for the sender client certificate.

3. Create a service instance and service key in the SAP BTP Cockpit by choosing the subaccount that houses your SAP Cloud Integration virtual environment. You can check creation of Service key in Chapter 2.

4. See Chapter 2 (Section 2.7.6) to the assign the correct user to the tenant.

Table 3-8 shows the different options that are available and the different data available when creating the service keys.

Table 3-8. *Configuration of Service Keys for Client Certificate Authentication*

| Option (Certificate Type) | Service Instance | | Service Key | | | |
	Roles	Grant Types	Key Type	External Certificate	Validity	Key Size
SAP certificate	Maintain the default ESBMessaging role. Send or apply a few custom roles.	Client Credentials	Certificate	N/A	Specify validity in days.	Specify key size.
External certificate	Maintain the default ESBMessaging role. Send or apply a few custom roles.	Client Credentials	External Certificate	Add PEM-encoded X.509 certificate.	N/A	N/A

The SAP Service Key with the key type set to Client Certificate is shown in Figure 3-33.

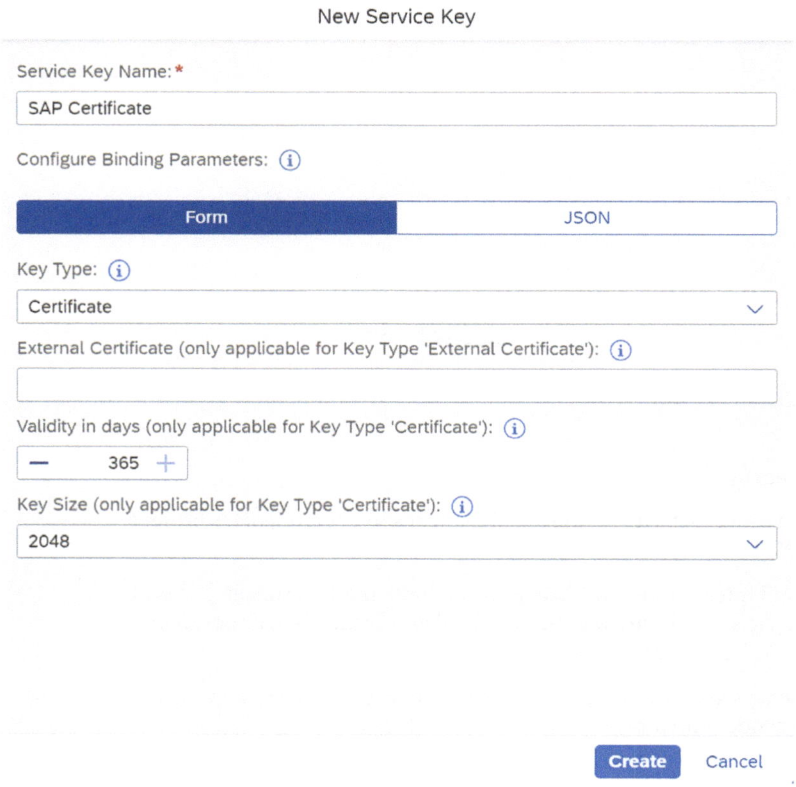

Figure 3-33. *The SAP Certificate Service Key*

The External Certificate Key with the key type set to External Certificate is shown in Figure 3-34.

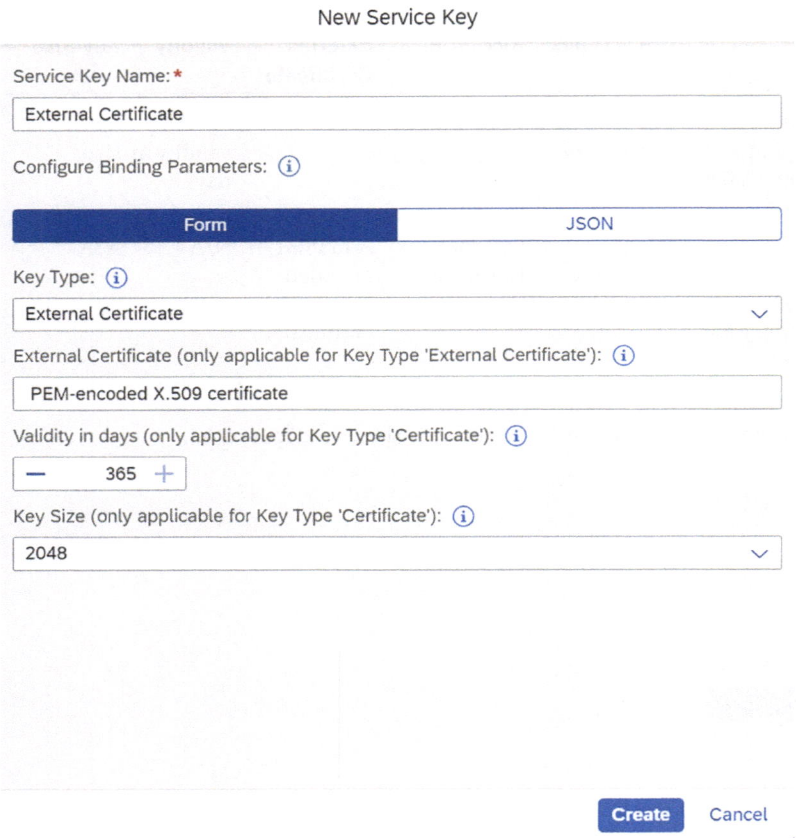

New Service Key

Service Key Name: *

| External Certificate |

Configure Binding Parameters: ⓘ

| Form | JSON |

Key Type: ⓘ

| External Certificate | ⌄ |

External Certificate (only applicable for Key Type 'External Certificate'): ⓘ

| PEM-encoded X.509 certificate |

Validity in days (only applicable for Key Type 'Certificate'): ⓘ

| — 365 + |

Key Size (only applicable for Key Type 'Certificate'): ⓘ

| 2048 | ⌄ |

Create   Cancel

Figure 3-34. External Certificate Service Key

1. Set the sender system up.

   - Ensure that the root certificate for the load balancer server certificate is present in the sender Keystore.

   - Utilize the Cloud Integration Connectivity Test to obtain this certificate. Choose the root certificate's `*.cer` file from the `downloaded.zip` file, then import it into the sender system's keystore.

   - A client certificate signed by one of the CAs that the load balancer supports is present in the sender keystore, as shown in Figure 3-35.

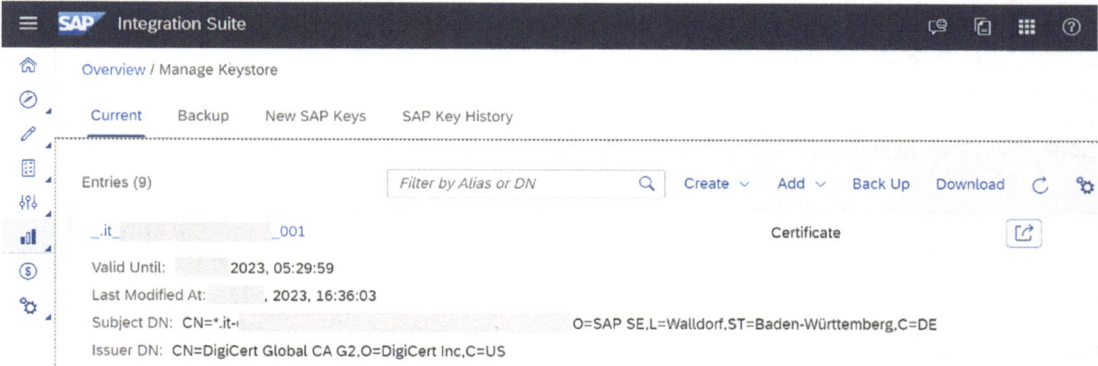

Figure 3-35. *Certificate in the keystore*

2. For the relevant integration flow, configure the inbound communication:

- For adapters that support client certificate-based authorization, authorization can be configured in the sender channel of the integration flow. The Connection tab of the channel includes the authorization configuration option. When you choose the client certificate, a table with space for adding client certificates is displayed. To include a new row in the table, select Add. You can access the uploaded dialog for a certificate in the row. The certificate file can be browsed and added to the channel using the Upload from the File System option, as shown in Figure 3-36.

Figure 3-36. *Client certificate in integration flow*

- Deploy the integration flow on the tenant once you set it up, depending on the necessary processing steps for your scenario.

- Save the integration flow, and then click Deploy.

You have seen how inbound HTTP communications can be set up using client certificate authentication. Next, you learn about the different option of setting up the HTTP connection using the OAuth with client credentials.

OAuth with Client Credentials Grant for Integration Flow Processing

For inbound calls from sender systems to the integration platform, OAuth authentication, in particular the Client Credentials Grant variation, is best. In this manner, OAuth authentication grants the sender (client) application access to the related worker node. Figure 3-37 shows an illustration of the OAuth Client Credentials Grant between the sender and SAP Cloud Integration.

Figure 3-37. OAuth Client Credentials Grant

Use the subsequent procedures to establish this authentication:

1. At the SAP BTP token server, the senders authenticate themselves. The token server can be accessed in one of two ways:

   - Using the service key's client ID and client secret.

   - By means of a client certificate obtained from the service key.

2. Access tokens are issued by a token server.

3. When calling the integration flow installed on the worker node, the senders authenticate themselves using an access token.

Follow these instructions to set up this authorization option:

1. Find the role that will give the sender permission to call the integration flow endpoint.

   - For the associated sender adapter of the integration flow to be invoked, this role must be supplied as the User Role argument. This could be a custom role or the default role called ESBMessaging.send.

   - Select the User Roles tile under Manage Security in the SAP Cloud Integration Monitor section to view the roles established for your tenant.

2. Create a service instance and service key in the SAP BTP Cockpit by choosing the subaccount that houses your SAP Cloud Integration virtual environment. You learned how to create service instances and security keys in Chapter 2 (Sections 2.8 and 2.9).

Table 3-9 lists the different options and details needed for each option when creating the service keys in the SAP BTP.

Table 3-9. *Configuration of Service Keys for OAuth with Client Credentials*

	Service Instance		Service Key			
Option (Certificate Type)	Roles	Grant Types	Key Type	External Certificate	Validity	Key Size
Client ID and client secret	Maintain the default ESBMessaging role. Send or apply a few custom roles.	Client Credentials	Client ID/ Secret	N/A	N/A	N/A
SAP certificate	Maintain the default ESBMessaging role. Send or apply a few custom roles.	Client Credentials	Certificate	N/A	Specify validity in days.	Specify key size.
External certificate	Maintain the default ESBMessaging role. Send or apply a few custom roles.	Client Credentials	External Certificate	Add PEM-encoded X.509 certificate.	N/A	N/A

You created the service key for the client certificate and the external client certificate, so now you'll create the client ID/secret service key, as shown in Figure 3-38.

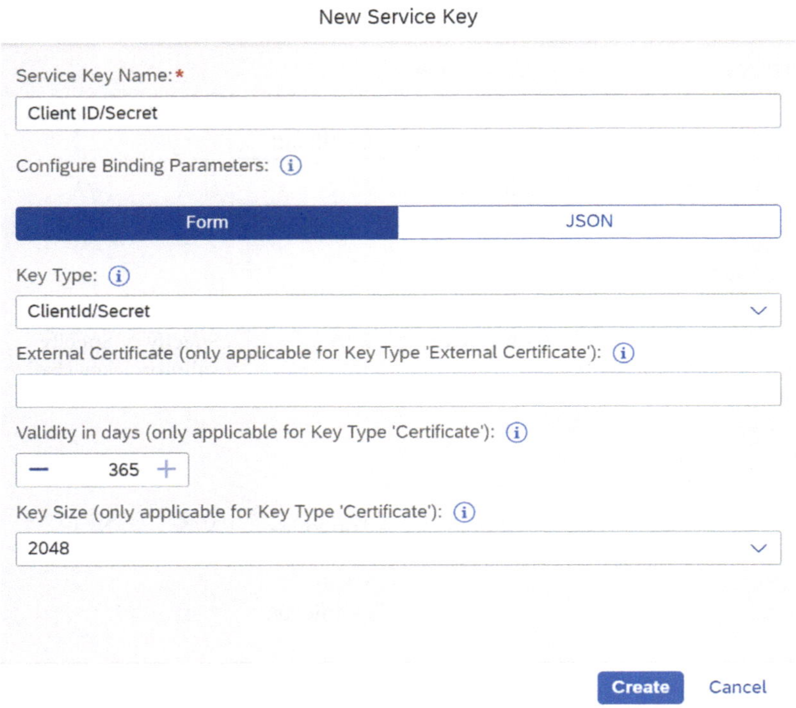

New Service Key

Service Key Name:*

Client ID/Secret

Configure Binding Parameters: ⓘ

Form	JSON

Key Type: ⓘ

ClientId/Secret ⌄

External Certificate (only applicable for Key Type 'External Certificate'): ⓘ

Validity in days (only applicable for Key Type 'Certificate'): ⓘ

— 365 +

Key Size (only applicable for Key Type 'Certificate'): ⓘ

2048 ⌄

Create   Cancel

Figure 3-38. *Client ID/secret service key*

1. Set the sender system up:

   - Ensure that the root certificate for the load balancer server certificate is present in the sender's keystore.

   - Utilize the SAP Cloud Integration Connectivity Test to obtain this certificate (pointing to the integration flow's endpoint address). The root certificate's *.cer file is chosen from the downloaded.zip file and then imported into the sender system keystore, as shown in Figure 3-30 and Figure 3-35.

2. Set up the appropriate inbound communication integration flow:

   - Edit the relevant integration flow by going to the SAP Cloud Integration Design section.

   - To use this authentication option, create a sender channel with the appropriate sender adapter type and then select the connection for that sender adapter.

   - Select User Role under Authorization and provide the role information. Keep the User Role's default entry of ESBmessaging.send. If you want to utilize a specific role to manage authorization for processing the integration flow, you can also choose a custom role.

- Deploy the integration flow on the tenant once you have completed its setup, depending on the processing steps in your scenario.

- Save the integration flow, and then click Deploy.

The next section shows the different options available for setting up the HTTP inbound communication (Basic Authentication) using the client ID and the client secret.

Basic Authentication Using Client ID and Client Secret

To access multiple APIs and online services, SAP Cloud Integration frequently uses basic authentication with the client ID and client secret. To access the service, the client must utilize a client ID and a client secret, which are created by the service provider and sent to the client as credentials. Figure 3-39 shows the authentication process using the client ID and client secret between the sender and SAP Cloud Integration.

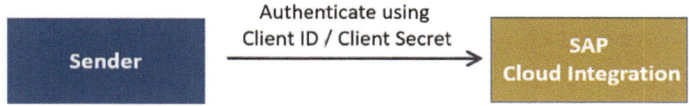

Figure 3-39. *Basic service key authentication*

Procedure:

1. Find the role that will give the sender permission to call the integration flow endpoint. For the associated sender adapter of the integration flow to be invoked, this role must be supplied as the User Role argument.

   - This could be a custom role or the default role called `ESBMessaging.send` (see Managing User Roles, Cloud Foundry Environment). Select the User Roles tile under Manage Security in the SAP Cloud Integration Monitor section to view the roles established for your tenant.

2. Create a service instance and service key in the SAP BTP Cockpit by choosing the subaccount that houses your SAP Cloud Integration virtual environment. However, no access token is collected from the token server during runtime.

The client ID and client secret values from the service key are used as user credentials to access the integration flow endpoint in place of an access token. The service instance and service key parameters should be specified as listed in Table 3-10.

Table 3-10. *Configuration of Service Key for Basic Authentication Using Client ID and Secret*

Service Instance		Service Key				
	Roles	Grant Types	Client ID/Secret	External Certificate	Validity	Key Size
Integration-flow	Keep standard role ESB-Messaging. send or use one or more custom roles.	Client Credentials	Client ID/Secret	N/A	N/A	N/A.

Create the service key with the key type set to client ID/secret in the SAP BTP Cockpit, as shown in Figure 3-40.

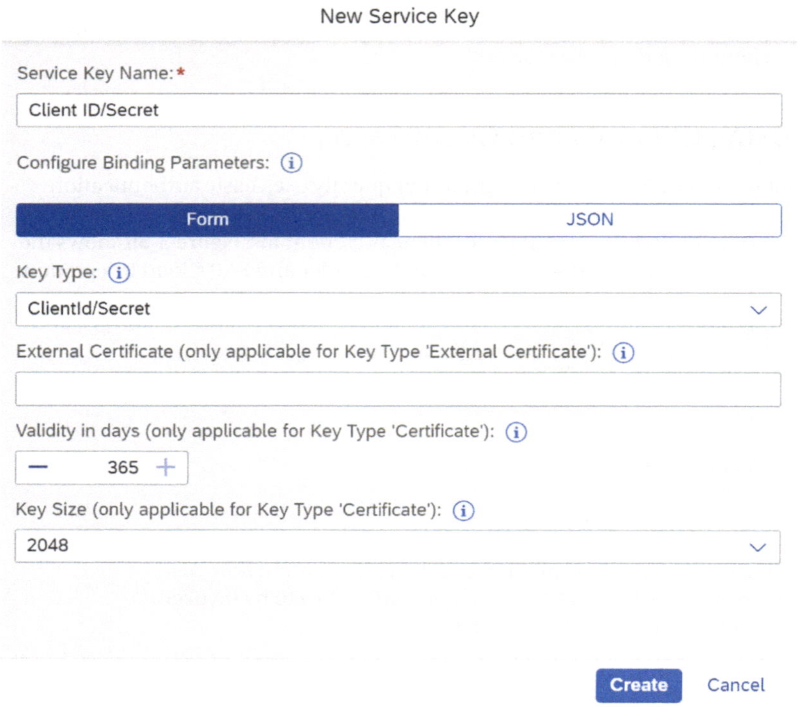

New Service Key

Service Key Name: *

Client ID/Secret

Configure Binding Parameters: ⓘ

Form	JSON

Key Type: ⓘ

ClientId/Secret ⌄

External Certificate (only applicable for Key Type 'External Certificate'): ⓘ

Validity in days (only applicable for Key Type 'Certificate'): ⓘ

— 365 +

Key Size (only applicable for Key Type 'Certificate'): ⓘ

2048 ⌄

Create   Cancel

Figure 3-40. *Client ID/secret service key*

1. Set the sender system up:

   - Ensure that the root certificate for the load balancer server certificate is present in the sender's keystore.

   - Utilize the SAP Cloud Integration Connectivity Test to obtain this certificate (pointing to the integration flow's endpoint address). Choose the root certificate's *.cer file from the downloaded.zip file and then import it into the sender system keystore, as shown in Figure 3-30 and Figure 3-35.

2. Set up the appropriate inbound communication integration flow:

   - Edit the relevant integration flow by going to the SAP Cloud Integration Design section.

   - To use this authentication option, create a sender channel with the appropriate sender adapter type and then select the connection for that sender adapter.

   - Select User Role under Authorization and provide the role information. Keep the User Role's default entry for the role name as ESBmessaging.send. If you want to utilize a specific role to manage authorization for processing the integration flow, you can also choose a custom role, as shown in Figure 3-41.

Figure 3-41. *Authorization in integration flow*

- Deploy the integration flow on the tenant once you have completed setting it up, based on the necessary processing steps for your scenario.

- Save the integration flow, and then click Deploy.

The next section covers another option, which is basic authentication using the IdP user to set up communication.

Basic Authentication of IdP User for Integration Flow Processing

Basic authentication can be used to authenticate an identity provider (IdP) user for integration flow processing. To authenticate an IdP user, the user's credentials must first be validated by SAP Cloud Integration using the basic authentication method.

The components and related security artifacts are depicted in Figure 3-42.

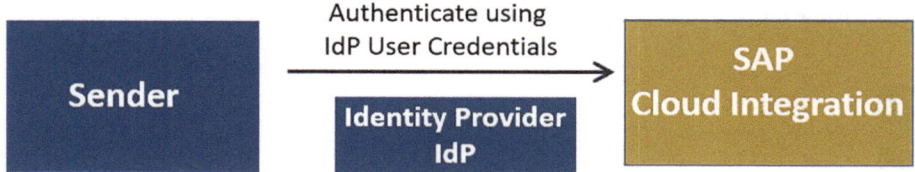

Figure 3-42. *Basic IdP authentication*

Table 3-11 highlights the necessary configuration procedures and gives you an overview of the necessary digital keys and their functions during the authentication process. Keep in mind that, to build secure communication across various systems, administrators connected to those systems often need to coordinate configuration chores and exchange public keys.

Table 3-11. *Security Artifacts for Authentication of IdP Users*

Security Artifact	Used To	Configuration Steps
Load balancer server root certificate	Make the load balancer trustworthy to the sender.	Manager of the sender: Utilize the Cloud Integration Connectivity Test to obtain a certificate (pointing to the endpoint address of the integration flow).
Load balancer server certificate (including certificate chain)	Make the load balancer a reliable component (for senders that like to connect to it).	No action is necessary because the cloud infrastructure operator maintains this artifact.

Assign the user a role in the SAP BTP Cockpit that will allow the sender to call the integration flow endpoint. Either a new role or the predefined ESBMessaging.send role can be assigned.

Procedure:

You learned about creating custom role collections and assigning roles and users to them in Chapter 2 (Sections 2.7.5 and 2.7.6).

You also saw how to create inbound communications for setting up the HTTP connection using the basic authentication of IdP user credentials. That means you have learned about all the possible methods for setting up HTTP connections. The next section explains the different methods available for SFTP connections.

Setting Up an Inbound SFTP Connection: Practical Example

In order for the tenant to read data from the SFTP server (a process referred to as *polling*), you must connect the SAP Cloud Integration tenant to the SFTP server using the sender SFTP adapter.

In other words, the tenant sends a request to the SFTP server, as depicted in Figure 3-43, but the data flow is in the other direction—from the SFTP server to the tenant. The direction of the request is shown by the arrow in Figure 3-43, while the direction of the connection arrow indicates the direction of the data flow.

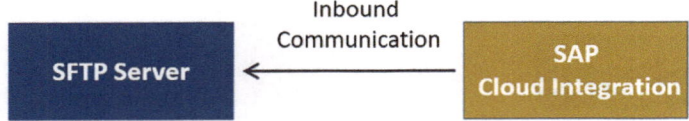

Figure 3-43. *Inbound communication for SFTP connections*

The SFTP adapter occasionally enables you to choose one of the following choices for SFTP connectivity authentication:

- Username/password
- Public key

If you choose the public key, take the following actions:

1. Create an SSH known host file in the Manage Keystore option, under the Monitor Section, as shown in Figure 3-44.

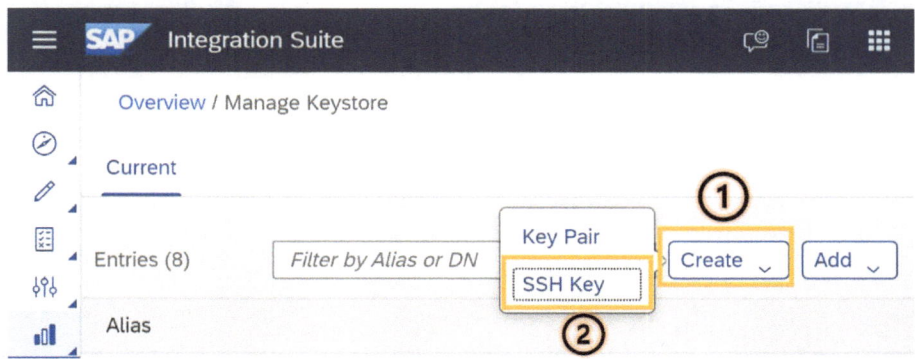

Figure 3-44. *Create an SSH Key*

2. Provide the necessary details, as shown in Figure 3-45.

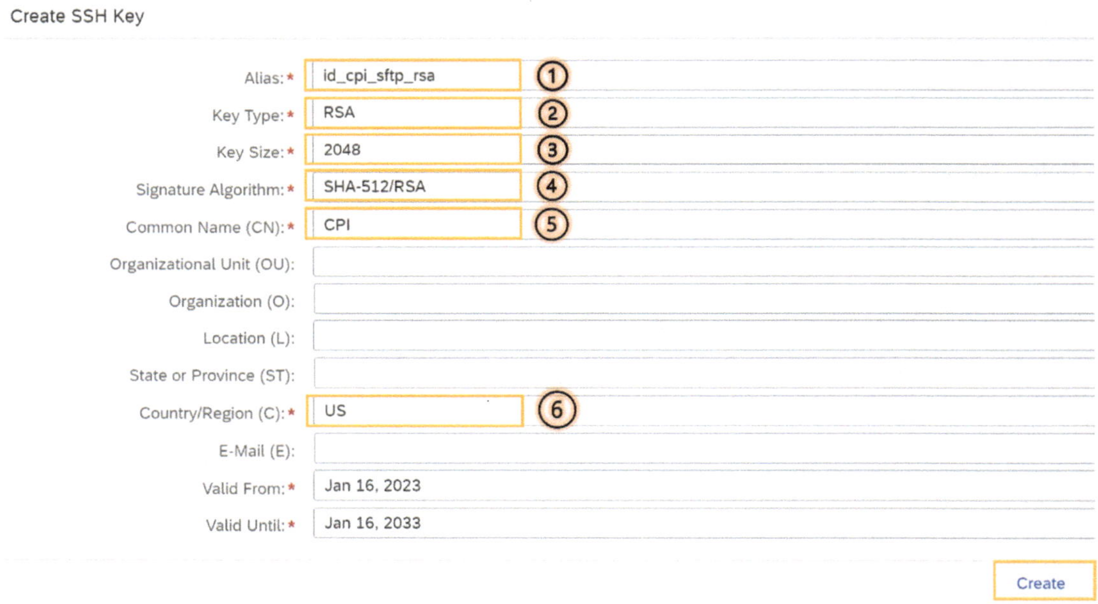

Create SSH Key

Alias: *	id_cpi_sftp_rsa ①
Key Type: *	RSA ②
Key Size: *	2048 ③
Signature Algorithm: *	SHA-512/RSA ④
Common Name (CN): *	CPI ⑤
Organizational Unit (OU):	
Organization (O):	
Location (L):	
State or Province (ST):	
Country/Region (C): *	US ⑥
E-Mail (E):	
Valid From: *	Jan 16, 2023
Valid Until: *	Jan 16, 2033

Create

Figure 3-45. *Provide the necessary details for the SSH key*

3. Click the key created in the keystore and download the public OpenSSH key, as shown in Figure 3-46.

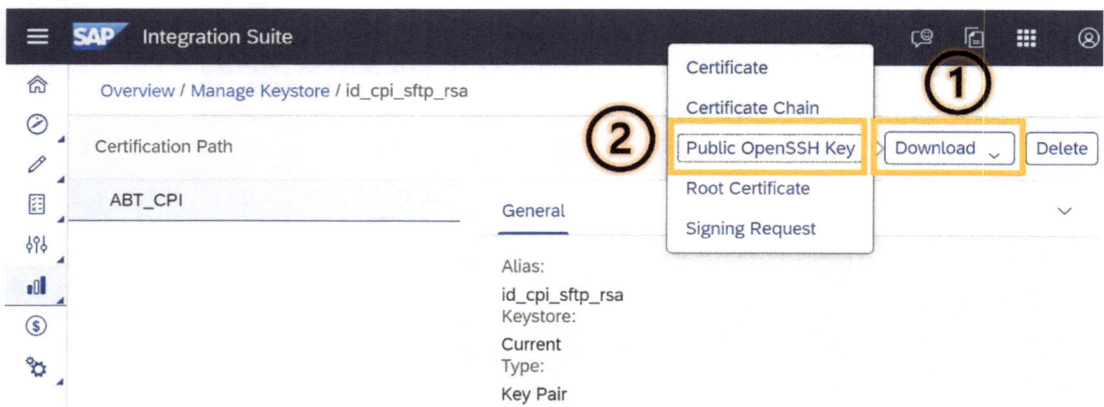

Figure 3-46. *Download the public OpenSSH key*

4. Import the public OpenSSH key to the SFTP server. You can set up your SFTP connection using the SFTP server as Rebex and SFTP client as FileZilla.

Test Connectivity

In the Connectivity Test section of SAP Integration Suite, when you enter all the required information, you obtain a successful connectivity test result, as shown in Figure 3-47.

■ **Note**   The username is the SFTP server username.

| TLS | SSH | FTP | SMTP | IMAP | POP3 | AMQP | Kafka | Cloud Connector |

Request

Host: *	my-sftp-virtual-host
Port: *	2222
Proxy Type:	On-Premise
Location ID:	POC
Timeout (in ms): *	10000
Authentication:	○ None
	● Public Key
	○ User Credentials
	○ Dual
User Name: *	tester
Private Key Alias:	id_cpi_sftp_rsa
Host Key Verification:	● Against Tenant
	○ Against Partner Directory
	○ Off
	☐ Check Directory Access

Response   2023-01-16 12:18:42 UTC

✓ Successfully reached host at my-sftp-virtual-host:2222

Host Key Type: ssh-rsa

Host Key Fingerprint: 14:59:a6:66:c7:8d:2e:a5:d8:6f:08:0b:40:a1:1a:36

Copy Host Key

Send

Figure 3-47. *Public key connectivity test is successful*

If you choose the username/password method to authenticate SFTP connectivity, perform the following actions:

1. Open the SAP BTP Integration Suite. The Monitor section navigates to the Manage Security Material pane. Select Create User Credentials, as shown in Figure 3-48.

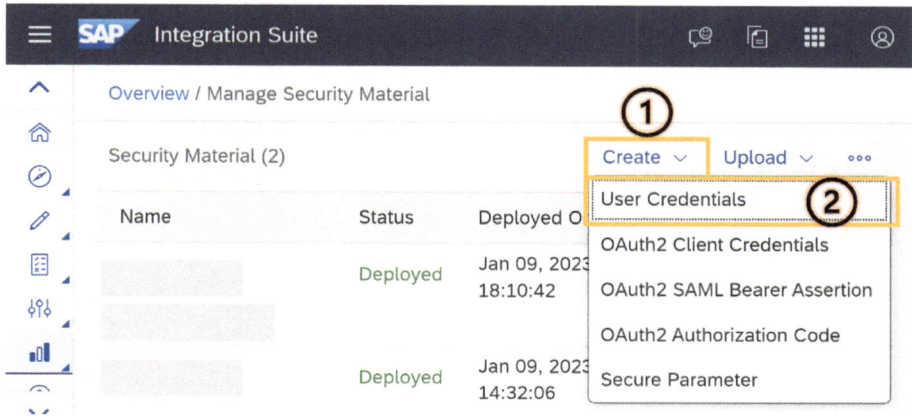

Figure 3-48. *Create the user credentials*

2. Enter the desired name. Choose the User Credential type. Specify the user and password of the SFTP tenant. Click Deploy.

3. You can also upload the SSH known host key in the Security Material pane, as shown in Figure 3-49.

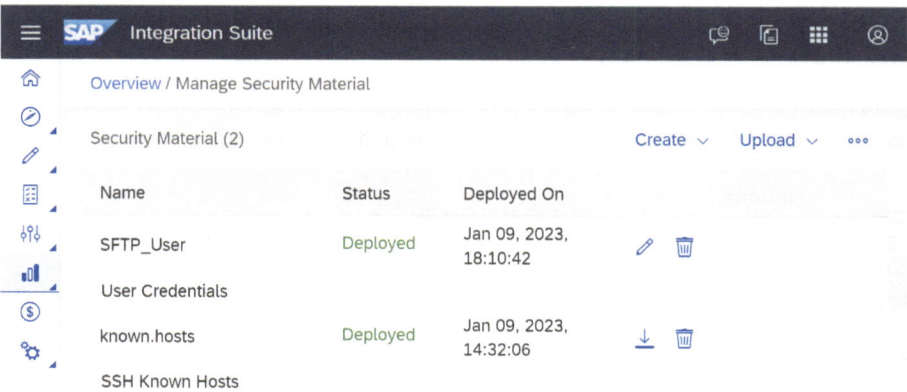

Figure 3-49. *Configuring the user credentials and SSH key*

Test Connectivity

When you test the connection of the inbound SFTP connection, you should obtain a successful result, as shown in Figure 3-50.

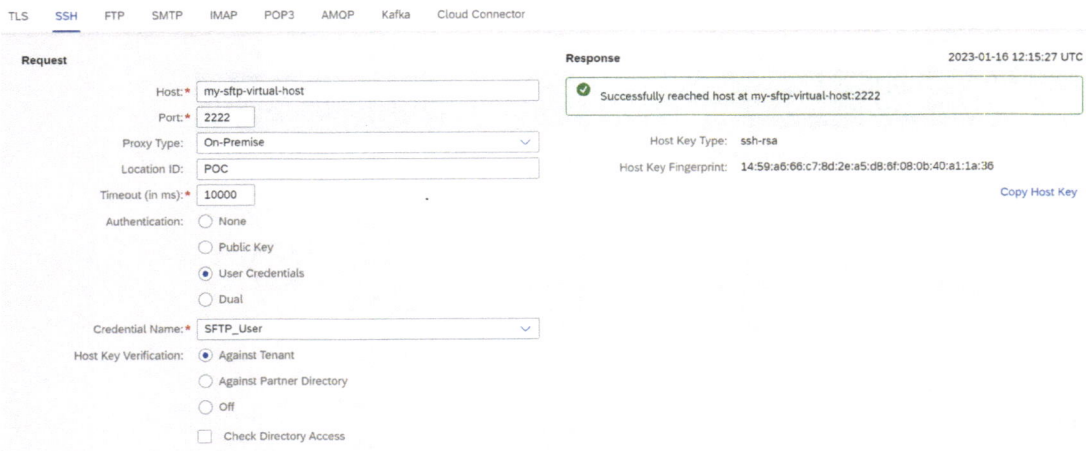

Figure 3-50. *User credentials are successfully connected*

You have successfully set up the inbound SFTP connection and observed successful test results for the UPublic key and the username and password. The next section explains how to set up the inbound connection for the Mail adapter.

Setting Up Inbound Mail Connections

You'll connect the tenant to an email server using the mail sender adapter so that the tenant can read data from the email server (in a process referred to as *polling*).

To put it another way, the tenant sends a request to the email server, but the data flow is in the opposite direction—from the email server to the tenant. In Figure 3-51, the arrow next to the R notation denotes the direction of the request, and the direction of the connection arrow denotes the direction of the data flow.

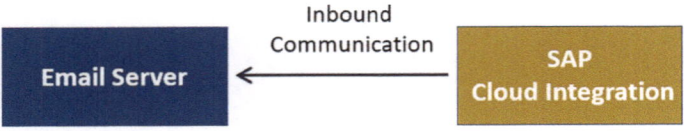

Figure 3-51. *Email server inbound communication*

The IMAP and POP3 protocols can be used to connect to mail servers when using the sender mail adaptor. The possibilities for creating secure connections for the various protocols are listed in Table 3-12. You can use this table as a connection setup checklist.

Table 3-12. *Authentication Options for Mail Connection*

Authentication	Description	How to Configure
Encrypted user/password	The mail sender adapter can download emails from email servers and access both the email body and attachments.	Create and deploy a user credentials artifact that contains the owner's login information (username and password).
Plain user/password	The mail sender adapter can download emails from email servers and access both the email body and attachments. The password and username are transmitted in plain text (only used together with SSL or TLS).	Set the mail adapter settings in the integration flow's mail receiver adapter. Name the User Credentials artifact to use for this connection, specifically, as Credential Name.

You have briefly explored inbound communication in SAP Cloud Integration, and it is time to turn your attention to outbound communication. As organizations increasingly rely on a diverse array of enterprise applications and external systems to support their operations, the ability to enable seamless communication between these systems is essential. With SAP Cloud Integration, you can leverage a range of powerful tools and features to streamline outbound communication and ensure that your enterprise applications are speaking the same language as the external systems they interact with. The next section dives into the details of outbound communication in SAP Cloud Integration, exploring the key features and best practices that can help you achieve seamless integration across your enterprise.

3.4.7 Outbound Communications

The term "outbound communication" describes the transfer of messages from the integration platform to a distant system (where the integration platform is the client).

Setting up a distant receiver system's connection with the integration platform is referred to as configuring outbound communication. You will now see the different steps and options available for setting up the outbound HTTP connection.

Setting Up an Outbound HTTP Connection

To link the tenant to a receiver system using the HTTP protocol, you can use a variety of receiver adapters (for example, the SOAP adapters, the IDoc adapters, and the HTTP adapters). Figure 3-52 shows an illustration of the outbound communication between SAP Cloud Integration and the sender.

Figure 3-52. *Outbound communication*

For connections such as these, you must make sure that the on-premises business systems connecting to the cloud are not immediately accessible to the Internet (when SAP Cloud Integration delivers a message to the on-premises system).

To protect the on-premises system from outside calls, an additional component is linked between the on-premises system and the integration platform in the SAP Cloud (from the Internet).

For this section, there are two alternatives:

- SAP Cloud Connector.

- Reverse proxy (for example, SAP Web Dispatcher).

You have seen the process for inbound HTTP connections in SAP Cloud Integration. Now you will see the different options available for outbound HTTP connections.

Setting Up SAP Cloud Connector with a SAP Account

A feature of the SAP Cloud Platform called SAP Cloud Connector enables a safe and dependable connection between on-premise systems and cloud-based applications. Follow these steps:

1. Install SAP Cloud Connector on your on-premises system.

2. SAP Cloud Connector works as a reverse proxy, and to use it with the SAP BTP, it is necessary to define which subaccount it will be connected to. SAP Cloud Connector requires a subaccount to which it can be connected to Cloud Integration. First, create the subaccount.

3. Copy the subaccount ID from the SAP BTP Cockpit, as shown in Figure 3-53.

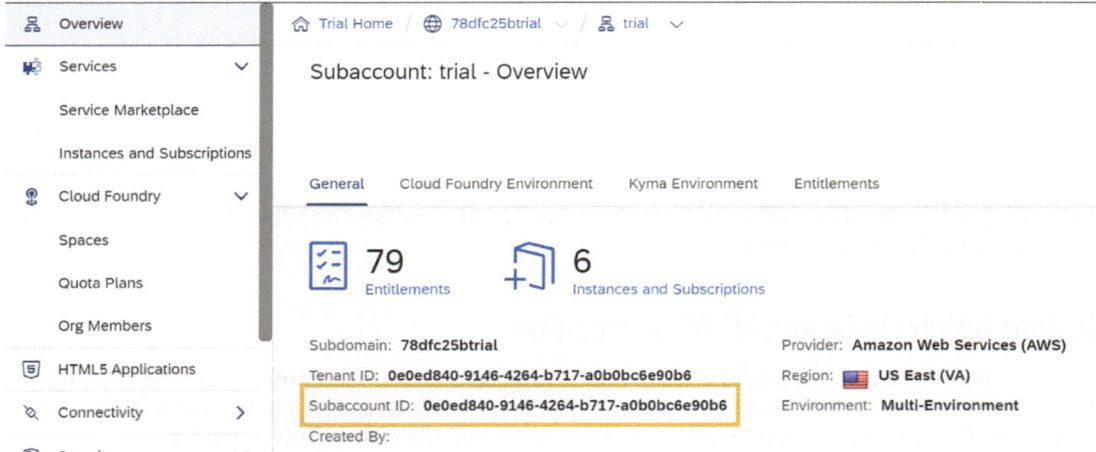

Figure 3-53. *Copy the subaccount ID from the SAP BTP Cockpit*

4. Enter the subaccount details and other details into the SAP Cloud Connector subaccount. The login email and password should be the same as those used for logging in to the SAP BTP Cockpit, as shown in Figure 3-54.

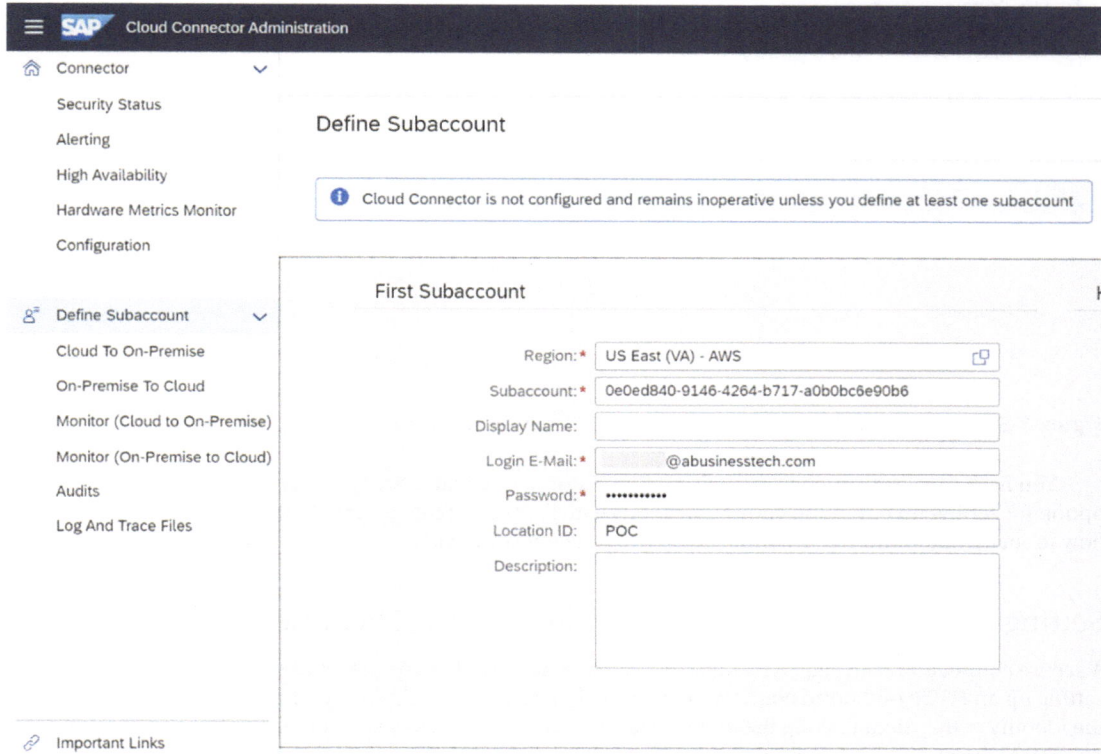

Figure 3-54. *Define the subaccount in SAP Cloud Connector*

5. In the BTP Cockpit, you can check whether the Cloud Connector is connected or not. Go to the Cloud Connector tab in the SAP BTP Cockpit, as shown in Figure 3-55.

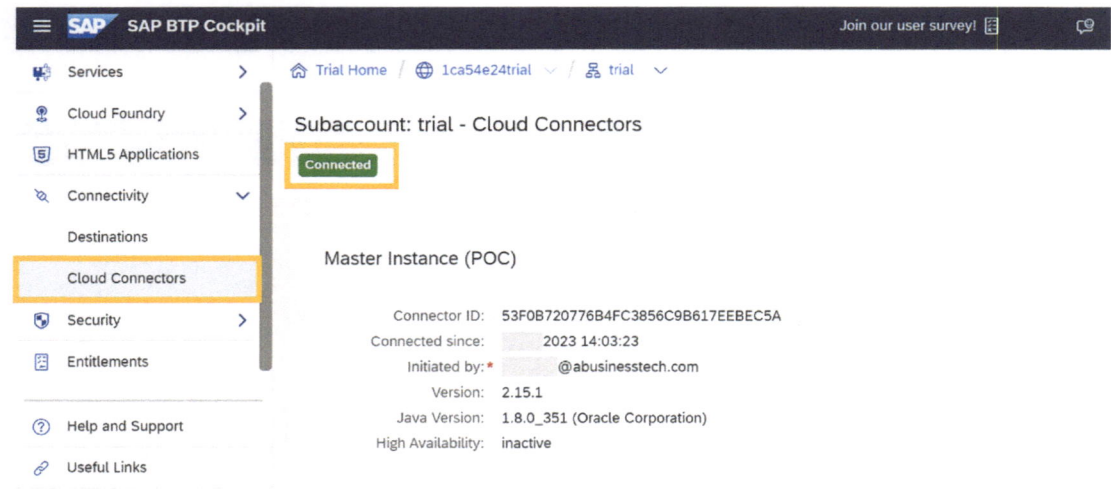

Figure 3-55. *SAP Cloud Connector connected via SAP Cloud Integration*

You have seen the setup process of SAP Cloud Connector with a SAP BTP account. This is the best option for setting up outbound communications with SAP Cloud Integration. In the next section, you will see how to set up outbound HTTP connections using a client certificate.

Setting Up HTTP Outbound Connections with a Client Certificate

A secure connection between SAP Cloud Integration and a third-party system or service is made possible by setting up an HTTP outbound connection with a client certificate in SAP Cloud Integration. By confirming the identity of the client making the request via client certificates, the security of the connection improves.

Procedure:

1. Maintain the tenant Keystore:

   - The tenant cannot self-authenticate as a client against the receiver without having a keystore with an installed client certificate.

   - The tenant that SAP initially provided has a keystore, which contains a foundational set of security artifacts. It can be suitable to establish the outbound connection using an existing key pair.

   - The keystore must also contain a certificate from the certification authority (CA) that authorizes the receiver system's server certificate, as shown in Figure 3-56.

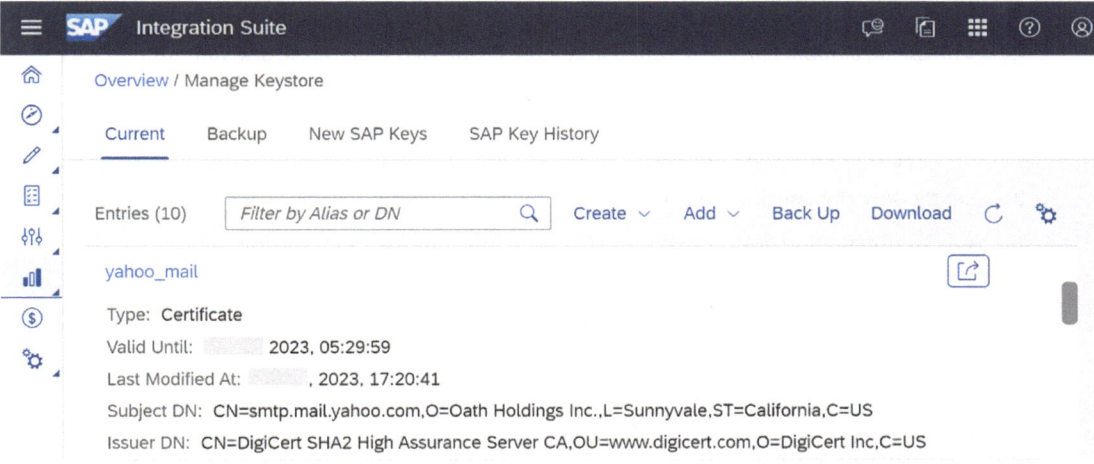

Figure 3-56. *Outbound certificate in keystore*

Import the tenant keystore with the receiving server root certificate. You have the following choices for obtaining this certificate:

- Get the certificate that the receiver administrator handed you.

- Use the receiver system as a test subject for the outbound connectivity test, as shown in Figure 3-57.

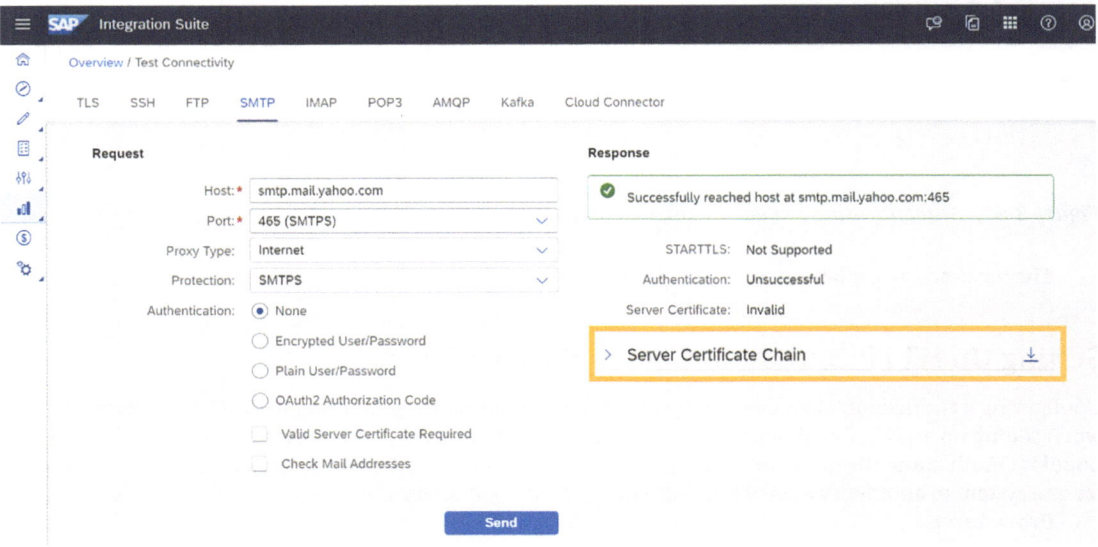

Figure 3-57. *Outbound certificate*

2. Make the receiver keystore appropriate:

- Create a public/private key pair and a certificate signing request, and have the certificate signed by a CA in the same manner as for the tenant keystore.

- Keep in mind that this must be the same CA from which the tenant keystore also acquired the root certificate.

3. Set the security-specific parameters in the relevant integration flow:

- For integration flows, open the SAP Cloud Integration design section.

- Go to the Design tab to construct and design integration flows.

- Open the associated receiver adapter (used to define the connection between the tenant and the receiver system) and select the client certificate for authentication.

- You can select a specific private key from the tenant keystore (tenant client certificate) to be utilized for this step by entering a Private Key Alias, as shown in Figure 3-58.

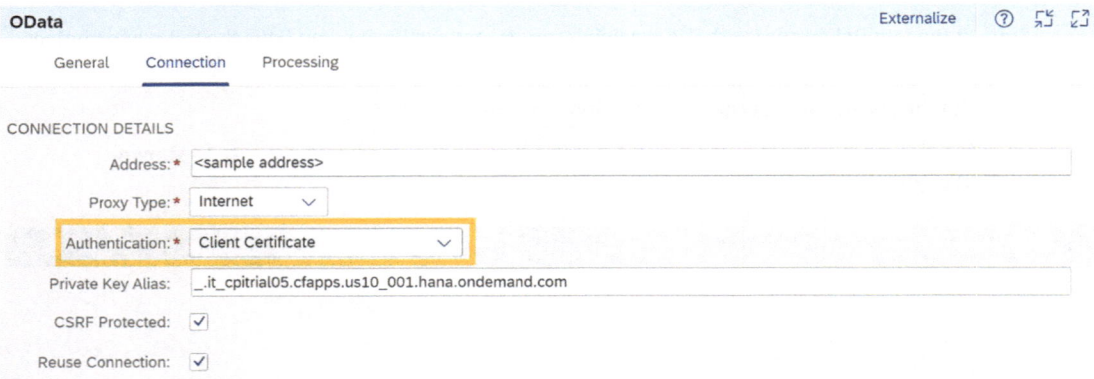

Figure 3-58. *Authenticating a client certificate*

The next section explains how to set up HTTP outbound connections using OAuth.

Setting Up HTTP Outbound Connections with OAuth

Configuring a connection to an external system that needs authentication using OAuth 2.0 is required when setting up an HTTP outbound connection using OAuth in SAP Cloud Integration. With the help of the popular OAuth authentication and authorization protocol, you can offer restricted access to your resources on one system to another system without disclosing your login information.

Procedure:

1. Obtain the OAuth connection information from the receiving system that has to be connected. Examples of this include the address of the token service, which issues the OAuth access token on behalf of the recipient.

2. Create one of the following artifacts, depending on the receiver adapter type and the OAuth grant type you want to use:

- Client credentials for OAuth2

- SAML Bearer Assertion OAuth2

- Authorization Code for OAuth2

3. Select the Security Material tile under Manage Security in the Monitor portion of the screen, as shown in Figure 3-59.

   Don't worry, you will learn how to create user credentials in the management security material in Chapter 5.

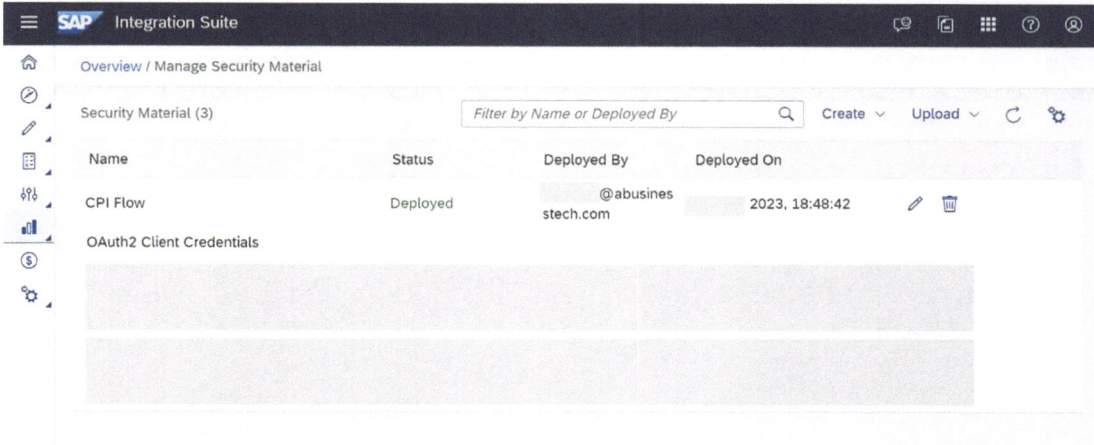

Figure 3-59. *OAuth2 client credentials*

4. Select the appropriate authentication option and enter the credential name in the receiver adapter of the associated integration flow, as shown in Figure 3-60.

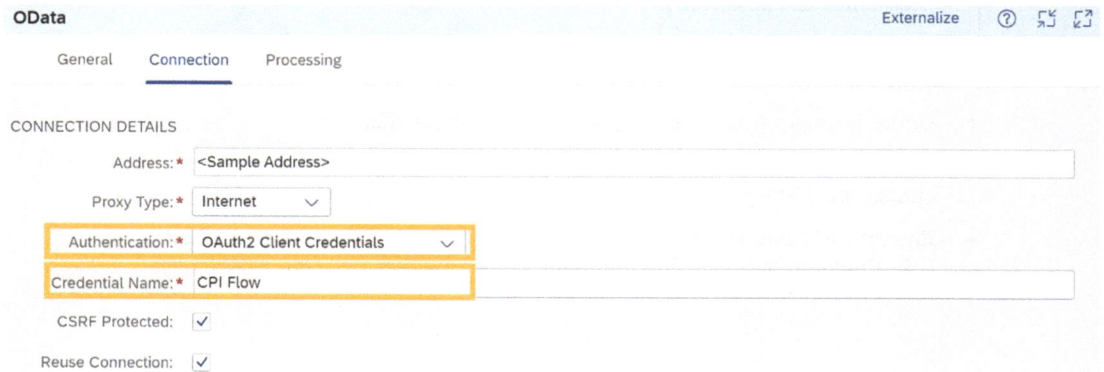

Figure 3-60. *OAuth2 client credentials in integration flow*

You have now successfully set up outbound HTTP connections using OAuth. In the next section, you learn to set up the HTTP outbound connections using basic authentication.

Setting Up HTTP Outbound Connections with Basic Authentication

You can securely communicate with an outside system or service that needs authentication by setting up an HTTP outbound connection with basic authentication in SAP Cloud Integration.

Procedure:

1. A tenant keystore containing the receiver server root certificate should be created and deployed. To identify (authenticate) the recipient system as a trusted server, this certificate is necessary. A tenant creates and deploys a keystore with the receiving server root certificate. This certificate is required to recognize (authenticate) the receiver system as a reliable server, as shown in Figure 3-61.

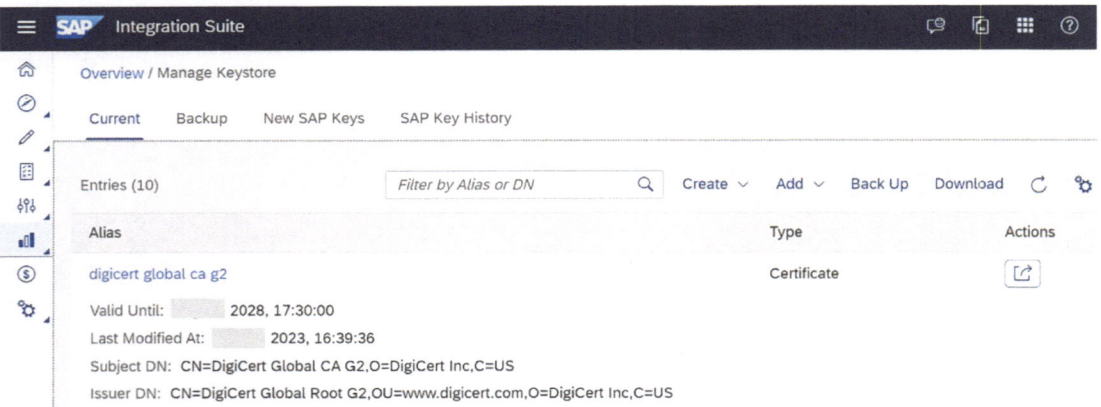

Figure 3-61. *Managing the keystore certificate*

2. Create and deploy the tenant's credentials.

   - The tenants calling the receiver system must be verified by using their usernames and passwords:

   - Select the Monitor section using the same URL as the integration flow design tool.

   - Under Manage Security, select the Security Material tile.

   - Select Add to add a new user credentials artifact to the tenant's collection or change an existing one.

   - Enter the attributes (Credential Name, User Type, and Password) to the Add User Credentials screen, and then click Deploy, as shown in Figure 3-62.

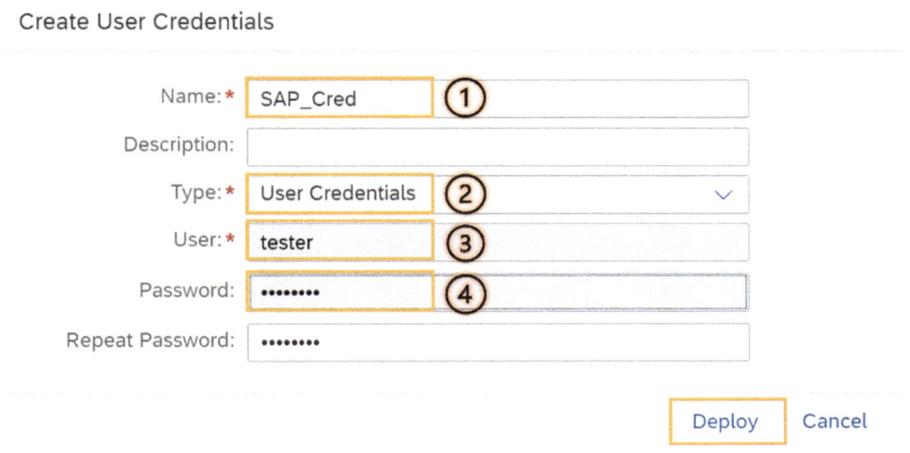

Create User Credentials

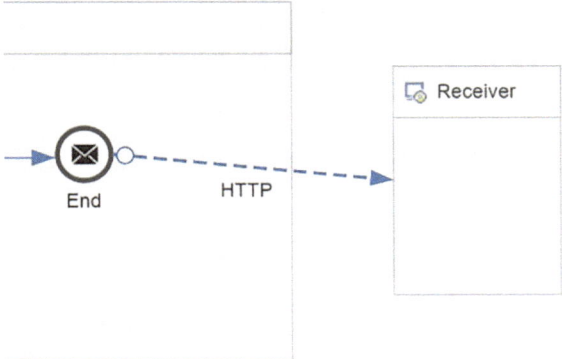

Figure 3-62. *Creating user credentials*

3. Set the security-specific parameters in the relevant integration flow:

    - For integration flows, open the SAP Cloud Integration design section.

    - Go to the design area to construct and design integration flows.

    - Open the associated receiver adapter (HTTP), as shown in Figure 3-63.

Figure 3-63. *HTTP receiver adapter*

    - Select Basic under Authentication, then input the credential name to specify the relationship between the tenant and the receiver system.

    - The User Credentials artifact that you previously deployed on the tenant has this name, as shown in Figure 3-64.

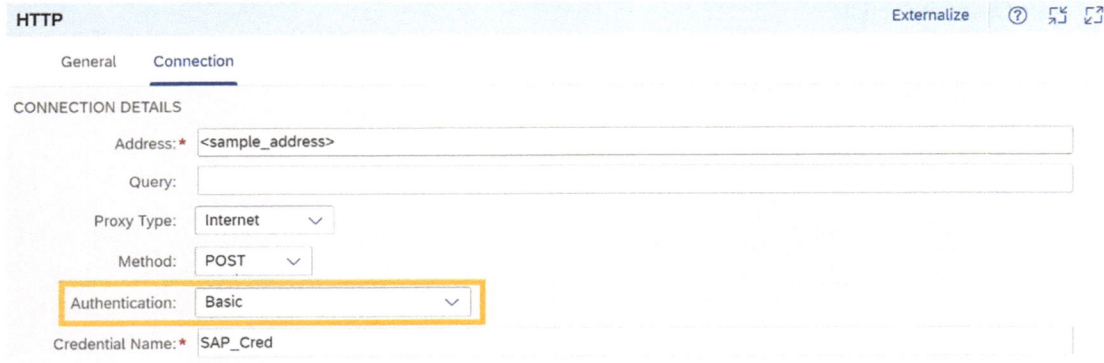

Figure 3-64. Basic authentication

You have now successfully set up HTTP outbound communications using basic authentication. You have seen the HTTP connection and various steps available for setting up this outbound communication. The next section explains the outbound connection for SFTP and explores the world of SFTP communication.

Setting Up an Outbound SFTP Connection in Cloud Integration

With the help of the SFTP receiver adapter, you connect the tenant with an SFTP server so that they can upload files to the SFTP server.

To put it another way, the tenant contacts the SFTP server, as shown in Figure 3-65, and data travels in the same path from the tenant to the SFTP server. The direction of the connection arrow indicates the flow of data.

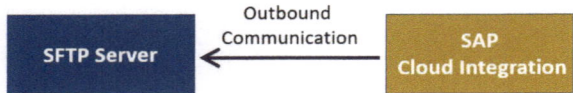

Figure 3-65. Outbound communication for SFTP connections

The SFTP adapter occasionally enables you to choose one of the following for SFTP connectivity authentication:

- Username/password
- Public key

If you choose public key, take the following actions:

1. Maintain a private key pair in the tenant keystore (Create or Add). Provide an alias and use it again in the next steps. (Check out Section 3.4.6.2 to learn how to add the private key pair.)

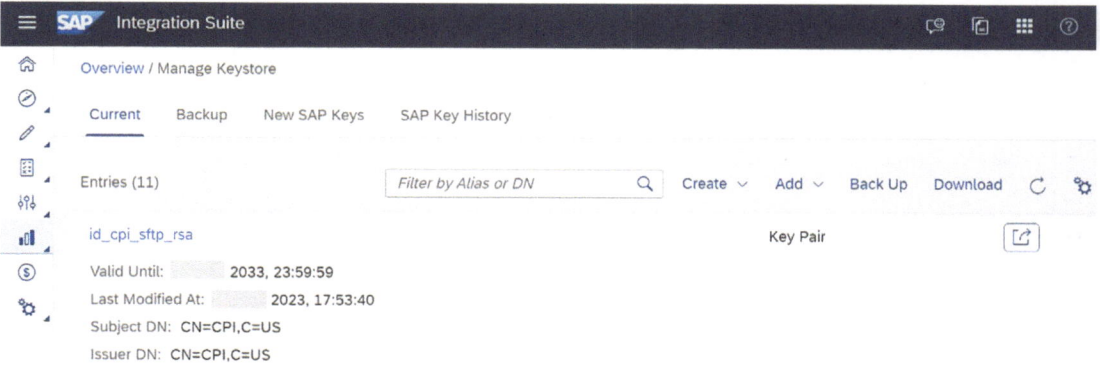

Figure 3-66. *Key pair for SFTP*

2. Add the host key for the SFTP server to the tenant's known host file (as an SSH known host artifact).

3. Select Public Key in the integration flow for the SFTP receiver adapter, then enter the alias for the key in the keystore, and enter the user (which is defined in the SFTP server and provided by the server admin), as shown in Figure 3-67.

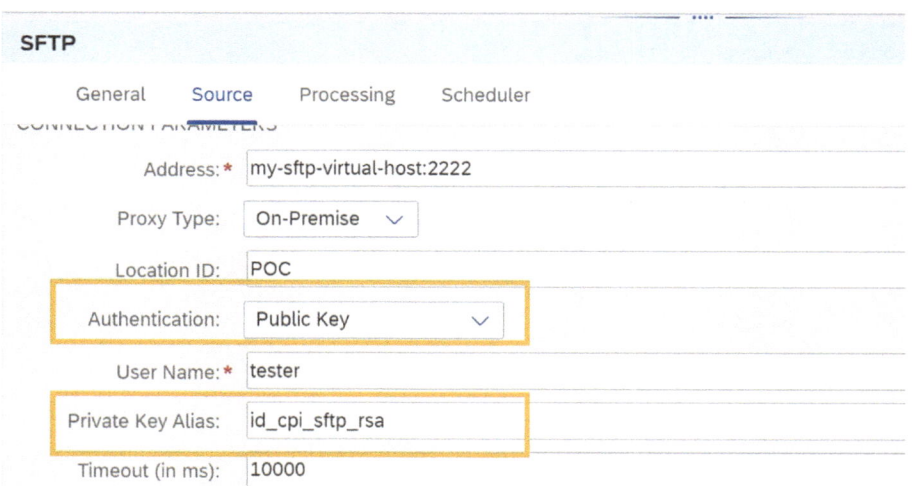

Figure 3-67. *Public key authentication in integration flow*

If you choose the username/password to authenticate SFTP connectivity, perform the following actions:

1. Create a User Credentials artifact, enter the username and password, and deploy the item to the tenancy, as shown in Figure 3-68.

Figure 3-68. *User credentials*

2. Install the known host file on the tenant and add the SFTP server's host key to it (as an SSH Known Hosts artifact).

3. Select User Name/Password authentication and the User Credentials artifact in the integration pipeline for the SFTP receiver adapter (and enter the credentials there), as shown in Figure 3-69.

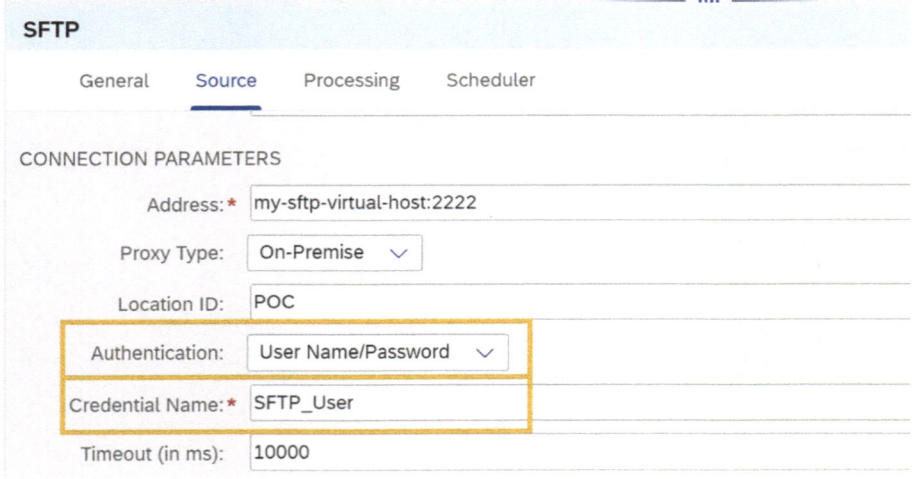

Figure 3-69. *Username/password authentication in integration flow*

You have successfully seen the outbound connection process for the SFTP. The next section explains the outbound communication process when using a mail connection.

Setting Up an Outbound Mail Connection in SAP Cloud Integration

With the mail receiver adapter, you connect the tenant to an email server so that the tenant can send emails to the server.

To put it another way, the tenant contacts the email server, as shown in Figure 3-70, and information travels in the same way from the tenant to the email server.

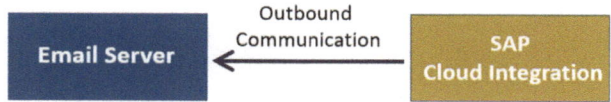

Figure 3-70. *Outbound communication with the mail adapter*

You can establish SMTP connections with mail servers by using the mail receiver adapter.

Table 3-13 provides a list of options for creating secure connections. You can use this table as a connection setup checklist.

Table 3-13. *Authentication for the Mail Connection*

Authentication	Description	How to Configure (Checklist) ...
Encrypted user/password	Prior to being transferred to the server, the username and password are hashed.	Create and deploy a User Credentials artifact that contains the owner's login information (username and password).
Plain user/password	The password and username are transmitted in plain text (only use together with SSL or TLS).	Set the mail adapter settings in the integration flow's mail receiver adapter. Name the User Credentials artifact to use for this connection, specifically, as Credential Name.

You have now set up the outbound mail connection successfully and seen the different options available to set up the outbound connection, whether it be HTTP or SFTP. The next section looks deeply into X.509 keys. You will learn how you can create such keys.

Creating X.509 Keys

You will need X.509 keys to configure communication when utilizing certificate-based authentication over HTTPS as well as PKCS#7 and XML Digital Signature security standards for digital message encryption and signing. Let's look at the different steps available for creating X.509 keys.

Generate a Key Pair

To create X.509 keys to use in protecting connections between various systems or applications in SAP Cloud Integration, it is important to first generate a key pair.

The procedure to generate a key pair for producing X.509 keys in SCI is as follows:

1. From the Manage Security section, select the Keystore tile, as shown in Figure 3-71.

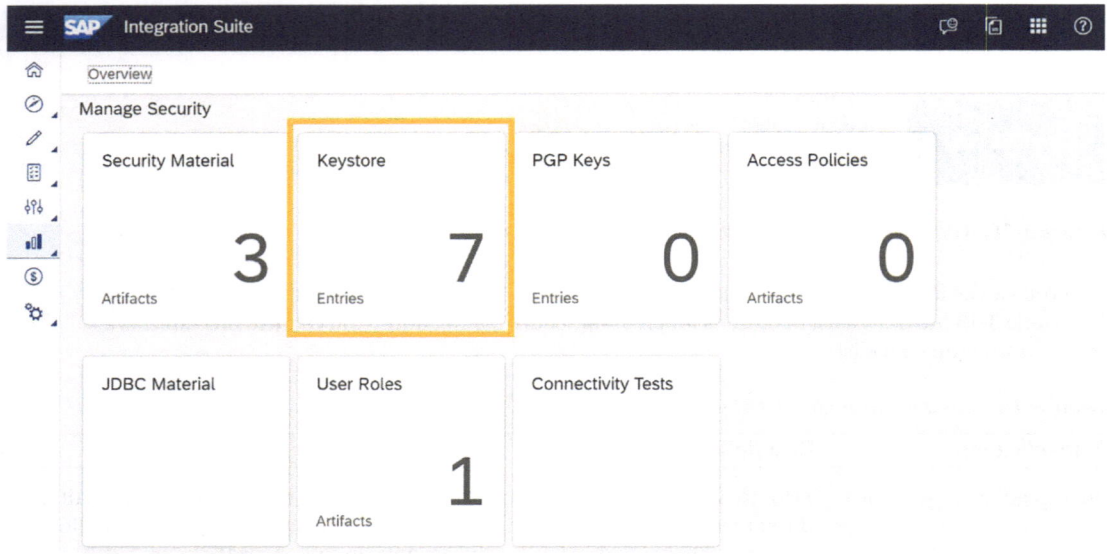

Figure 3-71. *Keystore tile*

2. Select Create from the Current tab. From the popup menu, select Key Pair, as shown in 3-72.

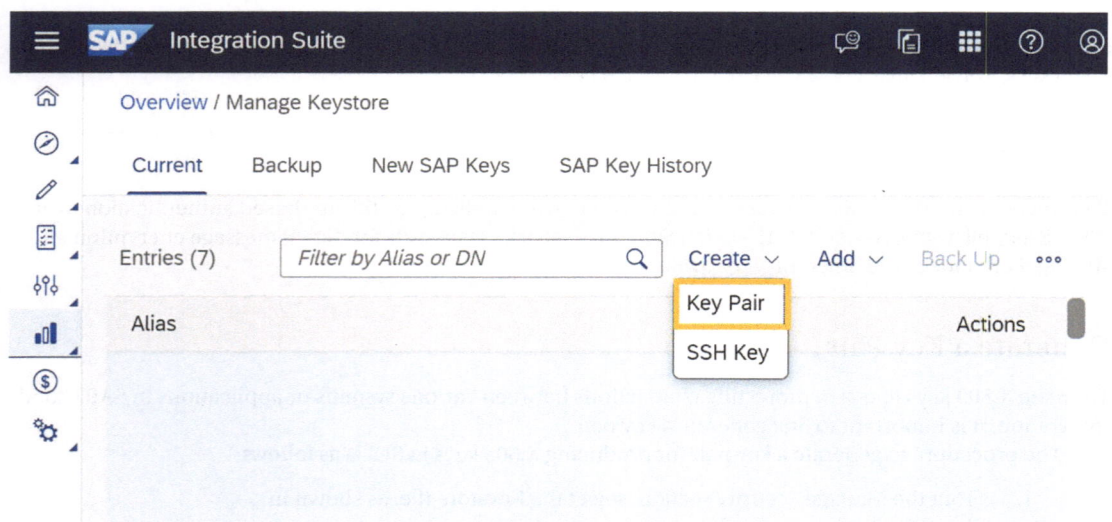

Figure 3-72. *Create a key pair*

3. Provide the necessary information by adding an alias. List the important values and designate the duration. Click Create, as shown in Figure 3-73.

Create Key Pair

Alias: *	abt_ci_091
Key Type: *	RSA ⌄
Key Size: *	2048 ⌄
Signature Algorithm: *	SHA-512/RSA ⌄
Common Name (CN): *	SAP_CPI
Organizational Unit (OU):	
Organization (O):	
Location (L):	
State or Province (ST):	
Country/Region (C): *	US
E-Mail (E):	
Valid From: *	Jan 24, 2023 📅
Valid Until: *	Jan 24, 2025 📅

Create   Cancel

Figure 3-73. *Creating an SSH key*

4. Save and Deploy.

You have now successfully generated a key pair. The next section moves on to the next step of downloading a certificate signing request.

Download a Certificate Signing Request

A certificate is initially created using a self-signed certificate. Before it can be utilized, it needs to be signed by a certification authority (CA). You must first download a certificate signing request (CSR) from the Keystore Monitor to obtain a certificate that has been signed by a CA.

Procedure:

1. Open the Cloud Integration Portal and navigate to the Monitor tab.

2. From the Manage Security section, select the Keystore tile.

3. Choose a key combination from the Current tab.

4. Select Download Signing Request after clicking the (Actions)icon. As an alternative, you can choose Download Signing Request option from the key pair details by clicking the key pair alias to access them, as shown in Figure 3-74.

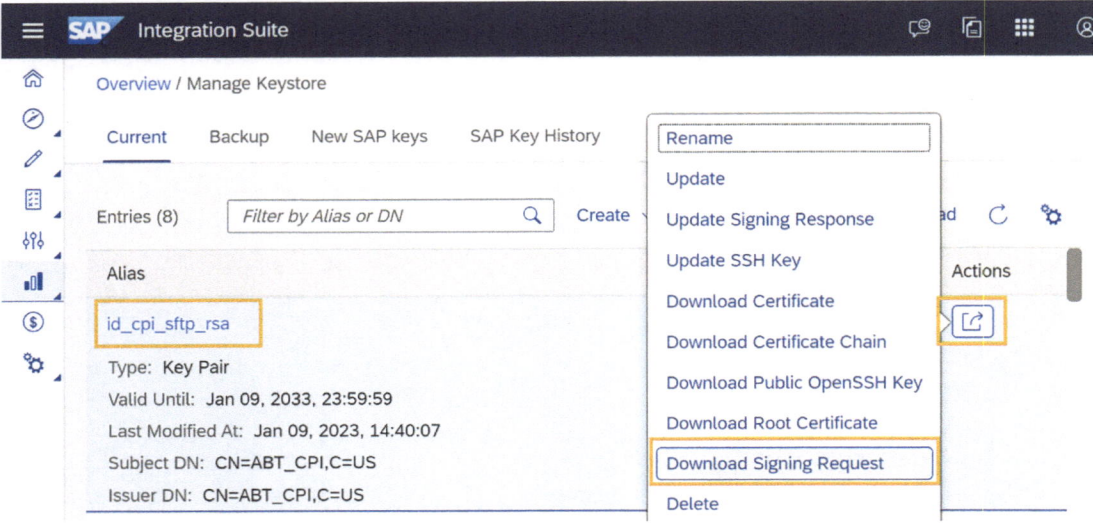

Figure 3-74. *Download Signing Request option*

    5.   This process downloads a file called `alias.csr`.

You have now downloaded a certificate signing request. The next section moves on to next step, i.e., you have to create a request for a signed certificate from a certification authority.

Request a Signed Certificate from a Certification Authority

The tenant cannot interact as a client with the customer system until a client certificate is imported into the tenant client keystore. This certificate must be signed by a certifying body (CA).

After requesting a signed certificate from a Certification Authority, you are ready, and your X.509 keys will be created successfully. In the outbound communication, the most crucial step is to exchange the key material, which you learn about in the next section.

Exchange the Key Material

In many cases, establishing a secure connection requires communication parties to exchange public keys.

Asymmetric (or public) key technology and private/public key pairs are used by communication partners to provide secure communication between software systems. At specific moments throughout the setup process, the partners need to exchange public keys.

When sharing essential materials, you must take specific precautions to guarantee that the security of your scenario is not compromised. In the next section, you learn about exchanging of the public keys.

Public Keys

When transferring public keys, you must ensure that they cannot be changed by a third party (for instance, X.509 certificates). You can choose from the following secure methods:

- To exchange keys, use a secure communication channel. For instance, you can utilize secure collaboration tools such as SAP Jam or email that has been encrypted and signed using PGP.

- Verify the sender and make sure they have permission to provide you with this important information (by using a signature for example).

- Make sure the material was not edited (usually using a signature).

You can verify the integrity of the keys using alternative methods, such as the ones listed here, if you are unable to use a secure communication channel:

- Verify the validity of the certificate and that a reliable certifying body issued it in the case of X.509 certificates (CA).

- Verify that the sender's fingerprint matches the fingerprint on the key using a different communication channel, such as a phone.

The next section explains the role of private keys.

Private Keys

Private keys are even more crucial than public keys. Your messages will be able to be signed with your signature and decrypted by others if they have access to your private key. Avoid sharing private key material wherever feasible.

If you must swap private keys in unusual circumstances, take one of the following precautions:

- Use a password-protected encrypted container (such as PKCS#12 or Java Keystore).

- Transmit the password through a different route of communication (for example, a phone).

- Use encrypted methods for communication. Never utilize plain HTTP or plain email.

- Receive a method for securely transferring keys from SAP.

You have seen the different steps for exchanging public and private keys. In outbound communication, you will see secure communication. You learn how SFTP works in the next section.

Secure Communication

Message exchange can be protected in a number of ways. By implementing particular authentication techniques and using the HTTPS or SFTP protocol, communication can be encrypted at the transport level. Additionally, you can set up procedures for digitally signing and validating messages, as well as for encrypting and decrypting the message content.

SFTP-Based Communication

SFTP (secure file transfer protocol) is a safe protocol for exchanging files across systems. The SFTP protocol is supported by SAP Cloud Integration, enabling users to transfer files securely between Cloud Integration and external systems.

How Does SFTP Work?

A tenant can connect to an SFTP server as an SFTP client (the latter either hosted at SAP or in the customer landscape). An SFTP sender adapter or SFTP receiver adapter is involved, depending on the direction of data flow (whether the tenant reads data from the SFTP server or writes data to it).

On the SFTP server, files are kept in designated folders, known as *mailboxes*. A user is provided a mailbox to manage access to the information.

The SFTP (sender or receiver) adapter offers the following authentication methods in some circumstances for SFTP connectivity:

- Username and password

- Private key

Username/Password

The tenant establishes a user-password connection to the server and performs SFTP server self-authentication. Prior to the connection setup process, a User Credentials artifact is deployed to the tenant and contains the user credentials (the username and password).

Public Key

A mix of symmetric and asymmetric keys establishes a secure connection between the SFTP client and server. Clients and servers that use SFTP to exchange data employ symmetric and asymmetric keys in the following ways:

1. A connection is made between the client and server.

2. The public key of the server is emailed to the client.

3. To determine whether the server is a trusted participant, the client examines a known host file on the client's side; if the server's public key is referenced there, the server's identity is confirmed.

4. The client generates a session key.

5. The client encrypts the session key using the public key of the server.

6. The client sends the encrypted session key from the server. As the public and private keys of one party are mathematically tied to one another, the server can use its private key to decrypt the session key.

7. The session can continue now that it has been encrypted.

8. As part of secure data transfer, the client sends the server its public key (using the session key exchanged in the preceding step).

9. The server assesses whether it is familiar with the client's public key by checking an authorized key file on the server side.

10. The server encrypts a random integer and sends it to the client after using the client's public key.

11. Using its private key, the client uses its private key to decode the random number and sends the server the unencrypted version. This is how the client authenticates on the server side.

3.5 Summary

This chapter explained the capabilities and connectivity choices offered by SAP Cloud Integration. It discussed the idea of cloud integration and highlighted the salient characteristics of SAP Cloud Integration. It included a general overview of the SAP Cloud Integration web UI and demonstrated how to construct an interface utilizing a variety of tools, including a Start timer, Content Modifier, outbound OData channel, Groovy scripts, and Deploy and Monitor features.

The chapter also covered the various adapters available in SAP Cloud Integration, namely, sender and receiver adapters. You learned how to configure the JDBC adapter with a practical example.

The chapter discussed the scenarios for inbound and outgoing communication, such as setting up connections for inbound and outbound HTTP, SFTP, and mail. The chapter explained how to generate X.509 keys, trade key material, and guarantee secure connections.

SAP Cloud Integration offers a number of inbound adapters, including HTTP, File, and JDBC, that let companies accept messages from different sources and translate them into a format that the target application can understand. On the other hand, outbound communication describes messages that are transmitted from the integration platform to a target application that is located outside the system. Outbound adapters from SAP Cloud Integration, including those for HTTP, SOAP, and SFTP, let companies send messages to a variety of target apps without any issues.

You should now have a thorough understanding of the many functions of SAP Cloud Integration and its range of adapters. This chapter discussed the concept of integration flow, but what does it mean and how is it created? In the next chapter, you learn how integration flows are developed using all of the adapters discussed in this chapter. You also learn about the many flow components that can be used during development.

CHAPTER 4

▪ ▪ ▪

SAP Cloud Integration: Development Part I

In this chapter, you learn about the numerous design object features and elements that are necessary for creating integration projects using SAP Cloud Integration.

You also see the development overview and go through some examples of packaging integration material, which involves building, importing, editing, and exporting integration packages. Together with looking at predelivered connection packages from SAP Partners, you will also learn about integration procedures for connecting SAP SuccessFactors with JIRA and SAP S/4HANA with ServiceNow API.

The chapter looks at various mapping methods, including Message Mapping and its testing and functionality as well as ID Mapping, Operation Mapping, XSLT Mapping, and Message Mapping. The chapter also examines the Cloud Integration Development design object elements and provides useful illustrations on how to use these *iFlow* (integration flow) development pallet functions.

4.1 Integration Content

Integration Content in SAP Cloud Integration refers to pre-built packages of integration artifacts that enable seamless connectivity and data exchange between different applications, systems, and processes. You should design integration content so that it indicates how messages will be passed between the connected components. With the tools and applications offered by SAP Cloud Integration, you can execute end-to-end processes on development and deployment, packaging and publishing, accessing, and editing the integration material. This topic covers the duties, organizational structure, and roles involved in managing integration content.

There are different roles and working environments, such as Integration Designer on the Eclipse platform or SAP Cloud Integration Web application on SAP UI5. To start with development in SAP Cloud Integration, you first learn about the integration packages.

4.1.1 Packaging Integration Content: Practical Example

Creating a package of integration artifacts, such as integration flows, mappings, and configurations, that can be readily shared and deployed across many environments or systems is referred to as packaging integration content in SAP Cloud Integration. This allows for easier management and maintenance of integration content, as well as the ability to reuse integration content across multiple projects. The package can be created using the Cloud Integration Content Packager, which is a tool that allows users to select the integration content they want to include in the package and then export it as a file. Once created, the package can be imported into other systems or environments, making it easy to deploy and manage integration content across multiple locations.

© Jaspreet Bagga 2023
J. Bagga, *A Practical Guide to SAP Integration Suite*, https://doi.org/10.1007/978-1-4842-9337-9_4

There are different actions that you can perform with integration packages, such as Create, Import, Edit, and Export. The following sections describe each action that you can perform with integration packages, the prepackaged integration packages, and the features of the integration packages.

Creating an Integration Package

SAP Cloud Integration allows you to create an integration package for your tenant.
    Use these steps to create an integration package:

1. In the Design tab, choose Create.

2. In the Header section, provide all the mandatory details, like Name, Technical Name, and Short Description. You can also provide the rest of the information, as shown in Figure 4-1.

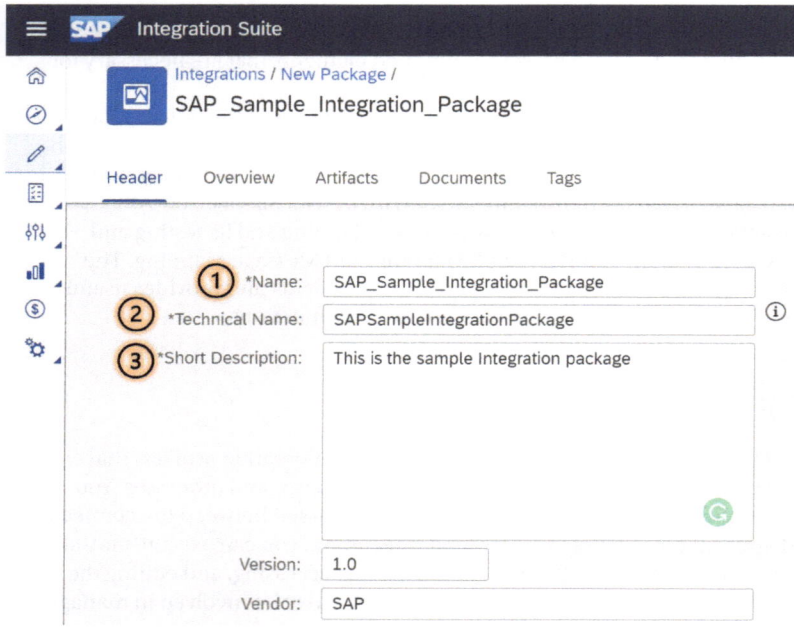

Figure 4-1. *Creating the integration package*

You have successfully created the integration package. But what if you have already created the integration package and you want to edit it further and add more artifacts to it? In such cases, you have to import the integration package, which you learn about in the next section.

Importing the Integration Package

You can quickly and easily deploy premade integration scenarios and templates into your Cloud Integration tenant by importing an integration package in SAP Cloud Integration (SCI). An integration package is a collection of integration artifacts, including mappings, configurations, and flows, all of which have been specifically created to support a single integration use case.

Perform the following steps:

1. In the Design tab, select Import, as shown in Figure 4-2.

    The preset values of the externalized package parameters are not overwritten when a new version of the package is imported (via ZIP import).

Figure 4-2. *Importing an integration package*

2. To upload the package, browse to the integration package and click the OK button.

3. When a package contains one or more artifacts that are present in other packages of the tenant, the tenant experiences a uniqueness conflict, and the import of the package fails. For example, if there is an artifact say Integration flow A used in Integration package A, also the same Integration flow A is used in Integration Package B, then it will show error if you will try the import the integration package.

You have successfully imported the integration package into SAP Cloud Integration. At this point, you have learned how to create and import the integration packages. Do you know anything about integration packages? The next section covers the integration package features.

Features of the Integration Package

In SAP Cloud Integration, an integration package is a collection of integration artifacts that cooperate to carry out a particular integration scenario. One or more integration flows, APIs, mappings, and other artifacts can be included in an integration package. These artifacts are packaged together for easy administration and deployment.

The key features of the integration packages are listed here:

1. In the Design tab, select your integration package.

2. In the Artifact tab, you can perform various actions in accordance with the available artifacts in the package.

3. In the Action items, you can see various options that perform various actions according to their functionality, as shown in Figure 4-3.

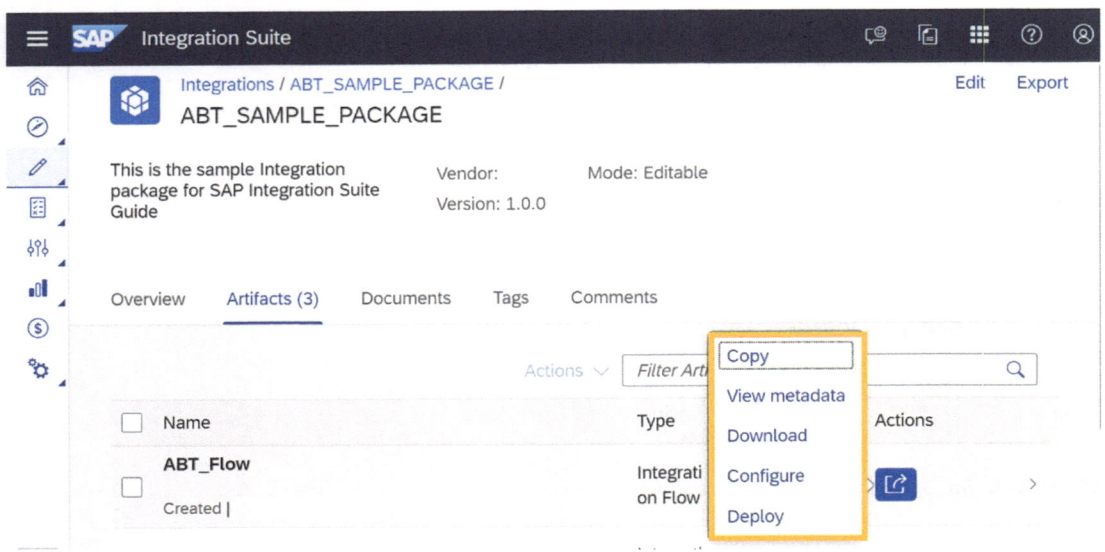

Figure 4-3. *Actions*

Table 4-1 lists the different options available in the Action tab along with their descriptions.

Table 4-1. *Options Available from the Action Tab*

Action	Description
View Metadata	This makes it easier to view the artifacts' metadata.
Deploy	This helps deploy the artifact. However, only value mappings, integration flows, and data flows can be deployed. A number of value mappings can be deployed simultaneously.
Configure	This aids in setting up the artifact, but Cloud Integration enables you to set up a single integration flow or several integration flows simultaneously.
Download	This makes it easier to get the artifact; however, you can only download value mappings, integration flows, and files.

You have learned about the features of integration packages. You also learned how to create and import integration packages. What if you have to edit an integration package? The next section explains the process of editing integration packages in SAP Cloud Integration.

Editing an Integration Package

Add or remove content from the artifact, edit the header and overview sections, and delete the integration package by editing it.

Now, to edit the integration package, perform the following steps:

1. Select the integration package that you want to edit from the Design tab. Open the integration package in Edit mode, as shown in Figure 4-4.

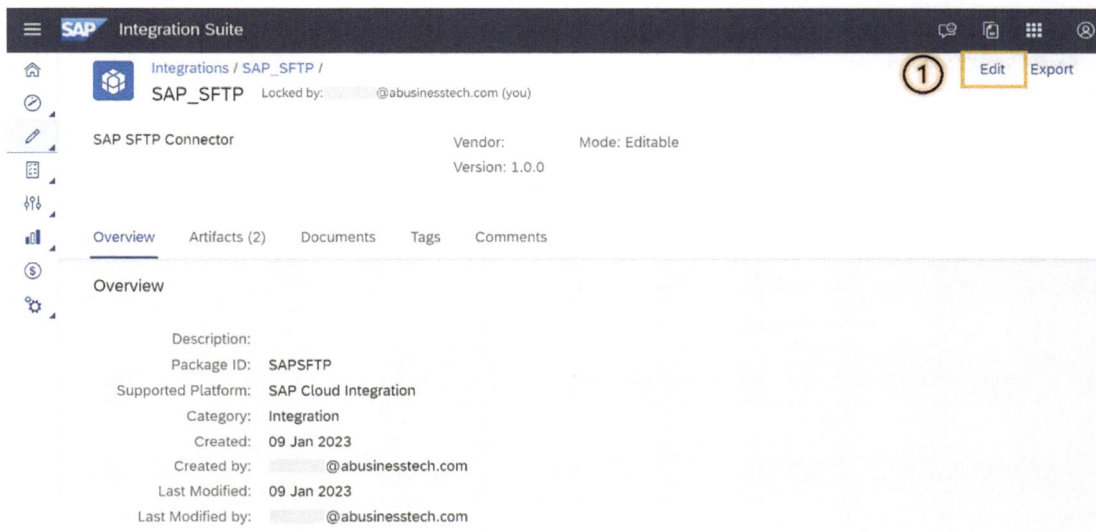

Figure 4-4. *Open the integration package in Edit mode*

2. From the Header table, you can edit the package name, a short description, a version number, and the package vendor, as shown in Figure 4-5.

    On top of the integration package, all the data is visible. Additionally, the name, a brief description, and the version are shown in the list of integration packages in your design workspace.

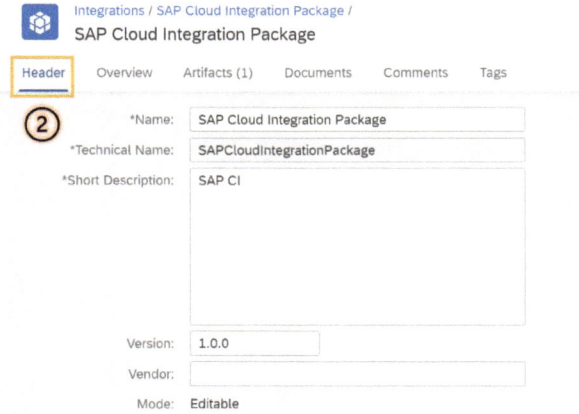

Figure 4-5. *Header of the integration package*

3. From the Overview tab, you can supply a more thorough description of the integration package, as shown in Figure 4-6. Links can also be included. You also have the choice of uploading a package image. The overview page's image is positioned at the top.

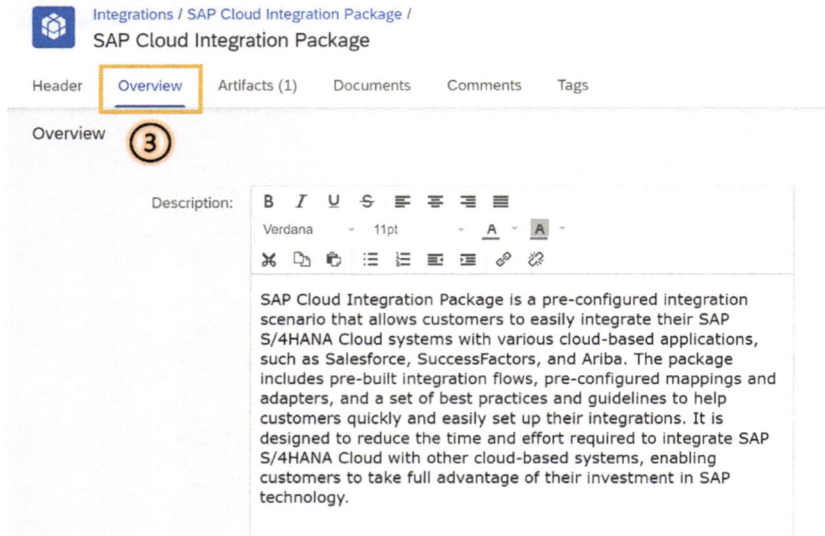

Figure 4-6. *Overview section of the integration package*

4. In the Artifacts tab, as shown in Figure 4-7, choose the Add option, whereby you can create the following functions:

- Integration Flow
- REST API
- Message Mapping
- SOAP API
- Value Mapping
- ODATA API
- Script Collection
- Integration Adapter

For various artifact types, a varied set of functions is available:

- Delete
- Copy
- View Metadata
- Download
- Configure (only for integration flows)
- Deploy (only for data flows and integration flows)

To edit the metadata of the artifact, you have to select the View Metadata option and select Edit. After giving the required information to the metadata, click Save. The changes will be updated in the Artifact metadata.

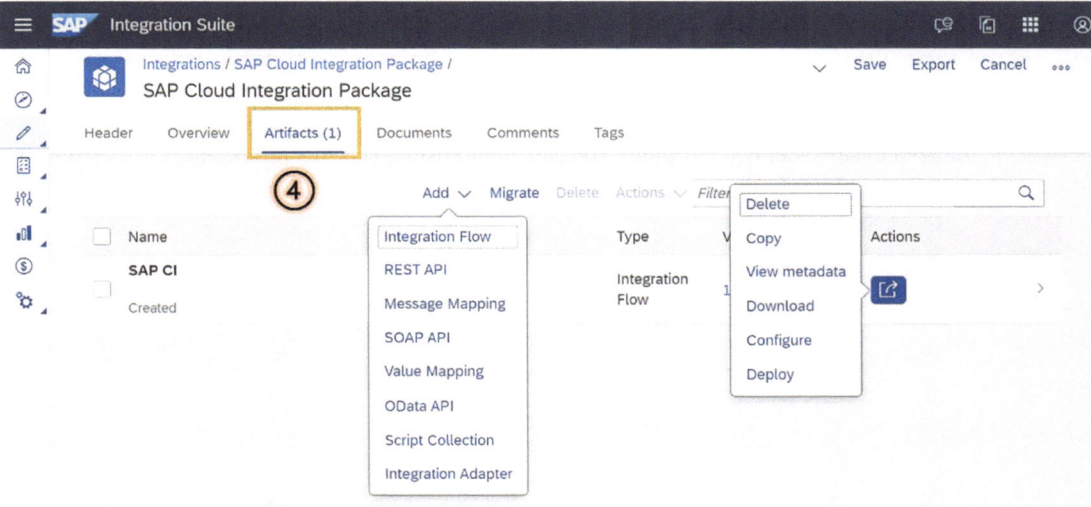

Figure 4-7. *Artifacts tab of the integration package*

5. From the Documents tab, you can add URLs and upload materials, such as configuration manuals, as shown in Figure 4-8.

Figure 4-8. *The Documents tab of the integration package*

6. From the Tags tab, as shown in Figure 4-9, you can maintain standard tags such as Product, Country/Region, Lines of Business, and Industry. Additionally, based on your custom-defined attributes, you can apply custom tags to identify and monitor your packages.

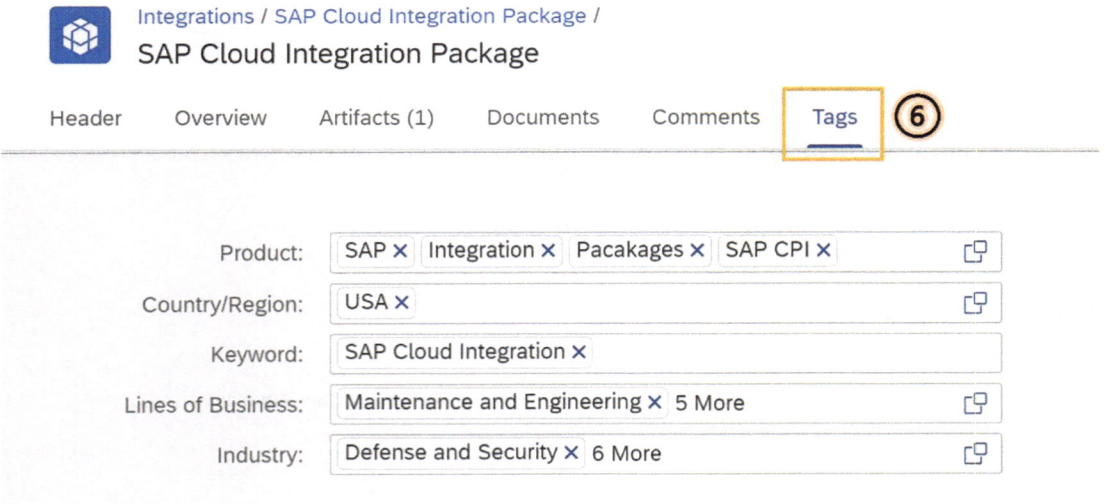

Figure 4-9. *Tags tab of the integration package*

7. You can save or cancel the changes you made to the integration package.

You have now seen all the detailed versions of the integration packages. You also created an integration package. The next section explains how to send/export that package to your development team for changes or approval.

Exporting an Integration Package

Users can create a package that comprises all the artifacts and parameters necessary for a specific integration scenario by exporting an integration package in SAP Cloud Integration. To reuse the package or create a backup, you can import it into another Cloud Integration tenant.

To share or download the integration package, perform the following steps to export an integration package:

1. Choose the integration package from the Design tab, which you have to export.

2. Open the integration package and click Export, as shown in Figure 4-10.

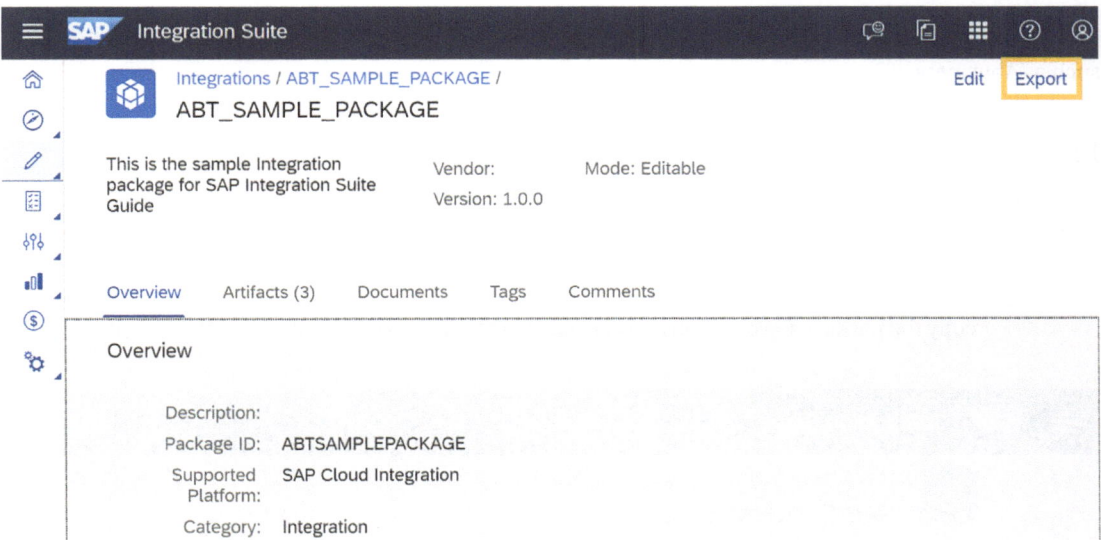

Figure 4-10. *Exporting the integration package*

3. The integration package will be downloaded to the local system with the name of the integration package by default as a ZIP file.

You can also back up and restore your artifact instead of downloading the entire integration package. Perform these steps to do so:

1. From the Design tab, choose the integration package.

2. Select the integration artifact you want to download from the Artifact tab.

3. In the Actions section, Select the download button and the integration artifact is downloaded as a ZIP file.

4. You can now follow the import procedure to upload the artifacts to another package or tenant.

You have exported the integration package successfully. Now, you will learn about the predelivered packages available in SAP Cloud integration from the Discover ➤ Integrations tab.

Predelivered Integration Packages (Interfaces)

Using SAP Cloud Integration to run business operations implies that participants will exchange data (messages). Integration content is created based on the needs of the business and specifies how messages are exchanged.

Integration flows are a crucial component of integration content because they show how SAP Cloud Integration handles messages delivered from one participant. In other words, particular integration patterns such as mapping or routing can be specified using integration flows.

For example, a set of integration flows states that, depending on the message's business content, SAP Cloud Integration will transmit a message delivered by participant A to three different recipients—B, C, and D. The endpoints of sender and receiver participants as well as data structure mappings between sender and receiver are specified in the integration flows.

The next section discusses the predelivered integration packages that are provided by SAP Partners, such as Abusiness.

Predelivered Integration Packages by SAP Partners (Abusiness)

Abusiness has also published SuccessFactors, ServiceNow, S/4HANA, and JIRA packages that can be freely used by the community. The Abusiness packages are listed here, and they are easily available in the SAP Cloud Integration Discover Section.

1. By selecting the copy on the screen's right side, as shown in Figure 4-11, you can copy the package to your integration package screen.

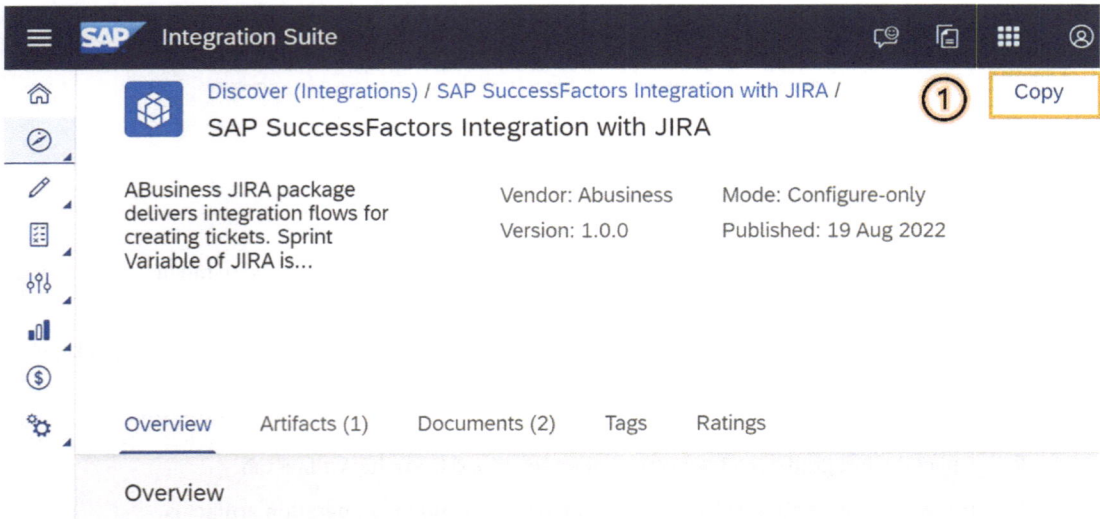

Figure 4-11. *Copy a predelivered integration package*

2. The integration package will be copied to your design screen, as shown in Figure 4-12.

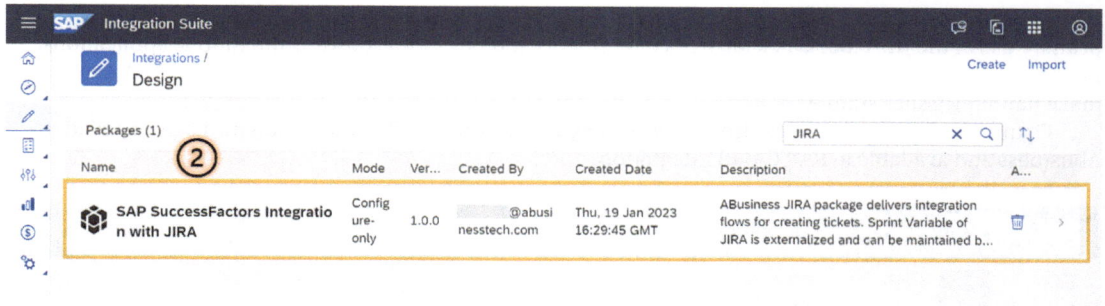

Figure 4-12. *Integration package was copied*

3. This predelivered content iflow can only be configured as suggested by the vendor, as shown in Figure 4-13.

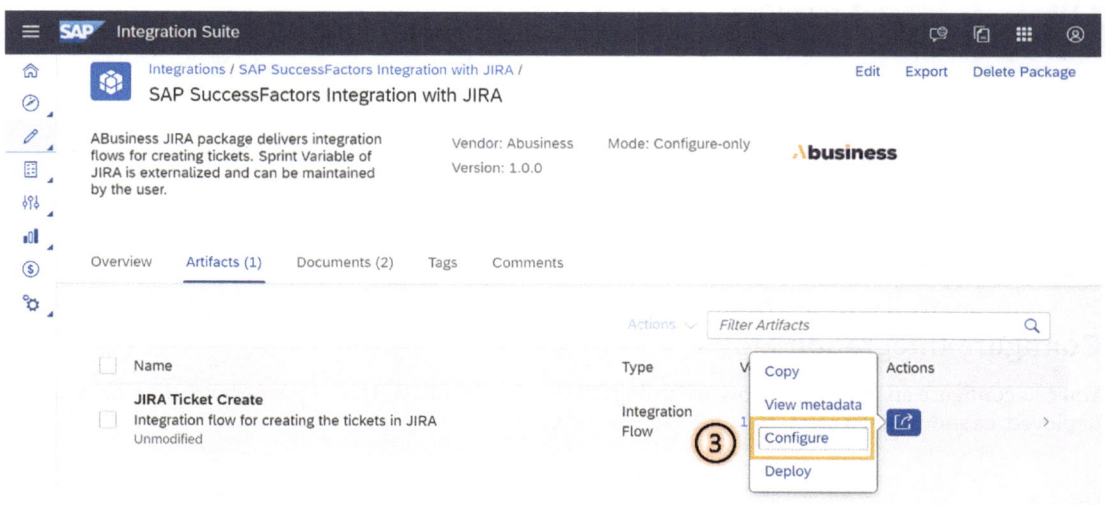

Figure 4-13. *Configuring a predelivered integration flow*

Different integration packages of Abusiness are available in SAP Cloud integration, which you learn about in the next section.

SAP SuccessFactors Integration with JIRA

To create tickets in JIRA, this package provides single sample integration flows.

This type of integration facilitates the comparison of planned, scheduled, or completed hours with SAP finances, automating report delivery, assisting in the monitoring of project timelines, or tying tasks to financial data points.

Project management, issue tracking, and bug tracking are all performed with JIRA software. The primary use of the JIRA tool is to keep track of problems and bugs with software and mobile applications. Project management is another application for it. Several practical tools and features on the JIRA dashboard make handling issues simple.

Figure 4-14 shows the SAP SuccessFactors Integration with the JIRA integration package provided by Abusiness and available in SAP Cloud Integration.

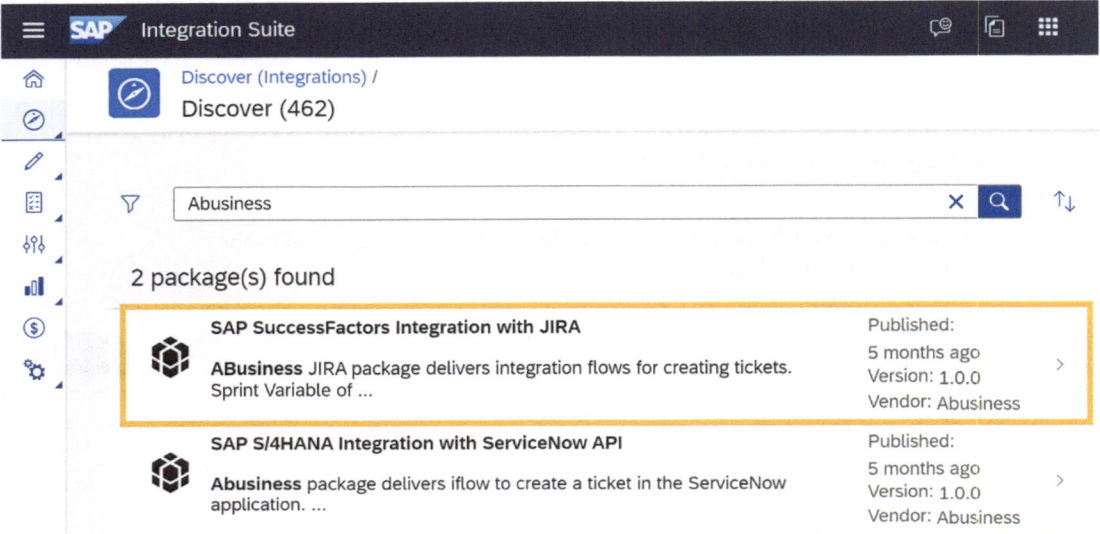

Figure 4-14. Abusiness Package 1

Configure Integration Flows

You can configure an integration flow according to your requirements. The parameters are then saved and deployed, as shown in Figure 4-15.

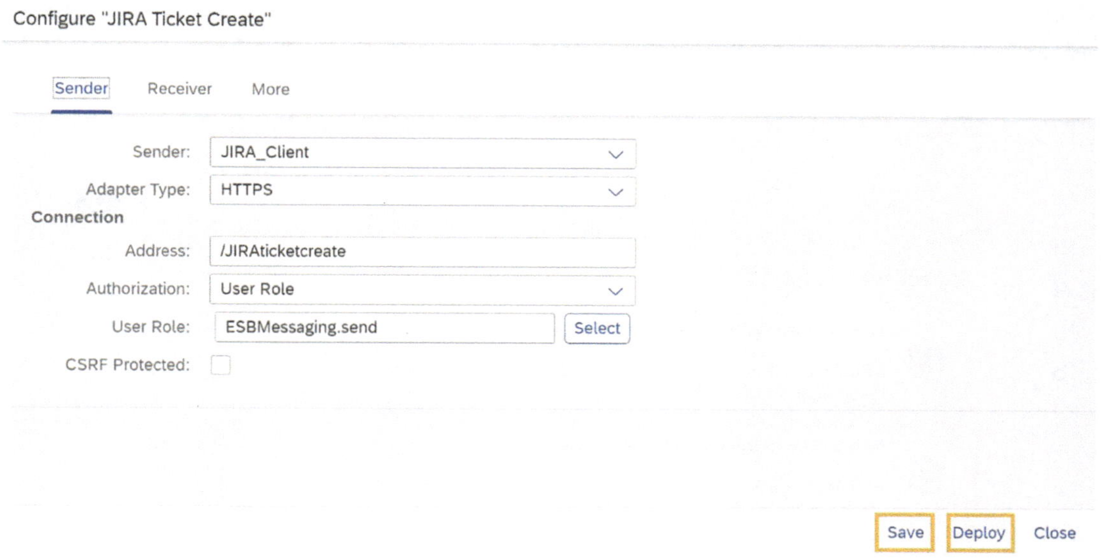

Configure "JIRA Ticket Create"

Sender	Receiver	More

Sender:	JIRA_Client ∨
Adapter Type:	HTTPS ∨
Connection	
Address:	/JIRAticketcreate
Authorization:	User Role ∨
User Role:	ESBMessaging.send [Select]
CSRF Protected:	☐

[Save] [Deploy] Close

Figure 4-15. *Configuration integration flow parameters*

Abusiness has also provided one more integration package as a predelivered integration package, which is based on SAP S/4HANA integration with ServiceNow API. You learn about this integration package in the next section.

SAP S/4HANA Integration with ServiceNow API

S/4HANA SAP ServiceNow Tickets are created with the help of an integration with the ServiceNow API integration flow using XML input from HTTP endpoints. Once the integration flows are installed in the tenant, the HTTP endpoints will be produced.

Enterprises can increase operational efficiencies by using ServiceNow, a cloud-based workflow automation technology, to stream and automate common work operations and create tickets. Any business situation or system that can call online services, such as DOST Add-on°, can use this interface.

Figure 4-16 shows the predelivered integration package of Abusiness (SAP S/4 HANA integration with ServiceNow API) available in SAP Cloud Integration.

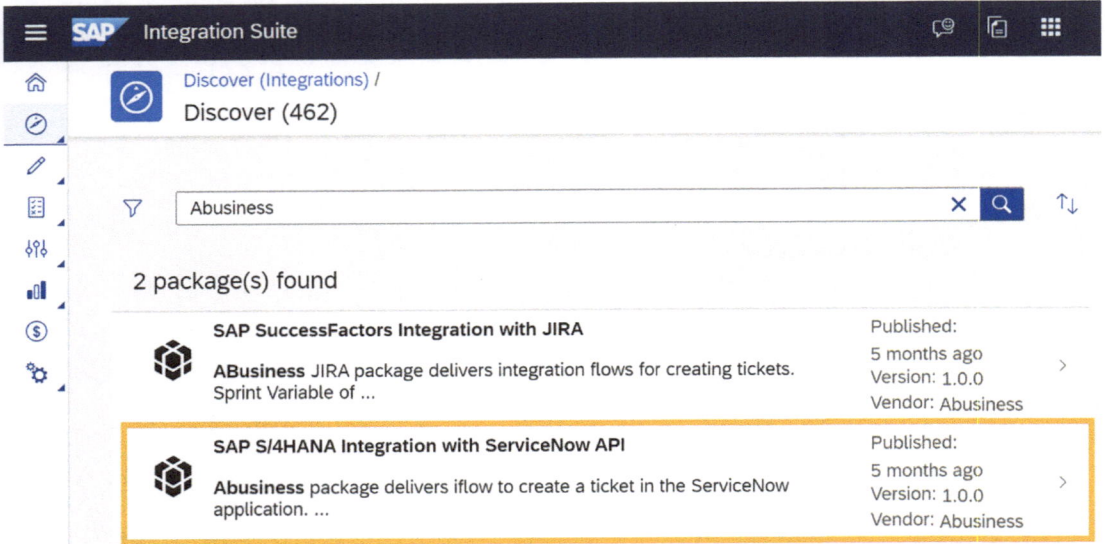

Figure 4-16. *Abusiness Integration Package 2*

Configure Integration Flows

You can configure the integration flow according to your requirements. Once the configuration is completed, the integration flow is saved and deployed, as shown in Figure 4-17.

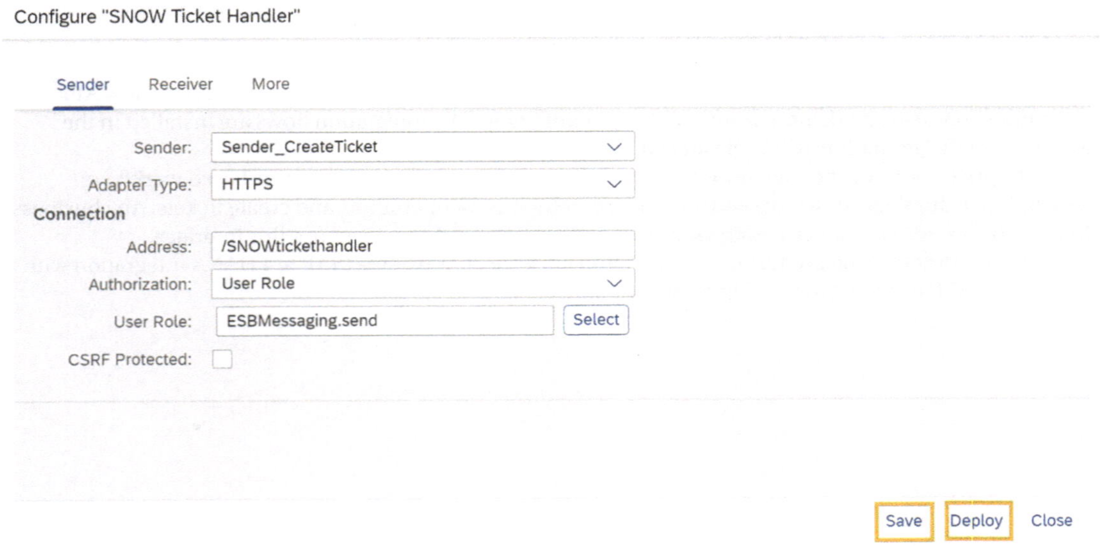

Configure "SNOW Ticket Handler"

Sender	Receiver	More

Sender:	Sender_CreateTicket
Adapter Type:	HTTPS

Connection

Address:	/SNOWtickethandler	
Authorization:	User Role	
User Role:	ESBMessaging.send	Select
CSRF Protected:	☐	

Save   Deploy   Close

Figure 4-17. *Configuration integration flow parameters*

You have learned a lot about integration packages. As you are learning about cloud integration, you will see the artifacts available in the SAP Integration packages. In the Artifacts tab, you have a very important option, called integration flow. You will learn how to create an integration flow in your integration packages in the next section.

4.1.2 Creating an Integration Flow Artifact

An integration flow is a prebuilt integration scenario in SAP Cloud Integration that specifies how data is shared between two or more systems. It is a visual representation of the integration procedure that incorporates message transformation, routing rules, and Message Mapping. Developers can create, edit, and test integration flows using a web-based integration designer tool because it does not require software to be installed locally. When an integration flow is installed, the SAP Cloud Integration platform allows it to run automatically and securely, facilitating seamless data interchange between diverse systems.

Cloud-to-cloud systems, cloud-to-on-premises systems, and even hybrid landscapes can all be integrated using integration flows. In situations such as data synchronization, data migration, and business process automation, integration flows are frequently employed. In general, SAP Cloud Integration's integration flows offer a strong and adaptable solution to link systems and automating business activities.

To create an integration flow in the integration artifact, follow these steps:

1. Open the integration package that you built in Section 4.1.1.

2. Navigate to the Artifacts tab of the integration package in Edit mode.

3. Choose Add ➤ Integration Flow, as shown in Figure 4-18.

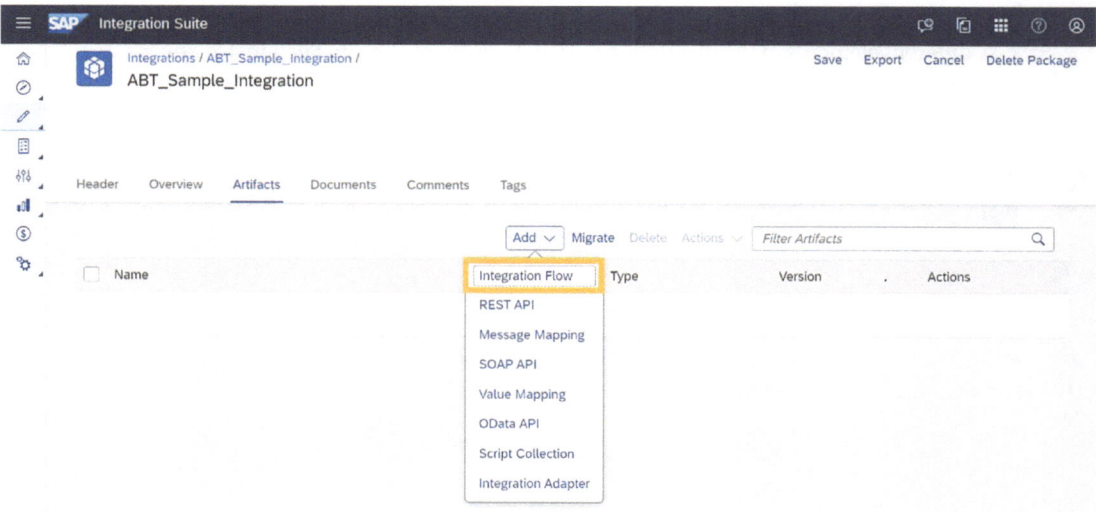

Figure 4-18. *Integration artifact*

4. Provide the name of the integration flow (such as My_First_Integration_Flow). Provide a suitable description of the integration flow. The ID will be automatically generated. Click OK, as shown in Figure 4-19.

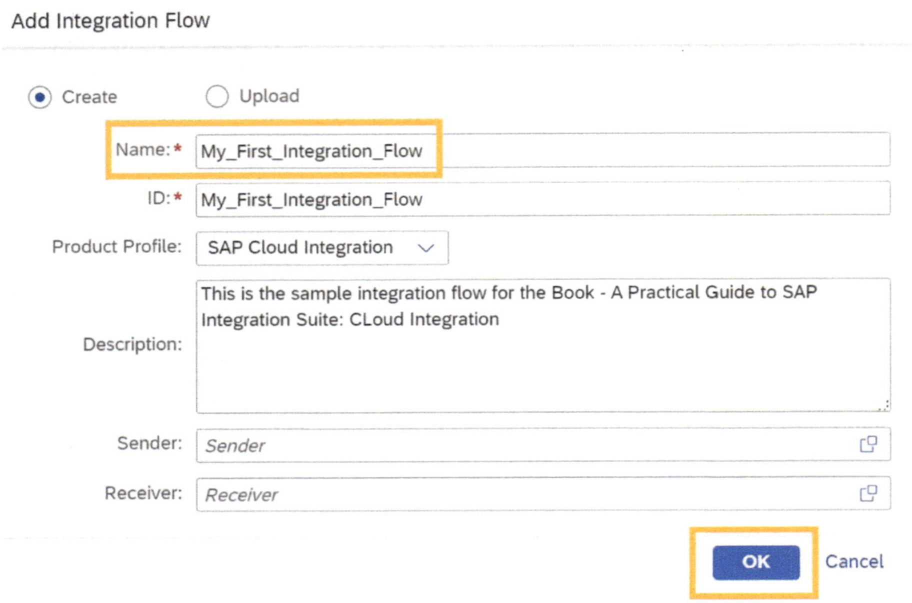

Add Integration Flow
Create
Upload
Name: *
My_First_Integration_Flow
ID: *
My_First_Integration_Flow
Product Profile:
SAP Cloud Integration
This is the sample integration flow for the Book - A Practical Guide to SAP Integration Suite: CLoud Integration
Description:
Sender:
Sender
Receiver:
Receiver
OK
Cancel

Figure 4-19. *Configuration integration flow*

5. You will be directed to the Integration Flow Editor, where you will see the different elements of the iflow, which you will see thoroughly in Section 4.2. You can edit the iflow by clicking the Edit button in the top-right corner, as shown in Figure 4-20.

Integration Suite
Integrations / ABT_Sample_Integration / My_First_Integration_Flow /
My_First_Integration_Flow
Deployment Status: Not Deployed
Edit
Configure
Deploy
Delete
Search Step
Integration Process
Sender
Start
End
Receiver

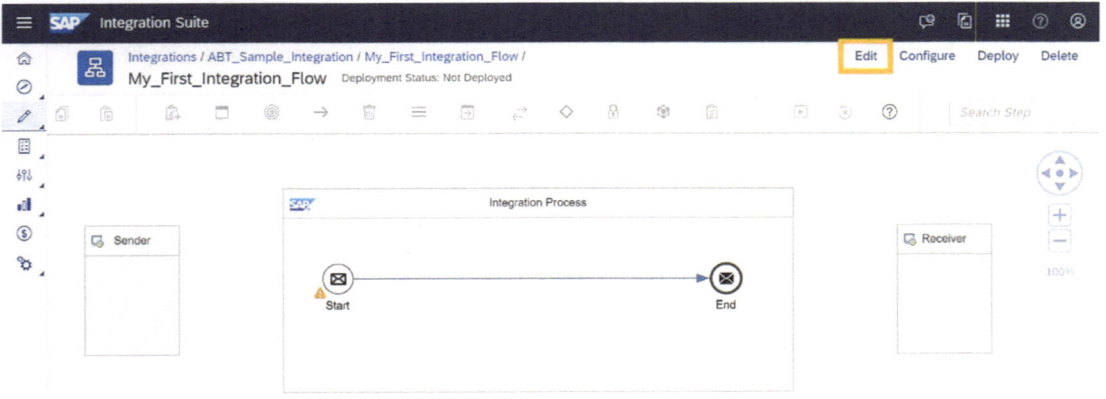

Figure 4-20. *Integration Flow Editor*

You have learned about the integration packages, including how to create, import, edit, and export them. Additionally, you read about the features of integration packages and saw examples of predelivered integration packages.

You also saw practical examples of creating integration packages and learned about the process of configuring integration flows. The chapter also provided examples of predelivered integration packages from SAP partners (Abusiness) and integration packages such as SAP SuccessFactors Integration with JIRA, as well as SAP S/4HANA Integration with ServiceNow API. You also learned about and created integration flow artifacts. In the next section, you learn about the different elements of SAP integration flow.

4.2 Iflow Design Object Elements

An integrated development environment (IDE) called Web IDE is a web-based IDE that SAP offers for creating and testing integration processes that link different applications, systems, and services. It offers a set of tools and templates for developing, deploying, and maintaining integrations in a cloud environment and enable developers to create, edit, and test integration flows using a visual design interface. Within a team, it also enables collaboration and version control for integration flows.

The following are the predelivered SAP palettes/functions/features that can be used in SAP Cloud Integration Development. You can also use Groovy/JavaScript if one of the predelivered functions does not meet your requirements.

■ **Note** WebIDE is still used in SAP Integration Suite, although SAP created the Business Application Studio, which is available in the SAP BTP.

You will see the different sections of the elements available in SAP Integration Flow. Let's first look at the participants.

4.2.1 Participant

In the Participant tile are the Receiver and Sender adapters, which allow you to assign your source and target, as shown in Figure 4-21.

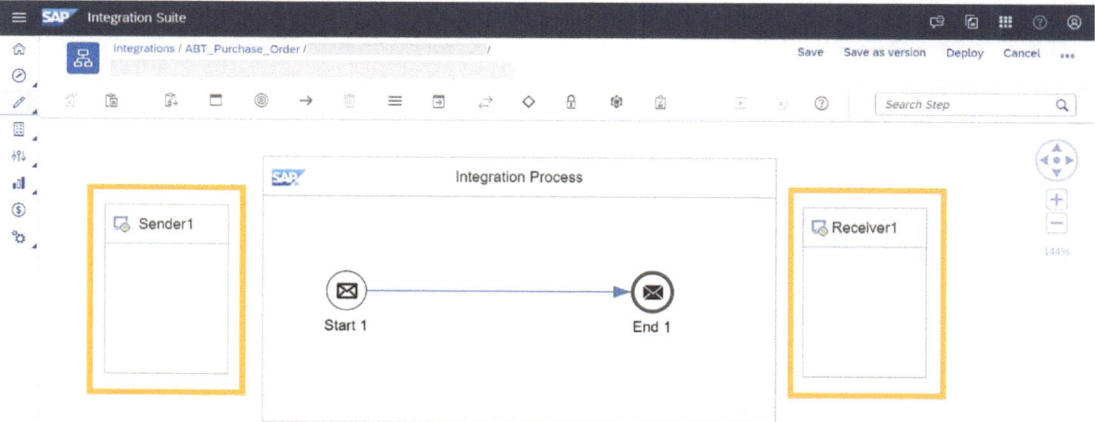

Figure 4-21. *Participant in the integration flow*

You finish this assignment by including the receiver and sender participants in the integration flow. To allow the sender participant to connect to the tenant, you must either provide the client certificates or log in using the SDN username and password.

To simulate different systems connected to your integration flow (either as message sender or receiver), utilize the sender and receiver components. To add the receiver or sender, navigate to the Participant section in the palette, as shown in Figure 4-22.

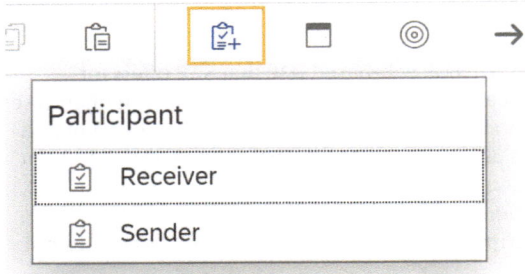

Figure 4-22. *Participant in the SAP Integration Flow Editor*

4.2.2 Process

Process defines the different types of processes that lead to communication from the sender to the receiver: Exception Subprocess, Integration Process, and Local Integration Process. To add the process step to the Integration Flow Editor, select the Process tab from the palette, as shown in Figure 4-23.

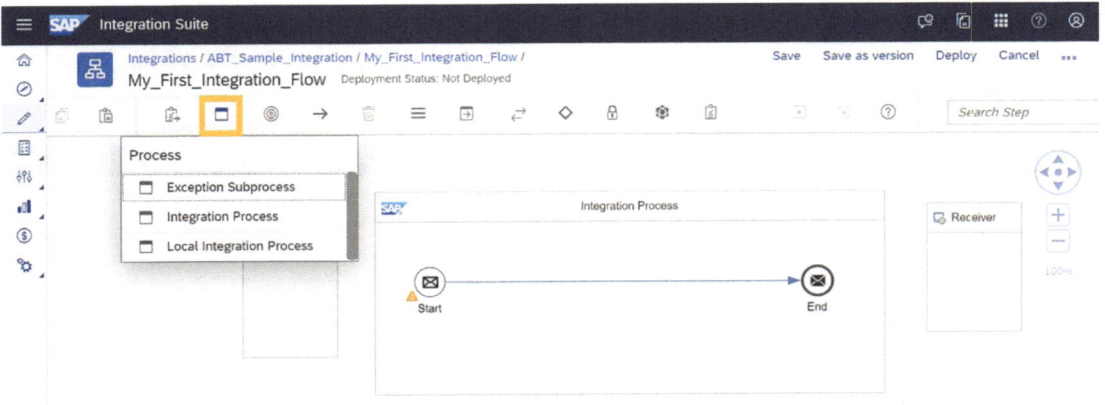

Figure 4-23. *Process tile in the SAP Integration Flow Editor*

You will learn about the three processes available in SAP Cloud Integration one by one. The next section starts with the Exception Subprocess.

Exception Subprocess (Error Handling): Practical Example

This component is used to handle any exceptions that are thrown during the integration process.

1. Open the editor and select any integration flow that you created in the Integration package in Section 4.1.7.

2. Select Process Exception Subprocess from the palette to include an exception subprocess in the integration flow. The subprocess does not need to be connected to any of the integration flow's components and can be dropped into the integration process.

3. Always begin the process with the Error Start event.

4. Use the End Message, Error End, or Escalation events to end the process. Between the Start and end Events, additional flow components can be added, as shown in Figure 4-24.

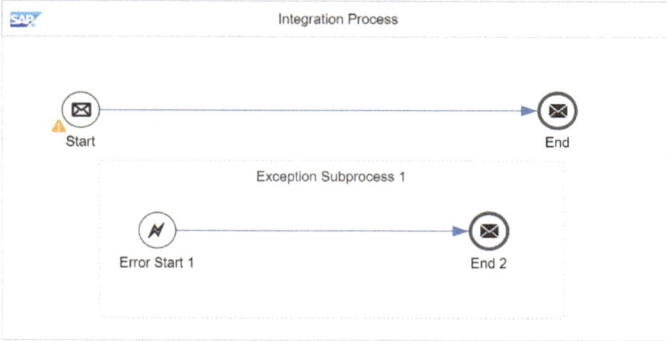

Figure 4-24. *Exception subprocess*

5. Responding to serious mistakes that occur while processing messages is known as *exception handling.* An exception can occur during mapping execution error, context router, and so on. The exception subprocess can be used to address such situations.

6. Consider an example of the integration flow in which we used the SOAP adapter or any other adapter, Before the exception handler, note the SOAP message, as shown in Figure 4-25.

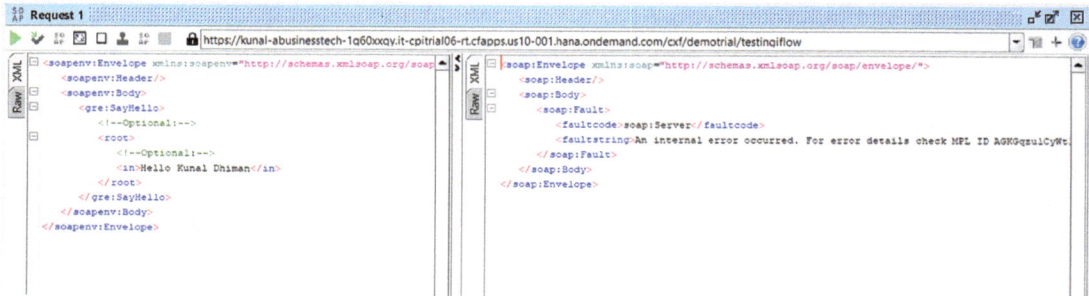

Figure 4-25. *Exception message from SOAPUI 5.6.7*

7. The error details can be seen in the Monitor Message Processing tab, as shown in Figure 4-26.

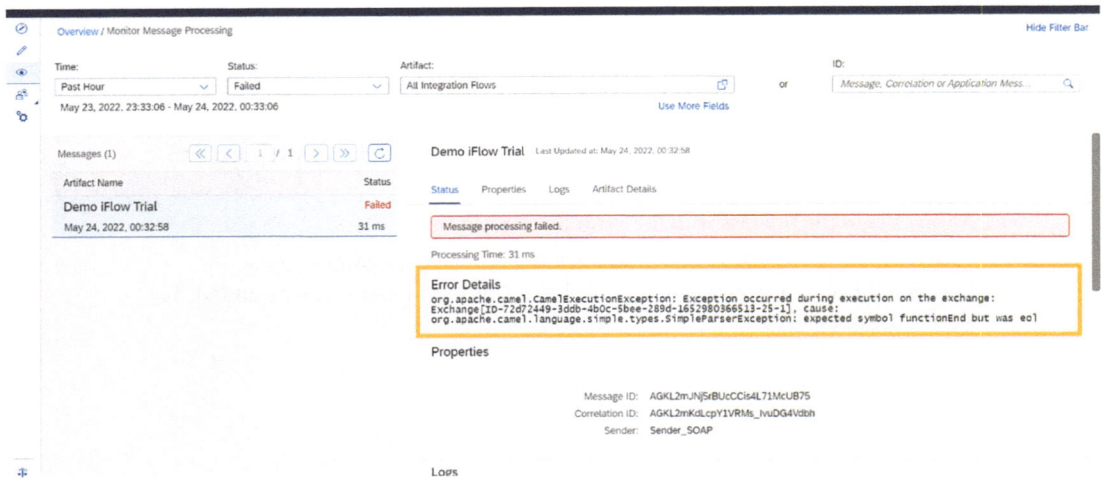

Figure 4-26. *Error details in the Monitor tab*

To see the error while you run your endpoints, you have to use the exception subprocess. Follow these steps to do so:

Error Handling Solution

1. Go to your package and then choose your artifact and select Edit.

2. Add a new Exception Process (in Process) in your integration process and add a new Content Modifier, as shown in Figure 4-27.

3. Choose Process ➤ Exception Subprocess, as shown in Figure 4-27.

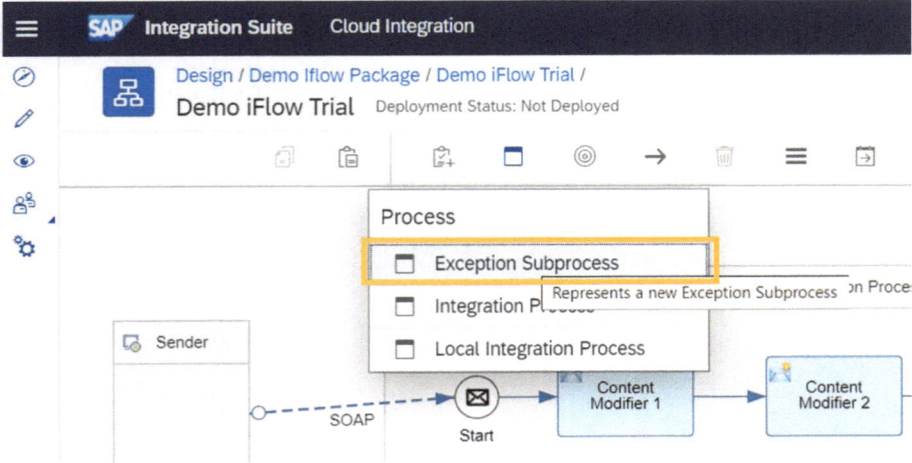

Figure 4-27. *Exception subprocess*

4. The Content Modifier is added in the Exception subprocess, as shown in Figure 4-28.

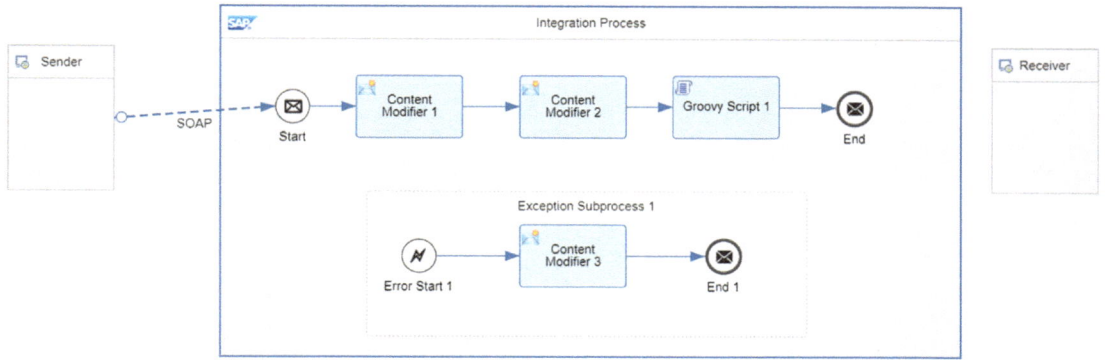

Figure 4-28. *Content Modifier in the Exception subprocess*

5. In the message body (of the Content Modifier), add this logic:

```
<errorText>
You have issues related to simple message processing, please find below the
exception message and the stacktrace:
            ${header.var1}
            ${exception.message}
            ${exception.stacktrace}
</errorText>
```

■ **Note** ${exception.message}; ${exception.stacktrace} are the central variables used to catch exceptions in an integration flow.

---

6. Save the iflow and deploy it. You see that find the error that you were getting in the logs is now happening while testing the endpoint through SOAP UI, as shown in Figure 4-29.

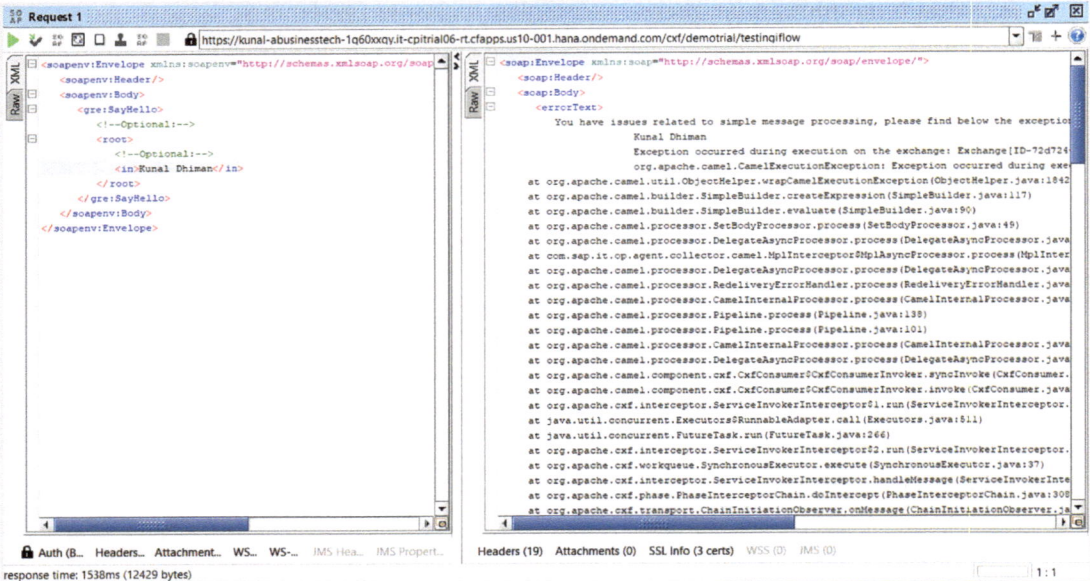

Figure 4-29. *Exception handling issue as the result*

You have seen the exception subprocess and learned about the error-handling solution using the exception handling subprocess. In the next section, you learn about the integration process in detail.

Integration Process

An integration process is used to establish the processes that handle message transfer between the sender and receiver systems.

An integration flow template is made up of the forms sender (which symbolizes your sender system), a receiver (which symbolizes a receiver system), and an integration process. The integration process has Start and End events, as shown in Figure 4-30.

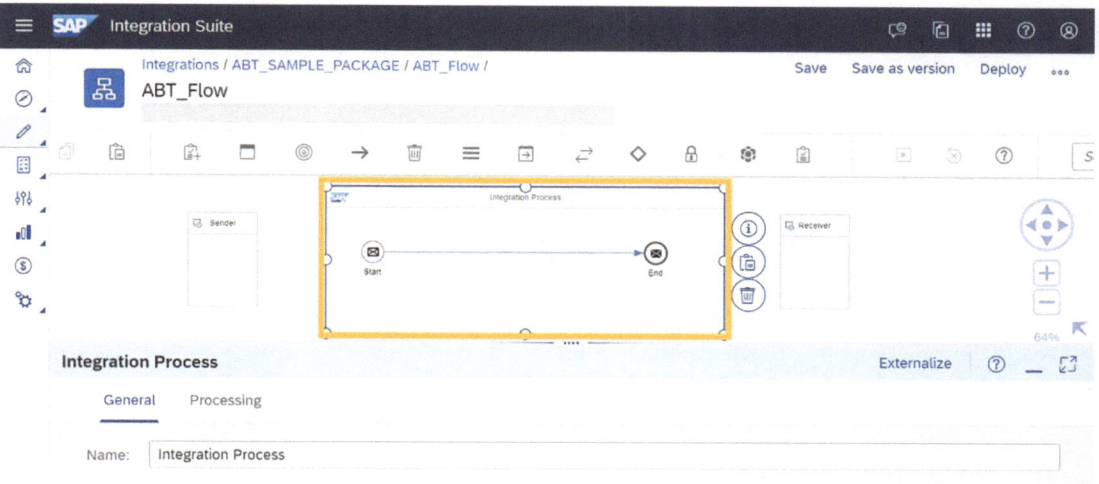

Figure 4-30. *Integration process*

The General tab shows the values listed here:

- **Name**—Enter a name in this field if you want to give the integration process a name. The integration process setting is the default.

The Processing tab shows the values listed here:

- **Transaction handling**—Choose the appropriate transactional database processing:
  - Required for JDBC
  - Required for JMS
  - Not Required
- **Timeout**—This is needed only for JMS or JDBC. Choose it from the Transaction Handling section, as shown in Figure 4-31.

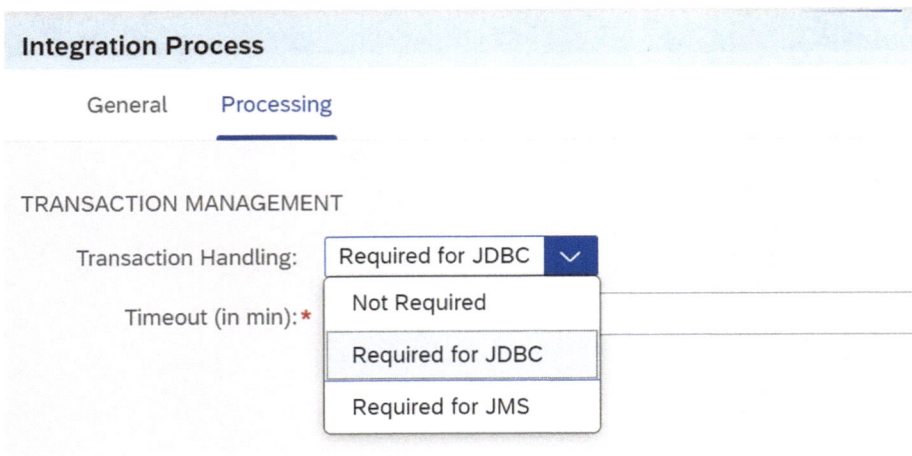

Figure 4-31. *Processing tab of the integration process*

Now that you have learned about the integration process, the next section explains the local integration process.

Local Integration Process

By utilizing the local integration process, the integration process in the integration flow becomes simpler. You can break up the main integration process into smaller pieces using local integration procedures. You put these elements together to finish your main integration procedure. Select the integration flow you created in Section 4.1.7.

1. Select the component of the local integration process that needs configuring.

2. If you want to give the local integration process element a name, enter a name in the Name box found under the General tab.

3. If you want to specify how the message will be handled during transactions, choose a value for the Transaction Handling field under the Processing tab.

4. If you want to include the local integration process in a process call element, the following substeps must be finished:

   a. Select the Integration Flow Editor's process call element.

   b. Select the Local Integration Process field.

   c. Choose the local integration process you want to ascribe to the process call in the Select Local Integration Process window. The local integration process is shown in Figure 4-32.

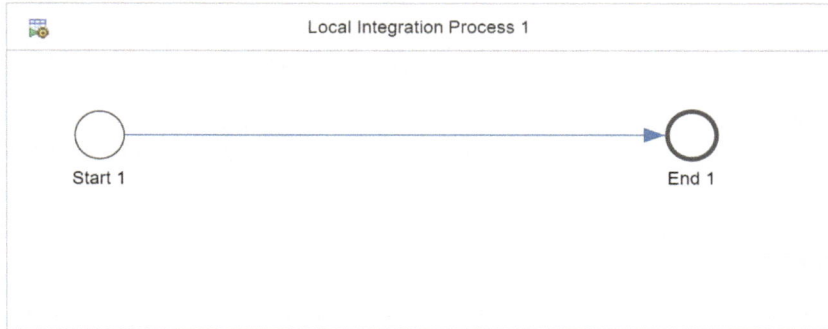

Figure 4-32. *Local integration process*

You have now seen the Process elements of the SAP Integration Flow Editor and learned about the three subparts of the process exception subprocess, integration process, and local integration process. In the next section, you learn about the events in the SAP Cloud Integration Flow.

4.2.3 Events

Events are the triggers that begin an integration flow's execution in the SAP Cloud Integration Flow Editor. An event can be any modification or activity that takes place in a networked system, including a message being received, a file being uploaded, or a timer expiring.

In the integration flow, you can find the event elements in the Event palette, as shown in Figure 4-33.

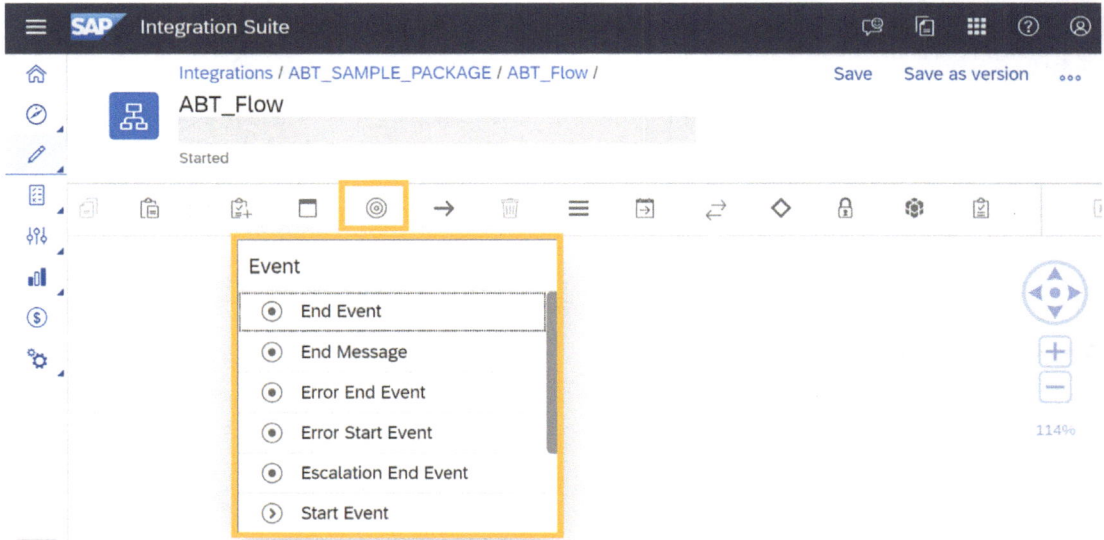

Figure 4-33. *Events in integration flow*

The events contain or subcategorize the elements shown in Figure 4-34. You will learn about these event elements one-by-one.

End Message Event

An End Message event marks the conclusion of a message processing sequence. The message status changes as a result of using an End Message event.

If an exception that has been dealt within an exception subprocess is thrown during the processing sequence, the message status displayed in the message processing log is *Failed*. When there is no error when handling the exceptions, the message status is *Finished* in the message processing log.

Terminating Message Event

The Terminate Message event ends the processing of a message. To prevent further processing of a message, use a Terminate Message event. If certain values were defined on the payload, for example, this might be the case. If the payload does not match certain values, the process is terminated.

If you want to set up your integration flow so that all messages have the Failed status (even if an exception occurs during the processing sequence and is successfully handled in an Exception subprocess), you can do so by selecting Terminate Message Events, as shown in Figure 4-34.

- Use the Error End event.

- Using an Escalation End Event in this situation changes the message status to Escalated.

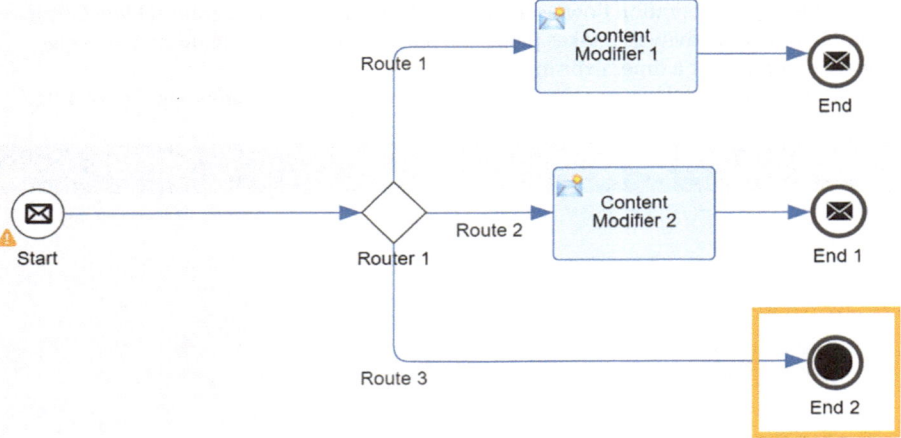

Figure 4-34. *Terminate Message event*

Start and End Message

The Start Message and End Message events are used when a message is received by SAP Cloud Integration from a sender and when it is delivered to a receiver. When you build an integration flow, the Start and End message events are made available by default, as shown in Figure 4-35.

Figure 4-35. *Start and End messages*

Error Start and Error End Message

Error Start and Error End are the additional message events that can only be used inside an exception subprocess, as shown in Figure 4-36.

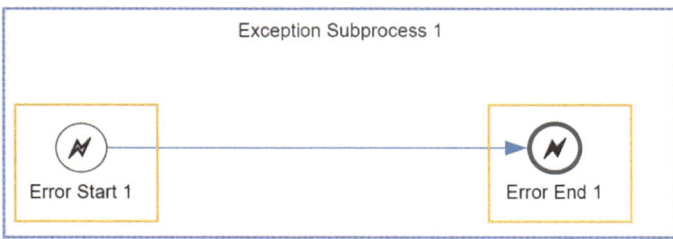

Figure 4-36. *Error Start and Error End messages*

Timer Start

When you need to pull data from systems or activate web services at specific times or intervals, a Timer Start event is extremely helpful. Because it is a pull-based adapter, you can presently only use it with a SuccessFactors adapter while polling (drawing data from systems).

A timer comes first, then Content Modifier, according to the standard usage pattern. The reason for this is that a timer does not add payload to the pipeline. You can create the request payload that will be submitted to the system using a Content Modifier, as shown in Figure 4-37.

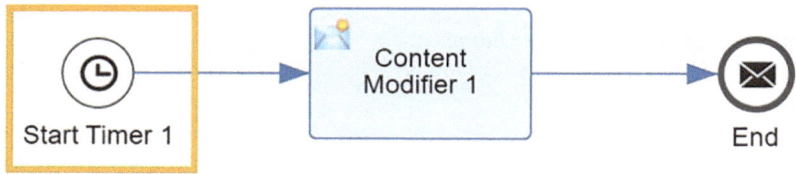

Figure 4-37. *Timer Start event*

Now that you have an understanding of events and learned about the different event options available to the integration flow, the next section explains the message transformations in iflow design object's palette.

4.2.4 Message Transformations

Message transformers change one message format into another. You can find the message transformations in the integration flow's palette, as shown in Figure 4-38.

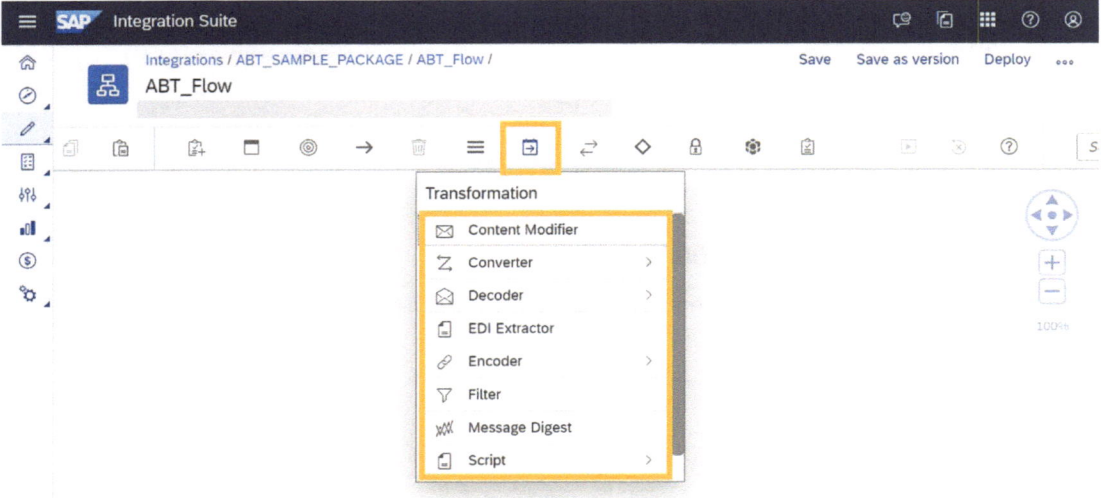

Figure 4-38. *Transformation options*

These transformations are further divided into elements, as shown in Figure 4-40.

The message transformations contain various other elements that can be used in iflow design. Let's start with the Content Modifier.

Content Modifier

You can modify the content of an incoming message by adding more information to the message's header or body using the Content Modifier step, as shown in Figure 4-39.

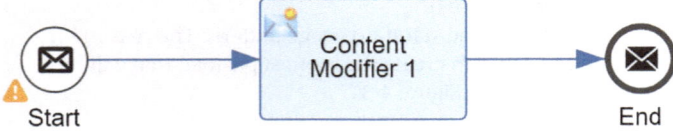

Figure 4-39. *Content Modifier in the SAP Integration Flow Editor*

The Content Modifier gives you the ability to edit a message's header, body, or exchange by changing the contents of the data containers used during message processing. Depending on which container you want to update, select the Message Header, the Message Body, or the Exchange Property tab. You can add the details you want to the message body by using an editor while modifying the message body. If you adjust the message header or an exchange property, you can learn how to access an incoming message's content. To alter the message header, choose XPath as the source type and add an XPath expression that corresponds to a particular element in the incoming message.

Procedure:

1. Select the Content Modifier step to alter the properties if it is included in the integration flow. Select the integration flow that you created in Section 4.1.7. See Figure 4-40.

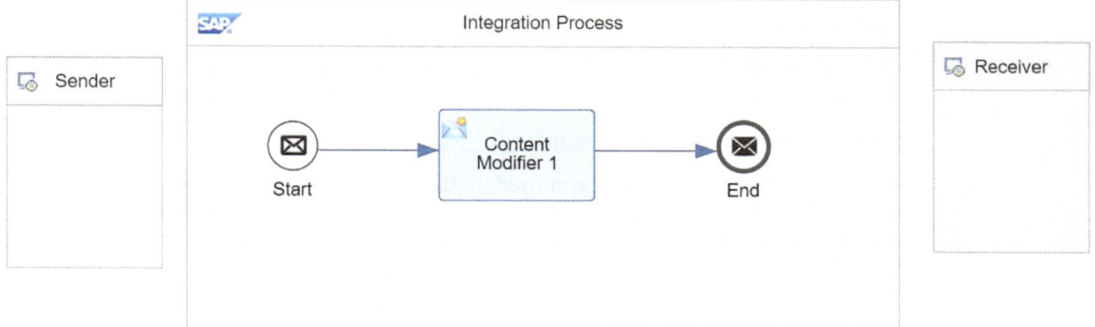

Figure 4-40. *Content Modifier in the integration flow*

2. Perform the following substeps if you want to add the Content Modifier step to the integration flow:

   • Select (Message Transformers) from the palette before selecting (Content Modifier).

   • In the integration process, place the Content Modifier step.

   • To define a new entry, select Add.

3. Define and specify the entry's parameters as shown in Figure 4-41.

   • **Action**—You can specify whether the header or property defined by the table row should be created or deleted by the Content Modifier.

   • **Name**—Name in which the chosen header or property data container must store the provided data.

   • **Source Type**—Select a header or property data container and then specify the type of data you want to use to modify its content.

      • **Constant**—Allows you to add a constant value to a property's data container or header.

      • **Header**—Enables you to specify a camel header's name.

      • **XPath**—Enables you to use XML Path Language to retrieve data from the incoming message (XPath).

191

- **Expression**—Permits you to enter a Camel Simple Expression Language expression. For instance, the word $exchangeId can be used to add the exchange ID, a unique identifier of the message exchange, to a data container.

- **Property**—Specifies an exchange property for you.

- **External Parameter**—Although Web UI does not have this option, you continue to support content that uses this field for compatibility reasons.

- **Local Variable**—Allows you to create local variables that can be used to write runtime values to property data containers or headers. The value can then be evaluated in the subsequent steps of the integration procedure.

- **Number Ranges**—Enables you to retrieve the value of a given number range (a unique identifier).

- **Global Variables**—Enables the definition of global variables and the runtime writing of their values to the header or property data container.

- **Source Value**—The value of the property.

- **Data Type**—Java data types are supported. In the Data Type field, only the XPath and Expression types are used. The data type can be present in any Java class. Use java.lang.String if you are addressing a string-type element.

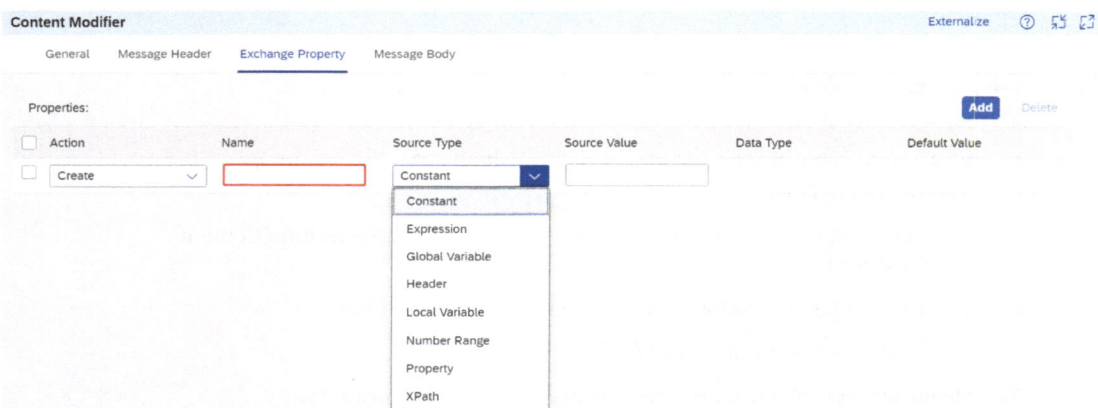

Figure 4-41. *The Exchange Property tab*

4. Go to the Message Body tab and define the values, as shown in Figure 4-42.

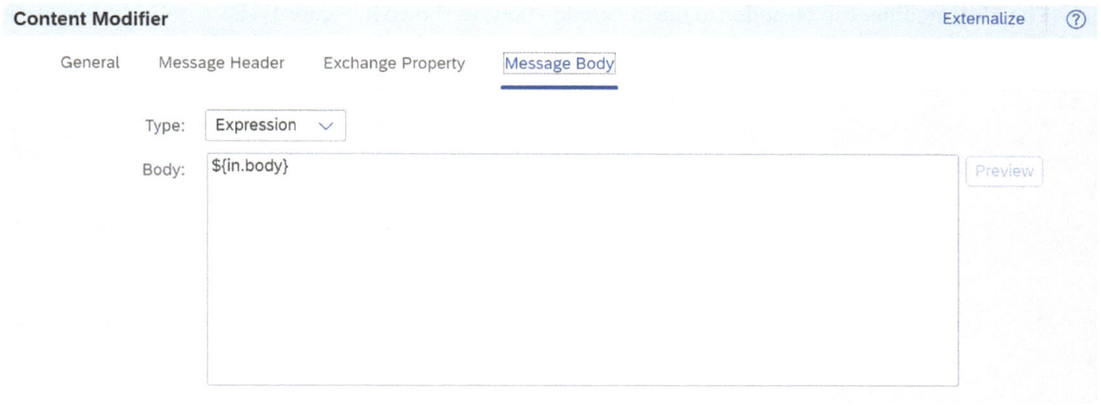

Content Modifier   Externalize   ⑦

  General   Message Header   Exchange Property   Message Body

        Type: Expression ∨

        Body: ${in.body}   Preview

Figure 4-42. Message Body tab

5.   Save and deploy the changes.

The Content Modifier gives you the ability to edit a message's header, body, or exchange by changing the contents of the data containers used in message processing. You have learned about this Content Modifier thoroughly, so the next section goes through the XML modifier.

XML Modifier

The XML modifier is a type of step or activity that allows you to modify the structure and content of XML messages. It is a strong tool that enables you to modify XML communications between systems, add or delete nodes, update values, and execute other operations on XML data.

You can remove XML declarations and/or external DTDs using the XML modifier flow step. You can find the XML Modifier from the palette, by choosing Transformations ➤ XML Modifier, as shown in Figure 4-43.

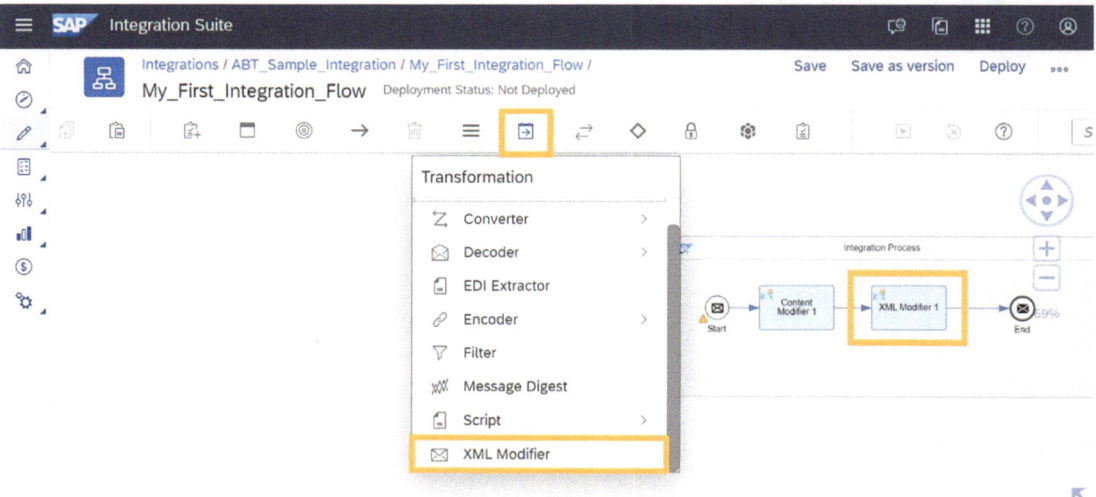

Figure 4-43. The XML modifier in the SAP Integration Flow Editor

The XML modifier can be added to the integration flow, as shown in Figure 4-44.

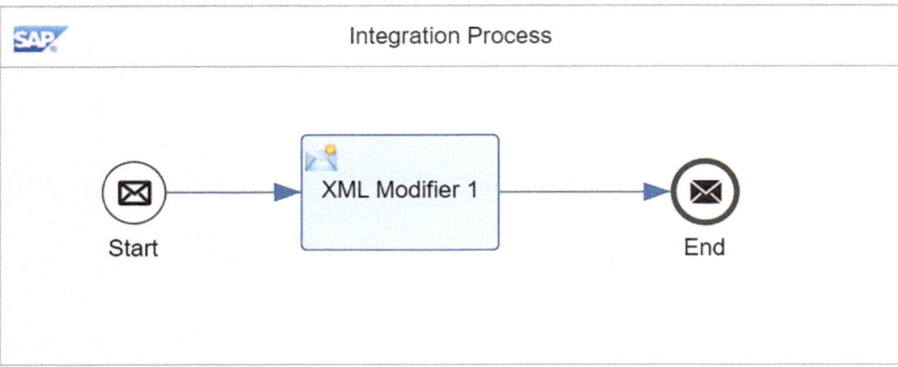

Figure 4-44. *The XML modifier in the Integration Flow*

With the help of XML modification, users can modify the incoming XML message. You can modify an XML message by removing the XML declaration and/or DTDs with external references using the XML modification to avoid issues with message processing since DTDs with external references are not supported.

In the next section, you learn about the different converters in the transformation element.

Converter

A converter is a component in SAP Cloud Integration that is used to change the data format so that it is compatible with the target system. Figure 4-45 shows the converter in the integration flow Web UI.

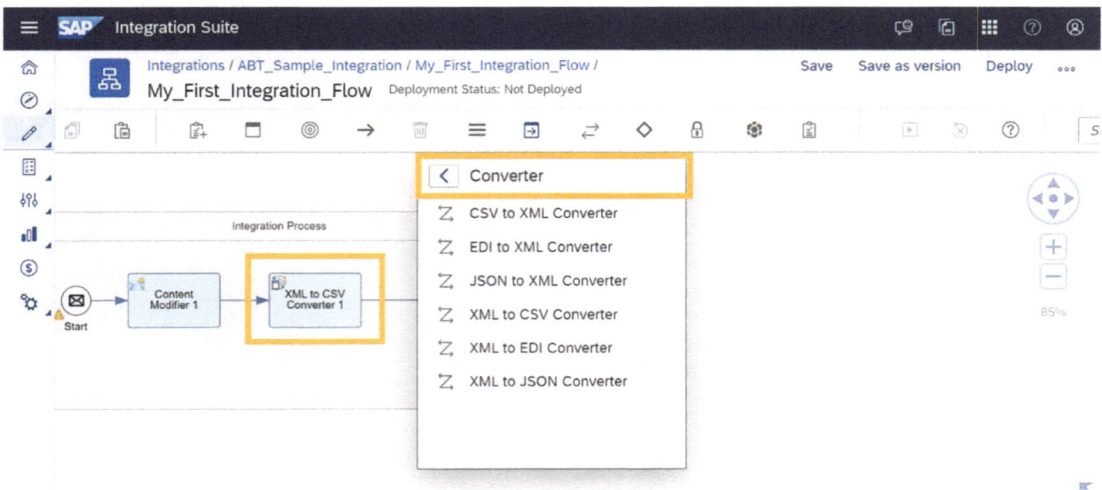

Figure 4-45. *The converter*

This can include tasks such as mapping fields, converting data types, and applying validation rules. Converters are typically used as part of an integration flow to ensure that data can be seamlessly exchanged between different systems. Navigate to the palette and choose Transformations ➤ Converter. Then select the converter of your choice, as shown in Figure 4-46.

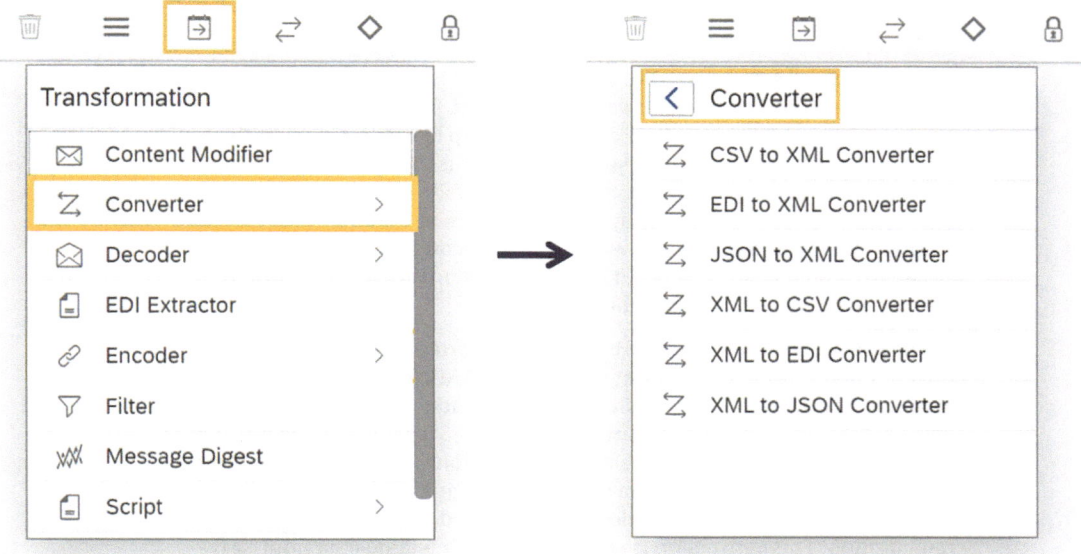

Figure 4-46. *Types of converters in the SAP Cloud Integration Flow Editor*

Here are the following types of converters.

1. **CSV to XML Converter**—A CSV to XML converter is a specific type of converter that is used to transform data from a CSV (comma separated value) format to an XML (extensible markup language) format. The converter takes the input data in CSV format and converts it into an XML document, following the rules and structure defined in the converter's configuration.

2. **EDI to XML Converter**—An electronic data interchange (EDI) to XML converter is a specific type of converter that is used to transform data from an EDI format to an XML format. A common format for electronically transmitting commercial documents, including orders, invoices, and shipping notifications between various systems, it is known as EDI. XML is a widely used markup language for structuring data, which makes it a popular choice for integrating data from EDI into other systems. This conversion can be performed using different middleware platforms, such as SAP PI/PO, MuleSoft, Boomi, and so on.

3. **JSON to XML Converter**—A JSON to XML converter is a specific type of converter that is used to transform data from a JSON (JavaScript object notation) format to an XML (extensible markup language) format. JSON is a lightweight, human-readable format for structuring data that is commonly used in web applications, while XML is a widely used markup language for structuring data. The converter takes the input data in JSON format and converts it into an XML document, following the rules and structure defined in the converter's configuration.

4. **XML to CSV Converter**—An XML to CSV converter is a specific type of converter that is used to transform data from an XML (extensible markup language) format to a CSV (comma separated values) format. XML is a widely used markup language for structuring data, while CSV is a simple, plain-text format for storing tabular data. The converter takes the input data in XML format and converts it into a CSV file, following the rules and structure defined in the converter's configuration.

5. **XML to EDI Converter**—An XML to EDI converter is a particular kind of converter that is used to transfer data from an XML format to an EDI (electronic data interchange) format. XML (extensible markup language) is a popular option for integrating data from many systems because it is a widely used markup language for data organization. A common format for electronically transmitting commercial documents, including orders, invoices, and shipping notifications between various systems, it is known as EDI. The converter turns the XML-formatted input data into an EDI document by using the rules and structure specified in the configuration of the converter.

6. **XML to JSON Converter**—An XML to JSON converter is a specific type of converter that is used to transform data from an XML (extensible markup language) format to a JSON (JavaScript object notation) format. XML is a widely used markup language for structuring data, while JSON is a lightweight, human-readable format for structuring data that is commonly used in web applications. The converter takes the input data in XML format and converts it into a JSON document, following the rules and structure defined in the converter's configuration. This conversion is typically used to integrate data from XML-based systems into systems that support the JSON data format.

You will learn about converters by examining an example of a JSON to XML converter, which is discussed in the next section.

JSON to XML Converter: Practical Example

A component that converts JSON data into XML format is called the JSON to XML converter. This component is helpful for integrating systems that use several data formats and necessitate data conversion. To convert the JSON data to XML, follow these steps:

1. Create a new integration flow or open the integration flow created in Section 4.17. Open it in Edit mode.

2. Connect the Sender to the Start with HTTPS communication by dragging the arrow slider from Sender to Start. Select the HTTPS, as shown in Figure 4-47.

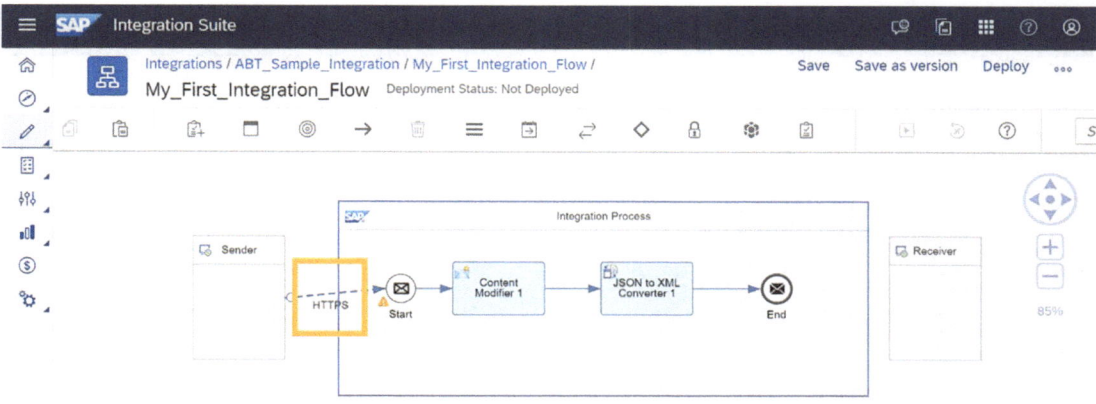

Figure 4-47. HTTPS inbound communication

3. The Connection tab provides the following details, as shown in Figure 4-48:

   ● Address—The URL Endpoint (for example, /JSON_to_XML).

   ● Authorization—User Role

   ● User Role—ESBMessageing.send

   ● CSRF Protected—Uncheck this field

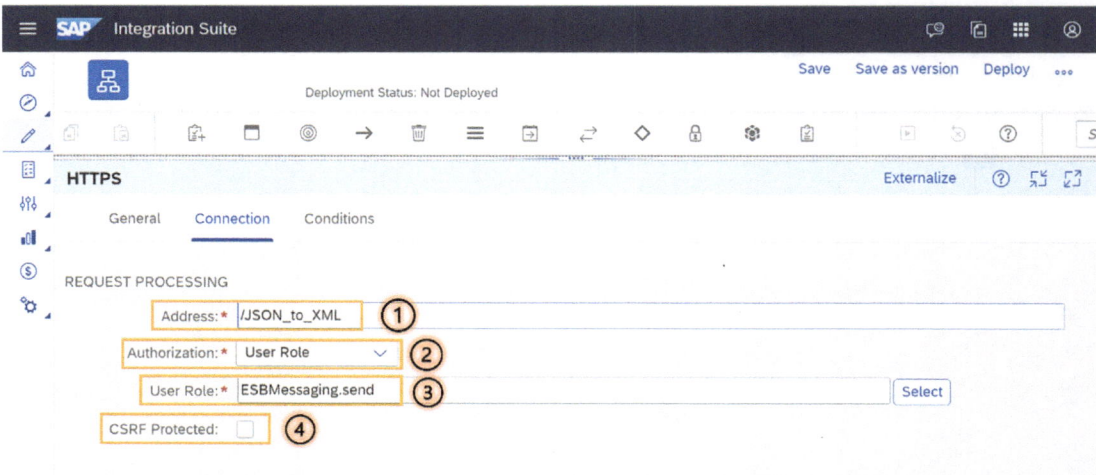

Figure 4-48. HTTPS communication

4. Click anywhere in the integration flow and, in Namespace Mapping under Runtime Configuration, enter the following address, as shown in Figure 4-49: xmlns:ns0=http://cpi.sap.com/demo;xmlns:ns1=http://sap.com/xi/XI/ SplitAndMerge.

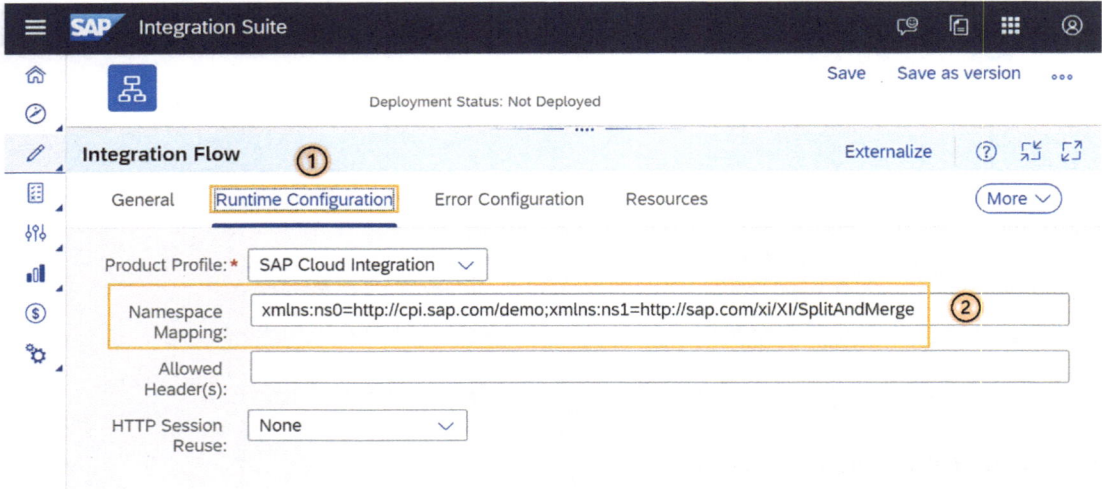

Figure 4-49. *Namespace mapping*

5. Add the JSON to XML Converter to the iflow.

6. In the Processing tab, the following fields are provided, as shown in Figure 4-50:

- Use Namespace Mapping—Uncheck the box

- Name—Provide the Name

- Namespace Mapping—Select the namespace using the Select button

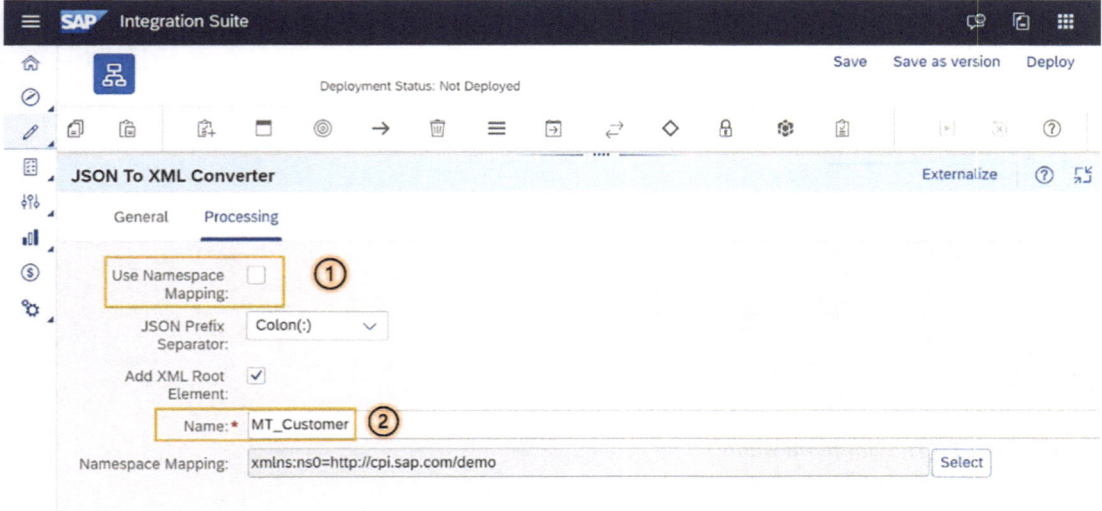

Figure 4-50. *JSON to XML configure*

7. Save the integration flow and then deploy it.

8. Navigate to the Monitor screen and open the Manage Integration Content section under Monitor Message Processing. Find your integration flow, and at the right side, you will find your endpoints, as shown in Figure 4-51.

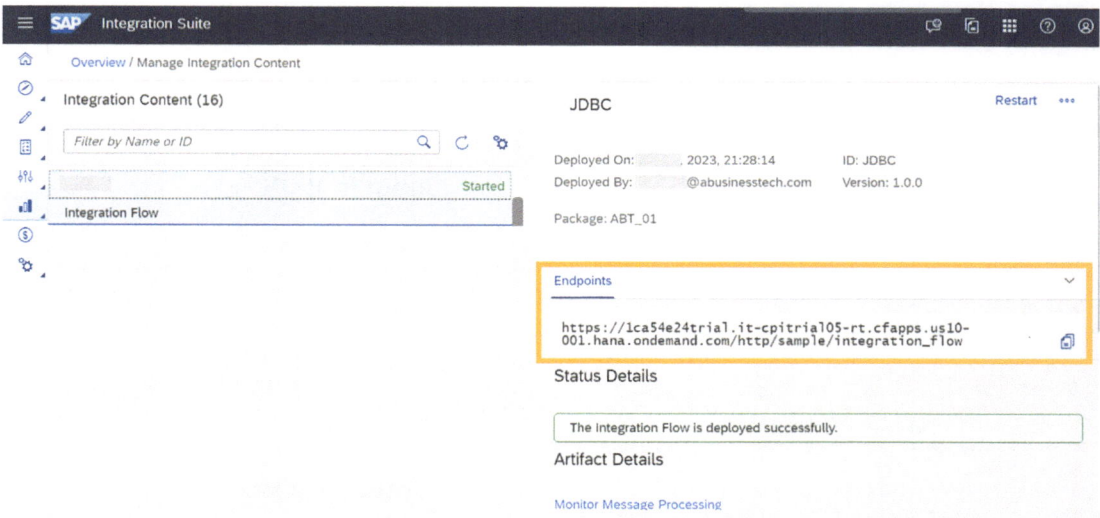

Figure 4-51. *Endpoints*

9. Copy the endpoint from the Monitor section. Run it through Postman, as shown in Figure 4-52.

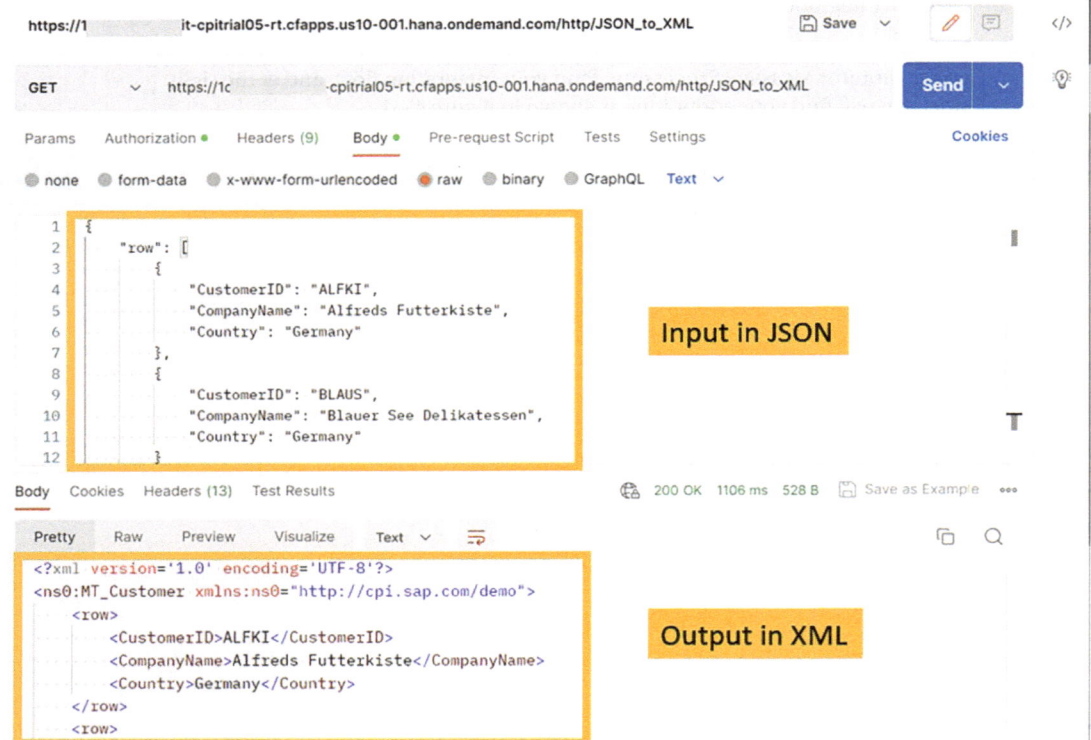

Figure 4-52. *Results in Postman*

10. In the body, provide the following code:

```
{
    "row": [
        {
            "CustomerID": "ALFKI",
            "CompanyName": "Alfreds Futterkiste",
            "Country": "Germany"
        },
        {
            "CustomerID": "BLAUS",
            "CompanyName": "Blauer See Delikatessen",
            "Country": "Germany"
        }
    ]
}
```

11. You will obtain the following result:

```xml
<?xml version='1.0' encoding='UTF-8'?>
<ns0:MT_Customer xmlns:ns0="http://cpi.sap.com/demo">
    <row>
        <CustomerID>ALFKI</CustomerID>
        <CompanyName>Alfreds Futterkiste</CompanyName>
        <Country>Germany</Country>
    </row>
    <row>
        <CustomerID>BLAUS</CustomerID>
        <CompanyName>Blauer See Delikatessen</CompanyName>
        <Country>Germany</Country>
    </row>
</ns0:MT_Customer>
```

12. The JSON code has been converted into XML through Cloud Integration.

Through this example, you have seen how JSON data is converted into XML data. You also learned about the converter with this example. The next section goes through the decoder.

Decoder

A decoder is used in SAP Cloud Integration to convert data from one format to another. Decoders are used to transform messages that are received in particular encodings or formats into versions that the integration flow can understand.

This operation allows the original data to be obtained by decoding the message that was transmitted over the network. Figure 4-53 shows the use of the decoder in the integration flow.

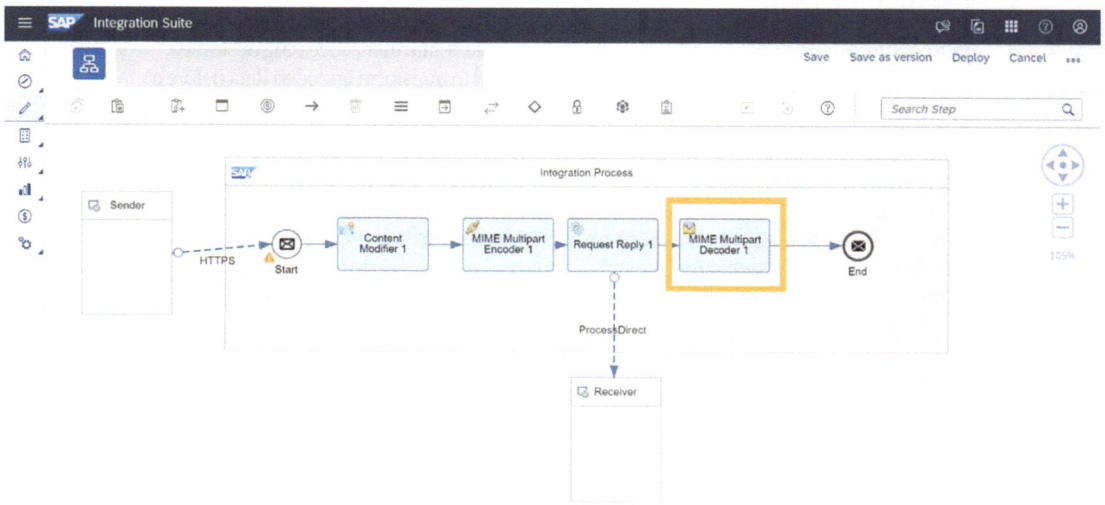

Figure 4-53. *Decoder in integration flow*

You can open the transformations from the palette and choose the decoder from it, as shown in Figure 4-54.

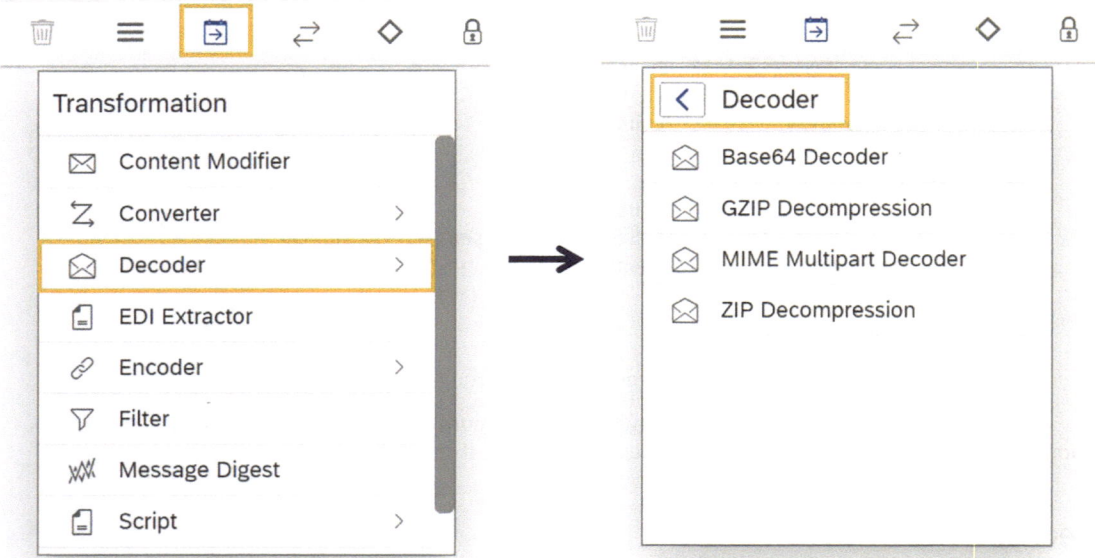

Figure 4-54. *Decoder in the SAP Cloud Integration Flow Editor*

There are four types of decoders that you can use in your integration flow:

1. **Base64 Decoder**—Base64 Decoder is a feature in SAP Cloud Integration that allows you to decode data encoded in Base64 format. This feature can be used to convert the encoded data back into its original format, such as into text or a file, for further processing or storage. The decoder can be used as a step in an integration flow to transform encoded data before it is passed to another system or application.

2. **GZIP Decompression**—GZIP Decompression is a feature in SAP Cloud Integration that allows you to decompress data that was compressed using the GZIP algorithm. This feature can be used to convert the compressed data back into its original format, such as into text or a file, for further processing or storage. The decompression step can be used as a part of an integration flow to transform the compressed data before it is passed to another system or application.

3. **MIME Multipart Decoder**—The MIME Multipart Decoder is a feature in SAP Cloud Integration that allows you to decode MIME multipart messages. Email communications can now include attachments of audio, video, photographs, and application programs in addition to text in character sets other than ASCII, thanks to the MIME (multipurpose internet mail extensions) standard. A MIME multipart message contains multiple parts, each of which can be in a different format. The MIME Multipart Decoder in SAP Cloud Integration can be used to decode such messages and extract the individual parts, which can then be processed or stored separately. This feature can be used as a step in integration flow to transform encoded data before it is passed to another system or application.

4. **ZIP Decompression**—ZIP Decompression is a feature in SAP Cloud Integration that allows you to decompress data that was compressed using the ZIP algorithm. This feature can be used to convert the compressed data back into its original format, such as into text or a file, for further processing or storage. The decompression step can be used as a part of an integration flow to transform the compressed data before it is passed to another system or application.

The ZIP software program is used to compress and decompress files. The ZIP algorithm is widely used for compressing files and folders, and it is the most common format for compressing and decompressing files on Windows, Mac, and Linux operating systems. The ZIP decompression feature in SAP Cloud Integration can be used to extract files and folders from a ZIP archive and decompress them, so they can be further processed or stored.

A decoder converts the data from one format to another and decodes the encoded data. You now know what a decoder is and understand how it can be used in the integration flow design artifact. In the next section, you learn about the encoder.

Encoder

Usually, when someone uses the word "encoder," they're referring to a part that transforms data between formats. A data-mapping procedure is used to transform data from the source system to the target system, and the encoder is a component of that process.

The encoder converts input data from one format, such as XML or JSON, into another format, such as EDI or a flat file. Figure 4-55 shows the iflow using the encoder.

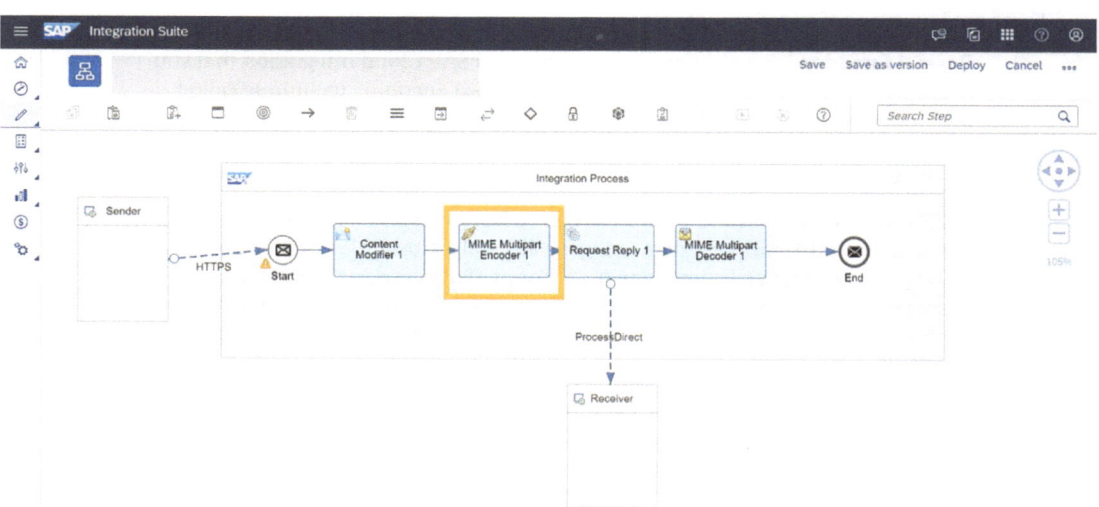

Figure 4-55. *Encoder in integration flow*

You use this task to encrypt communications using a technique to secure sensitive content while it is being transported over the network. Choose Transformation ➤ Encoder from the SAP Cloud Integration palette, as shown in Figure 4-56, to add the encoder to the integration flow, as shown in Figure 4-55.

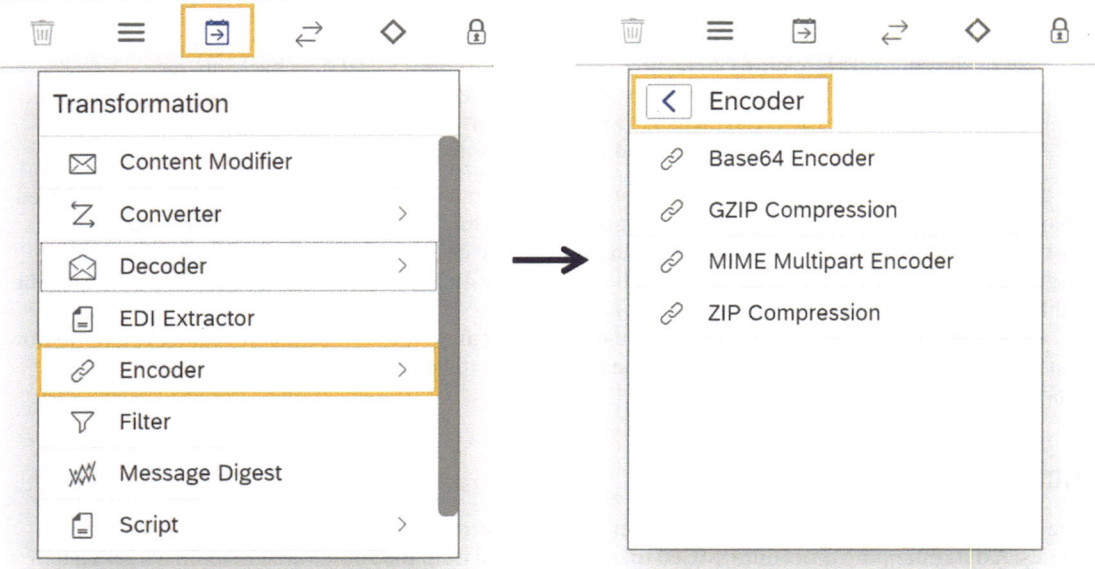

Figure 4-56. *Encoder in SAP Cloud Integration Editor*

There are four types of encoder that you can use in your integration flow:

1. **Base64 Encoder**—Base64 Encoder is a feature in SAP Cloud Integration that you to encode data into Base64 format. A binary-to-text encoding technique called Base64 represents binary data as a radix-64 representation before encoding it as an ASCII string. The encoded data is represented as a string of characters, which can be easily transported or stored.

2. **GZIP Compression**—GZIP Compression is a feature in SAP Cloud Integration that allows you to compress data using the GZIP algorithm. Data transit and storage can be sped up by lowering the amount of data needed to represent a given piece of information. The GZIP algorithm is a widely used method for compressing data, especially on the web, and it can compress data to a much smaller size than other algorithms such as ZIP.

   It is worth mentioning that GZIP compression is not suitable for all types of data, and some types of data, such as text or XML files, can be compressed with GZIP in such a way that the compression ratio is not good, so it is important to test before using it.

3. **MIME Multipart Encoder**—Several data portions can be encoded into a single MIME multipart message using the SAP Cloud Integration capability known as the MIME Multipart Encoder. Email communications can now include attachments of audio, video, photographs, and application programs in addition to text in character sets other than ASCII, thanks to the MIME (multipurpose internet mail extensions) standard.

4. **ZIP Compression**—ZIP compression is a feature in SAP Cloud Integration that allows you to compress data using the ZIP algorithm. Data transit and storage can be sped up by lowering the amount of data needed to represent a given piece of information. The ZIP algorithm is a widely used method for compressing data, especially on Windows, Mac, and Linux operating systems.

You now know what an encoder is and learned how it can be used in SAP Cloud Integration. You also went through the different types of encoders. The next section goes through the EDI Extractor.

EDI Extractor

EDI Extractor allows you to extract EDI headers and transform them into camel headers. Figure 4-57 shows the EDI Extractor used in the integration flow.

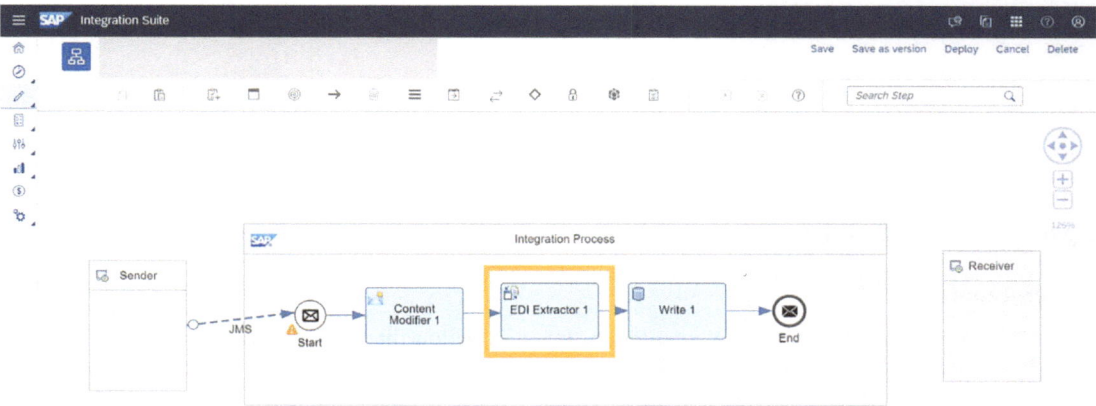

Figure 4-57. *EDI Extractor*

This component adds information to the exchange by extracting data from a single incoming EDI document to be used later in message processing. An EDI Extractor can read flat files and XML files in both formats. From the palette, choose Transformation ➤ EDI Extractor, as shown in Figure 4-58.

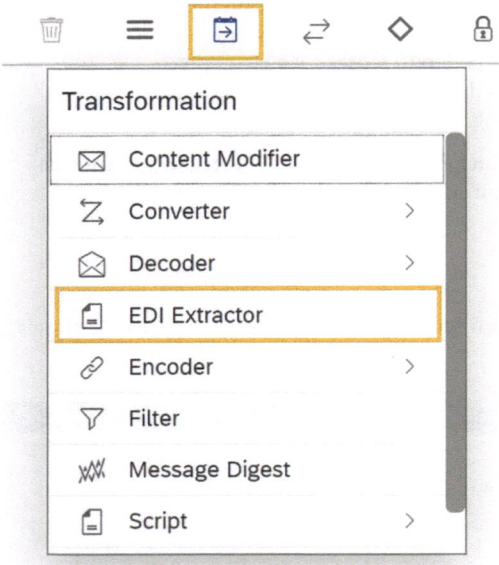

Figure 4-58. *The EDI Extractor in the SAP Cloud Integration Flow Editor*

Procedure:

1. Select the EDI Extractor element in the integration flow that you created in Section 4.1.7. Open it in Edit mode.

2. You can choose any integration flow that you created in the integration package.

3. Set up an extractor parameter.

4. Choose Save to continue modifying the integration package without exiting.

5. To save a copy of the current artifact, select Save as Version.

6. Before saving the package, select Cancel if you want to cancel its creation.

The next section goes through the next element of transformation, the filter.

Filter

To obtain data from an incoming message, use the Filter element. It allows you to filter out the message's unwanted components and just keep the information you need. Figure 4-59 shows the filter used in the integration flow.

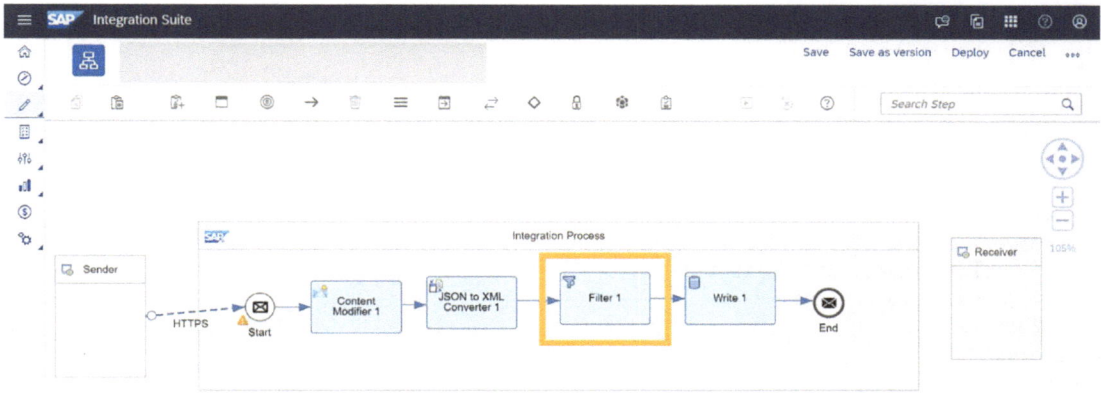

Figure 4-59. *Filter*

From the palette, you can select the Filter element by choosing Transformations ➤ Filter, as shown in Figure 4-60.

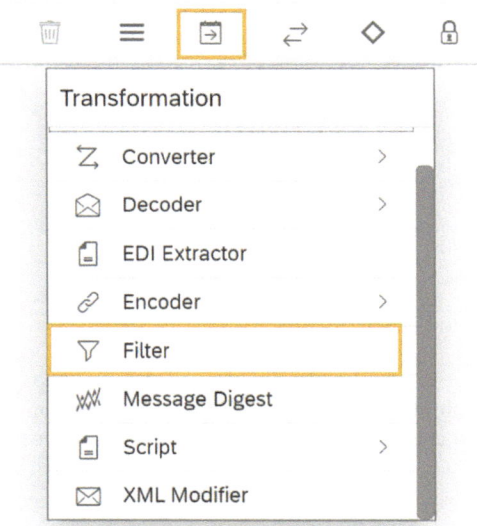

Figure 4-60. *Filter in the SAP Cloud Integration Flow Editor*

Here is an example of a message:

```
<Message>
<orders>
    <order>
        <clientId>ABT0012</clientId>
        <count>200</count>
    </order>
    <order>
        <clientId>ABT0013</clientId>
         <count>20</count>
    </order>
</orders>
</Message>
```

Suppose you are simply concerned with the count. You specify an XPath of the form /Message/orders/order/count/text() using the Filter, as shown in Figure 4-61.

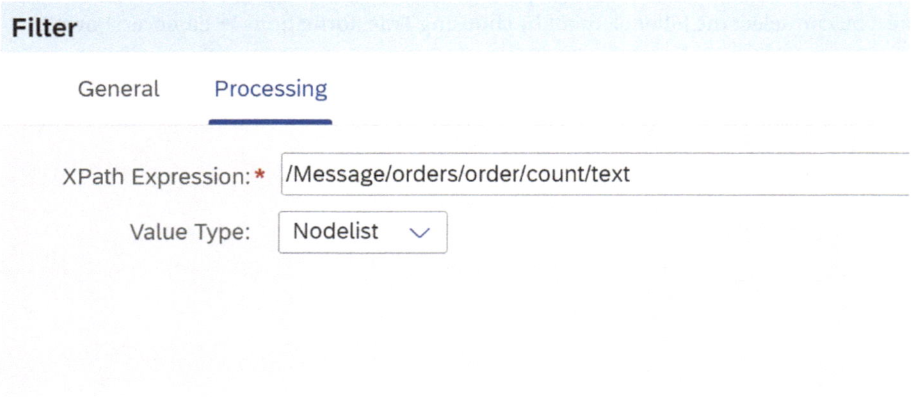

Filter

General   Processing

XPath Expression: * `/Message/orders/order/count/text`

Value Type: Nodelist

Figure 4-61. Filter

Data in the message's count fields are the output of the content filter. With the given XPath, the output in this example is 20020. The next section goes through the next element of transformation, message digest.

Message Digest

The payload or portions of it are digested in this integration flow stage, with the results being stored in the message header. Figure 4-62 shows the message digest used in the integration flow. A message is converted into a canonical XML document in the Message Digest integration flow step. A digest (hash value) is computed from this text and added to the message header.

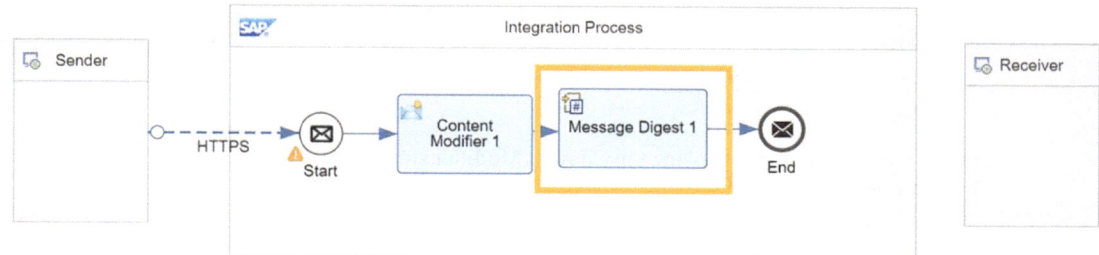

Figure 4-62. *Message Digest*

From the palette, you can select the Filter element by choosing Transformations ➤ Message Digest, as shown in Figure 4-63.

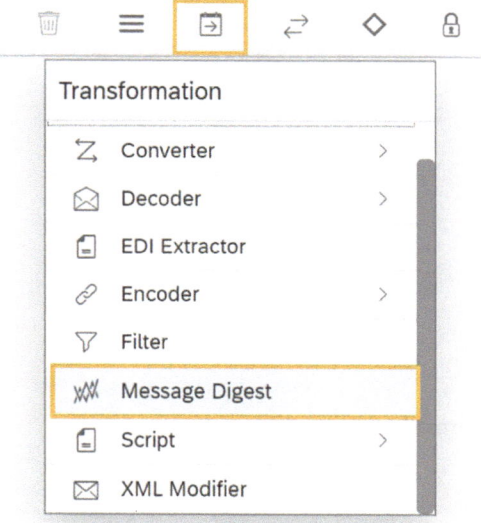

Figure 4-63. *Message digest in the SAP Cloud Integration Flow Editor*

This functionality can be used to implement situations such as the ones listed here:

1. You can transmit the digest along with a specific value in a message in place of a signature. By doing so, you give the recipient the ability to determine whether the value has changed throughout processing.

2. You can store its digest in place of a specific value. To determine whether a value has changed throughout message processing, look up the value's digest.

Procedure:

1. Select the Message Digest step to update the properties if it is included in the integration flow. You can choose the integration flow created in Section 4.1.7 or create a new integration flow.

2. If you want to incorporate a Message Digest step, the subsequent substeps must be finished:

    • Select (Message Transformers) from the palette before selecting (). (Message Digest).

    In the integration process, place the Content Modifier step, as shown in Figure 4-64.

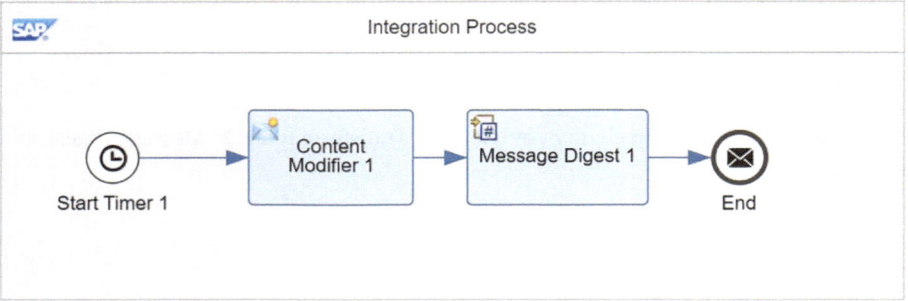

Figure 4-64. *Message Digest in integration flow*

3. Give the step the following characteristics.

    • Filter (XPath)

    • Canonicalization Method

    • Digest Algorithm

    • Target Header Name

The next section goes through the next element of Transformation, Script.

Script

In SAP Cloud Integration, a Script refers to a piece of custom code written in Groovy Script or JavaScript that can be executed within integration flows to perform specific operations and manipulations on data during the integration process. Scripts in SAP Cloud Integration are used to enhance the functionality and flexibility of integration flows by allowing users to add custom logic and transformations. They provide a way to extend the capabilities of the standard integration flow components and perform complex operations that cannot be achieved through configuration alone. You can utilize your own custom scripts in your integration artifacts when you need capabilities beyond what Cloud Integration's native functionality can provide. Figure 4-65 shows the use of Groovy Script in the integration flow.

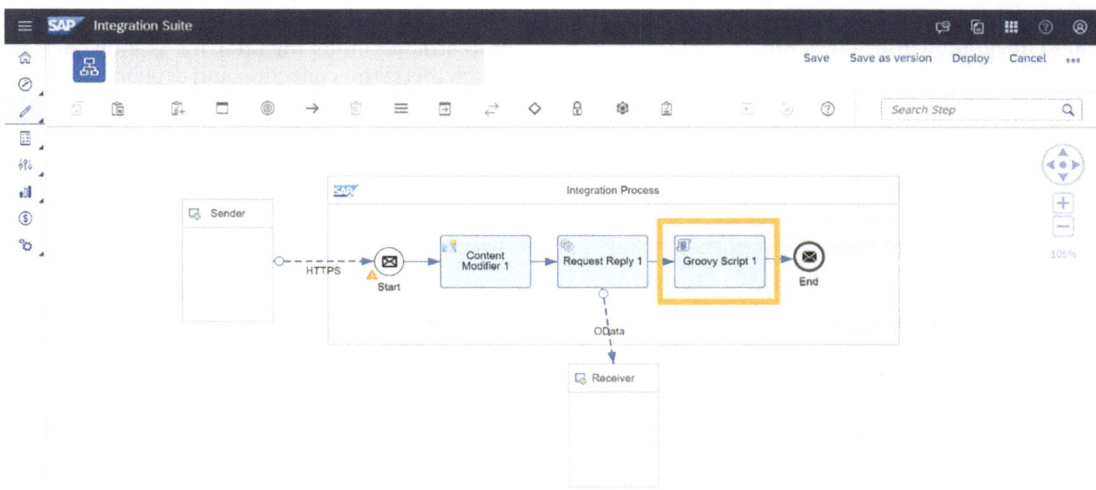

Figure 4-65. *Using Groovy Script*

For example, you can use scripts to handle exceptions, perform sophisticated transformations, or read or edit headers in adapters. There are two types of script that you can use in SAP Cloud Integration, as shown in Figure 4-66.

1. Groovy Script

2. JavaScript

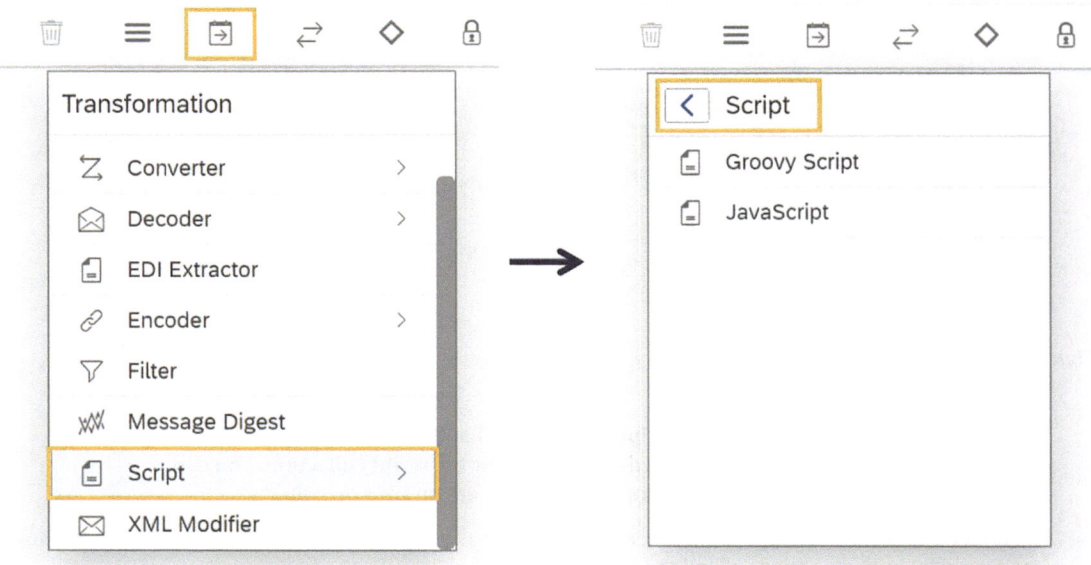

Figure 4-66. *Script and its types used in the SAP Cloud Integration Flow Editor*

You have learned about the script step, which can be added in the integration flow step. However, SAP Cloud Integration also allows you to create script collections as artifacts under the same integration package as you create the integration flow. The next section goes through this script collection and explains the different attributes and steps related to script collection.

Script Collection

An artifact in and of itself, script collection is a collection of scripts. To reuse a collection of scripts across any number of integration artifacts in an integration package, construct the collection within the integration package. JAR files, JavaScript, and Groovy Scripts are supported.

You learn how to create the script collection in the next section.

Create a Script Collection

Creating a script collection in the SAP Cloud Integration Flow Editor allows you to define and manage custom JavaScript or Groovy Scripts that can be used to enhance the functionality of integration flows. Follow these steps:

1. Open the integration package you created in Section 4.1.1 in Edit mode.

2. From the Artifacts tab, choose Add ➤ Script Collection, as shown in Figure 4-67.

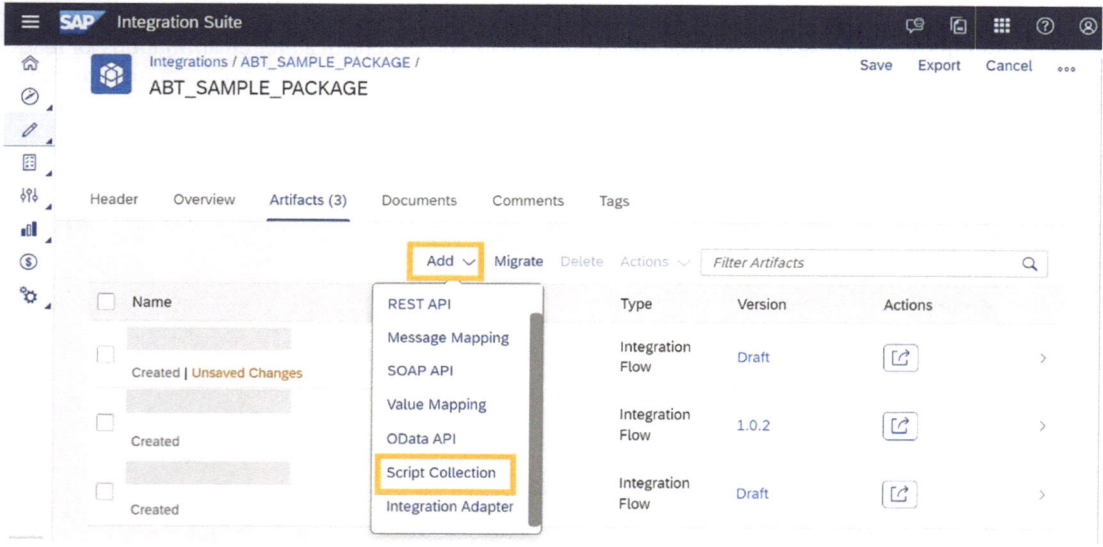

Figure 4-67. *Adding a script collection*

3. You can create or upload the script collection. To upload the script collection, upload the.ZIP file that contains the resources for the script collection. You can find this ZIP file from your developer, or you can create your own file.

As you have created the script collection as an artifact in the integration package, you will now see how to create the script in these script collections.

Create Script in Script Collection

You can add different scripts, such as Groovy Scripts and JavaScript scripts, to these script collections:

1. Open the script collection you created in previous step in Edit mode in your artifact.

2. Create a script:

   - From the Resources pane, choose Create ➤ Groovy Script or Create ➤ JavaScript, as shown in Figure 4-68.

   - Enter a name for the script and choose Save.

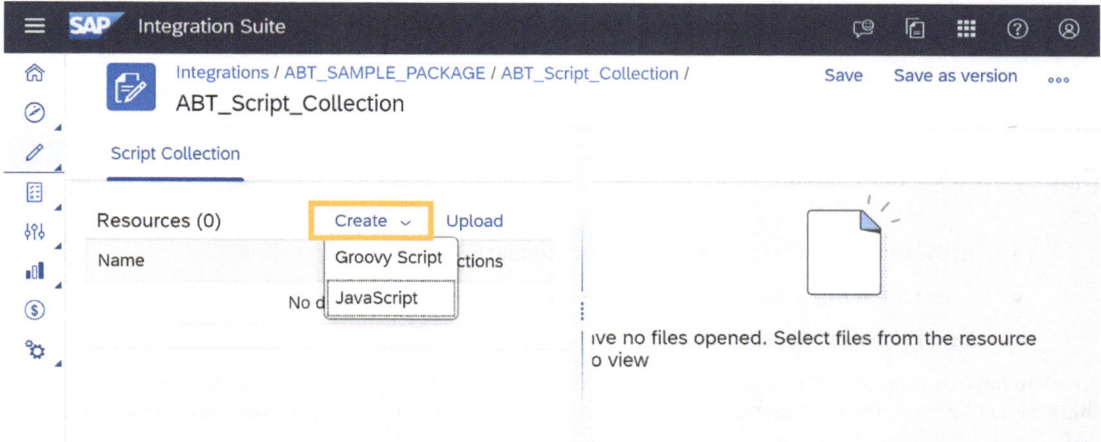

Figure 4-68. *Creating a script*

3. To upload a script, follow these steps:

   - In the Resources pane, choose Upload.

   - Select File System for source (upload a file containing the Groovy/JavaScript code from your local machine). Or select Integration Flow (upload a script file from integration flows in your tenant). See Figure 4-69.

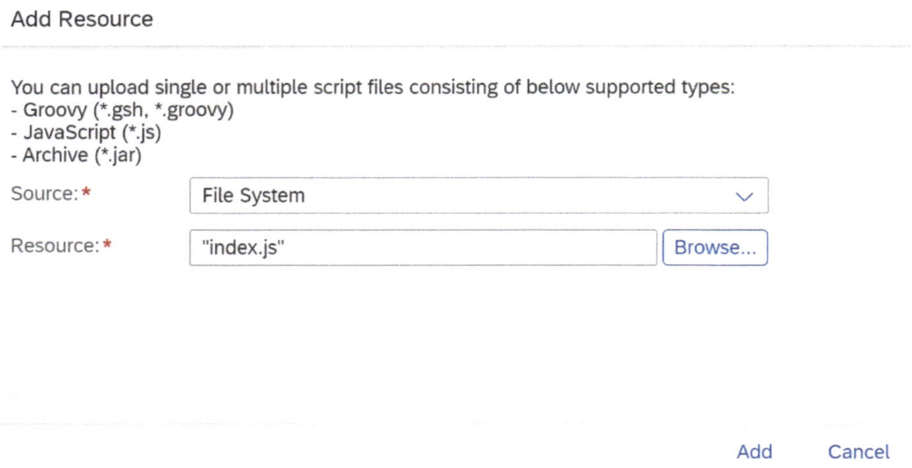

Add Resource

You can upload single or multiple script files consisting of below supported types:
- Groovy (*.gsh, *.groovy)
- JavaScript (*.js)
- Archive (*.jar)

Source: *	File System	⌄

Resource: *	"index.js"	Browse...

Add   Cancel

Figure 4-69. *Uploading a script*

- In the integration flow, select any integration flow.

- Select the script that you want to choose and click Add.

- Save the script collection artifact.

You have now created the script in the script collection and successfully saved it. In order to consume this script collection, the next section explains how to determine the consumption of script collection in the SAP Cloud Integration Flow.

Consuming Script Collection

Consuming a script collection refers to using a script from a script collection in an integration flow.
    A script collection must be consumed in two steps:

1. You establish a reference to the script collection in your integration artifact.

2. You establish a script step in your integration artifact at the required location and choose a script from the relevant script collection.

Deploying the Script Collection

1. Open the integration package in which you created the script collection in Section 4.2.4.9.1.

2. Select the script collection and choose Actions ➤ Deploy.

3. Choose Operations View.

4. Under Manage Integration Content, choose the All tile.

You learn about these script collections in depth with the help of an example, whereby you use the script collection created in the previous steps.

Script Collection: Practical Example

Deploying a script collection in the SAP Cloud Integration Flow Editor allows you to make your custom JavaScript or Groovy Scripts available for use in integration flows.

To deploy a script collection in the SAP Cloud Integration Flow Editor, follow these steps:

1. Open any desired integration flow or the one created in Section 4.1.7.

2. Choose Resources ➤ Reference ➤ Add Reference ➤ Script Collection in the Property sheet of the integration flow, as shown in Figure 4-70.

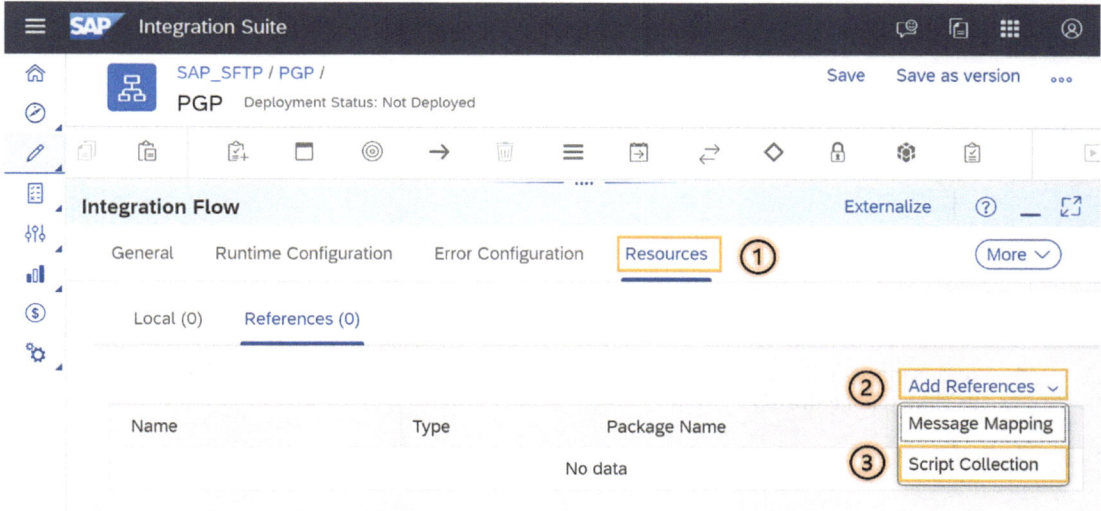

Figure 4-70. *Add script collection to the iflow*

3. Select the desired script collection and click OK, as shown in Figure 4-71.

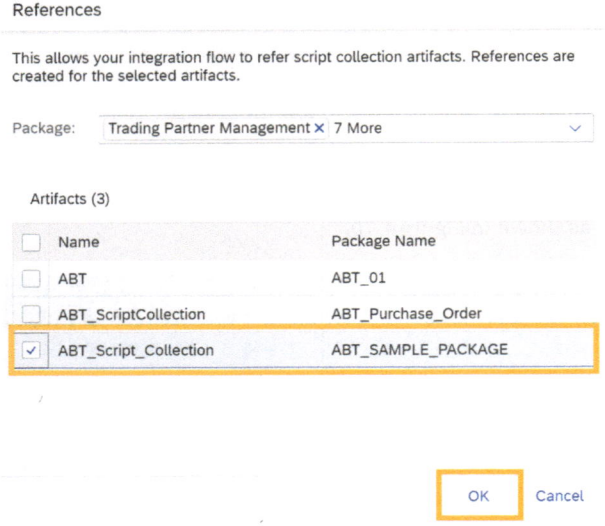

Figure 4-71. *Script collection*

4. In this manner, the script collection will be included in your integration artifact, as shown in Figure 4-72.

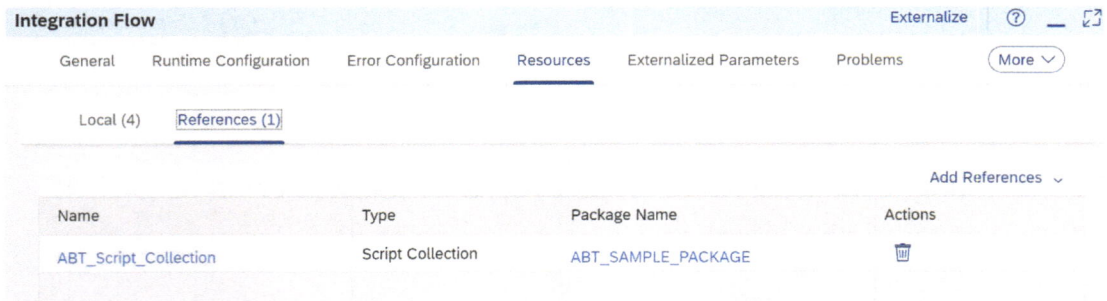

Figure 4-72. *Script collection in the integration flow*

5. The desired step can now be added to the script. Select the script by choosing Message Transformers ➤ Java Script, then add it to the integration flow.

6. In the Resources panel, select the desired script from the script collection, as shown in Figure 4-73.

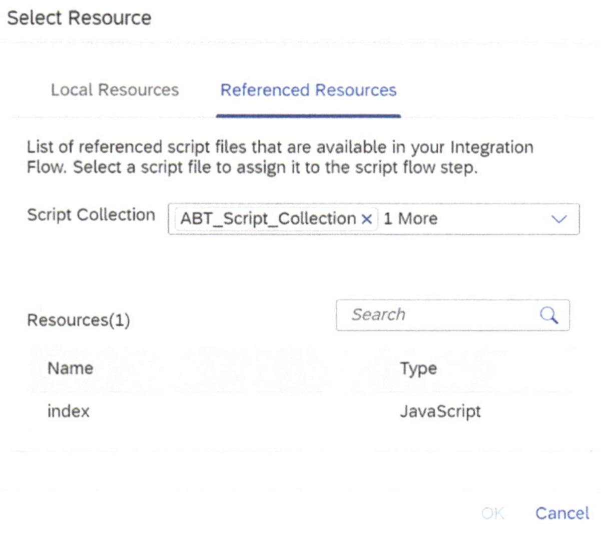

Select Resource
Local Resources
Referenced Resources
List of referenced script files that are available in your Integration Flow. Select a script file to assign it to the script flow step.
Script Collection
ABT_Script_Collection × 1 More
Resources(1)
Search
Name
Type
index
JavaScript
OK
Cancel

Figure 4-73. *Script collection*

7. After choosing the desired script, save the artifact.

You have gone through message transformations in which you learned about different tools and techniques, including the Content Modifier, XML modifier, converter, decoder, encoder, EDI Extractor, filter, message digest, and scripts. Each of these tools was explained in detail, with practical examples provided for JSON to XML conversion, creating a script collection, and deploying the script collection. In the next section, you learn about mapping as an integration flow design artifact.

4.2.5 Mapping

Data transformation from one structure or format to another is known as *mapping* in SAP Cloud Integration. It enables data to be tailored to the precise specifications of the intended system or application, allowing the system to understand and utilize the data. Figure 7-74 shows the use of mapping in the integration flow.

Integration Process
Message Mapping 1
XSLT Mapping 1
End

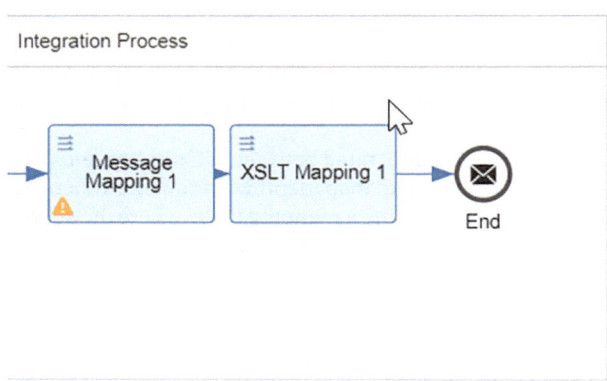

Figure 4-74. *Mapping in the integration flow*

Mapping in SAP Cloud Integration can be done using a graphical mapping tool, where you can drag-and-drop the source and target fields, and then use functions and expressions to transform the data. Mapping can be performed between different data formats, such as JSON, XML, CSV, and EDI.

Mapping can be used in different integration scenarios, such as data integration, application integration, and B2B integration, to adapt data to the specific requirements of the target system. It can also be used to enrich, filter, or validate data before it is passed to the target system. Additionally, it can be used to transform nonstructured data, such as text or images, into structured data, such as XML or JSON. Navigate to the Mapping section under the SAP Cloud Integration Flow Editor palette, as shown in Figure 4-75.

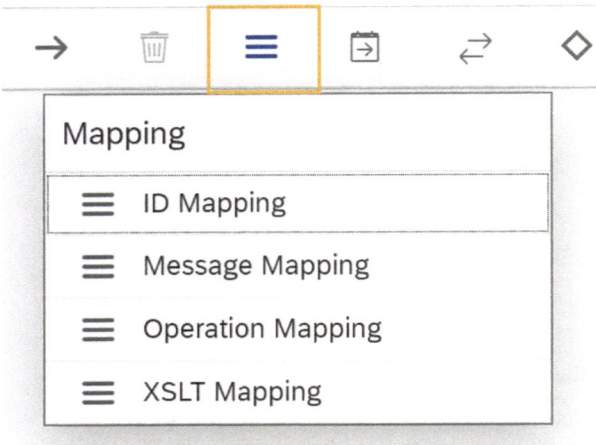

Figure 4-75. *Mapping in the SAP Cloud Integration Flow Editor*

There is a fourth option whereby you can use the mapping techniques in the integration flow. This is called ID Mapping, and is discussed in the next section.

ID Mapping

ID Mapping in SAP Cloud Integration refers to the process of mapping identifiers, such as primary keys or unique identifiers, between different systems or applications. It is used to ensure that data in the different systems can be linked or related correctly.

For example, in a system integration scenario, the source system can use a different identifier for a customer than the target system. ID Mapping allows you to map the customer identifier in the source system to the corresponding identifier in the target system so that the customer data can be linked correctly between the two systems.

ID Mapping can also be used in scenarios where data is being integrated between different systems, but the systems use different identifier schemes. ID Mapping can be used to map identifiers from one system to the other so that the data can be related correctly. This can be done using a mapping table, where the source and target identifiers are matched, or by using a function or expression to perform the mapping. Figure 4-76 shows the ID Mapping option in SAP Web UI.

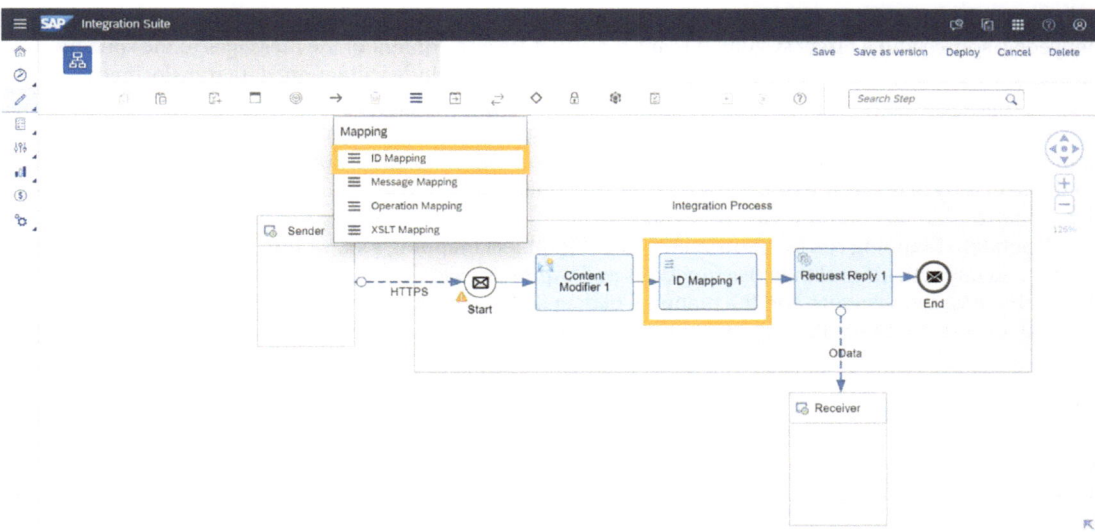

Figure 4-76. ID Mapping

The next section goes through operation mapping, which is another element of mapping.

Operation Mapping

Operation Mapping in SAP Cloud Integration is a specific type of mapping that is used to map operations and messages between different systems or applications. It allows you to adapt operations and messages to the specific requirements of the target system so that they can be understood and executed by the system. Figure 4-77 shows the Operation Mapping element in the Integration Flow Editor.

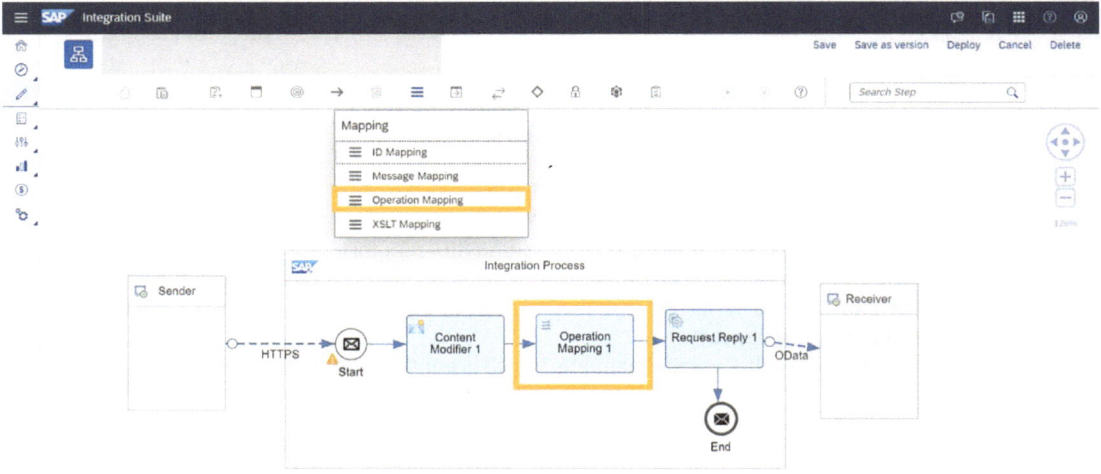

Figure 4-77. Operation Mapping

Operation Mapping defines the mapping of an operation and its request and response messages between a source and a target system. It maps the structure and content of the messages to the specific format and structure required by the target system. It can also include validation, data transformation, and routing rules.

Operation Mapping can be used in a variety of integration scenarios, including B2B integration, application integration, and service-oriented architecture (SOA), to modify operations and messages based on the demands of the destination system. It allows you to create a uniform and standard interface so that different systems can communicate with each other.

Operation Mapping can be performed using a graphical mapping tool, where you can drag-and-drop source and target fields and then use functions and expressions to transform the data. This allows you to define the message structure and the mapping of data fields between different systems.

The next section goes through the XSLT Mapping.

XSLT Mapping

XSLT (extensible stylesheet language transformations) Mapping is a method of transforming XML documents into other formats using an XSLT stylesheet. Figure 4-78 shows the XSLT mapping in the SAP Web UI.

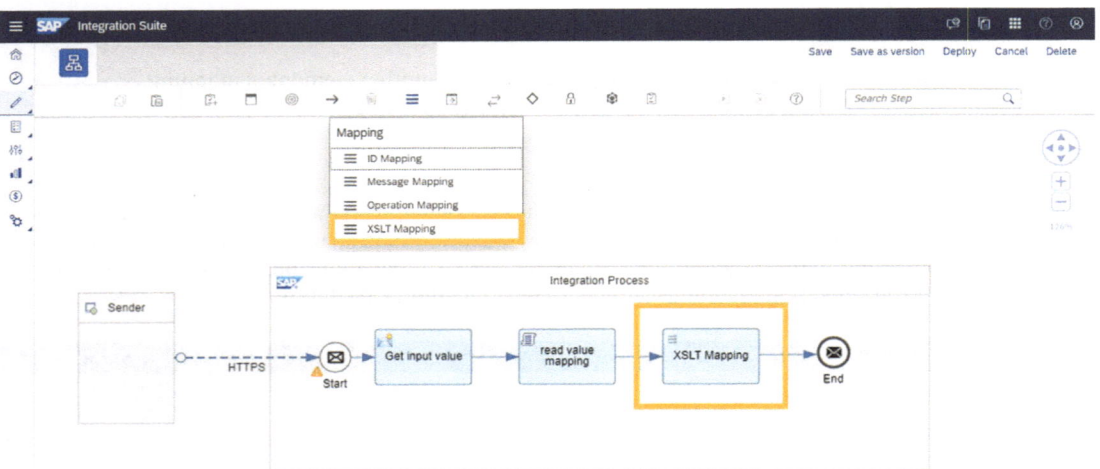

Figure 4-78. *XSLT Mapping*

SAP Cloud Integration is used to transform data between different formats, such as EDI to XML or XML to JSON, to enable communication between different systems. XSLT Mapping is done using a graphical mapping tool, which allows you to easily create and edit the XSLT stylesheet. This mapping can be used to map data from the source to the target system.

The next section goes through Message Mapping, which is another element of mapping.

Message Mapping

With the help of Message Mapping, you can provide a connection between message fields with various structures. Figure 4-79 shows the Message Mapping element used in the Integration Flow Editor.

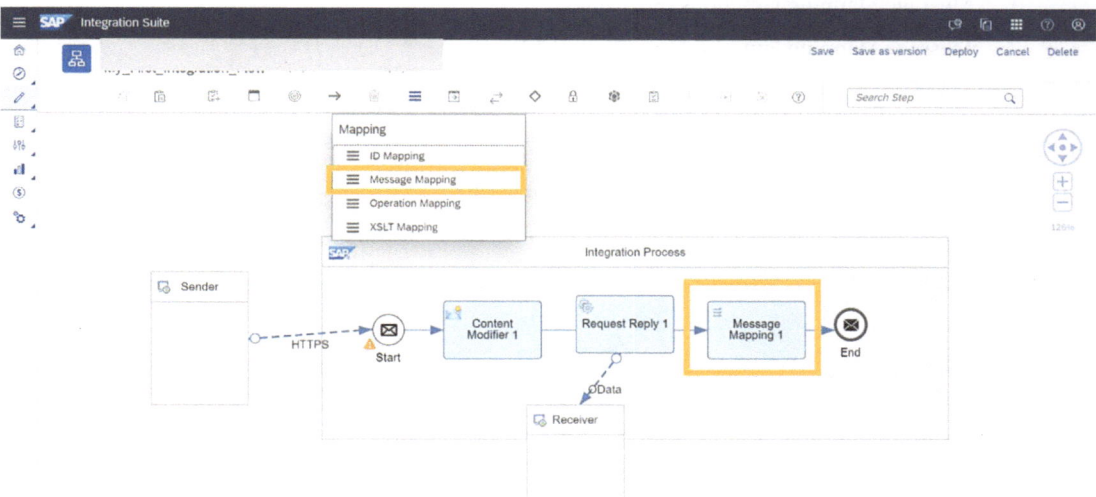

Figure 4-79. *Message Mapping*

When the receiving system takes the data in a format that differs from the sender system, Message Mapping is necessary. For example, purchase order replication from SAP ERP ➤ Salesforce. SAP Cloud Integration provides a single graphical editor for mapping. Message Mapping can be invoked from Mapping ➤ Message Mapping.

Message Mapping Example

Consider employee data that's in System A and System B. The data could be identical or different in both systems. However, in this scenario, assume that, in System A, the data is stored in different fields; in System B, the data is stored in another field, as shown in Figure 4-80. For example, say you want to convert the date of shipping in System A from its current DD-MM-YYYY format to the YYYY-MM-DD format.

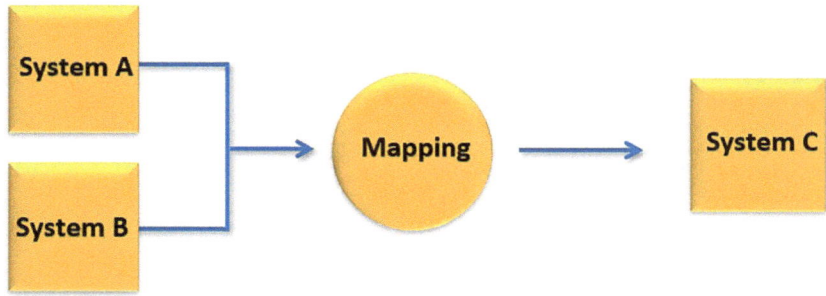

Figure 4-80. *Example of Message Mapping*

Message Mapping Functions

In Message Mapping, various functions can be used during the process of mapping. Here are some of the functions:

1. Message Mapping Functions (Fix Values)

2. Message Mapping Functions (Constant/Add)

3. Message Mapping Functions (Constant/Constant)

Fix Values

Fix Values functions in the message can be used to substitute the key/value combination from the source structure to the target structure, as shown in Figure 4-81.

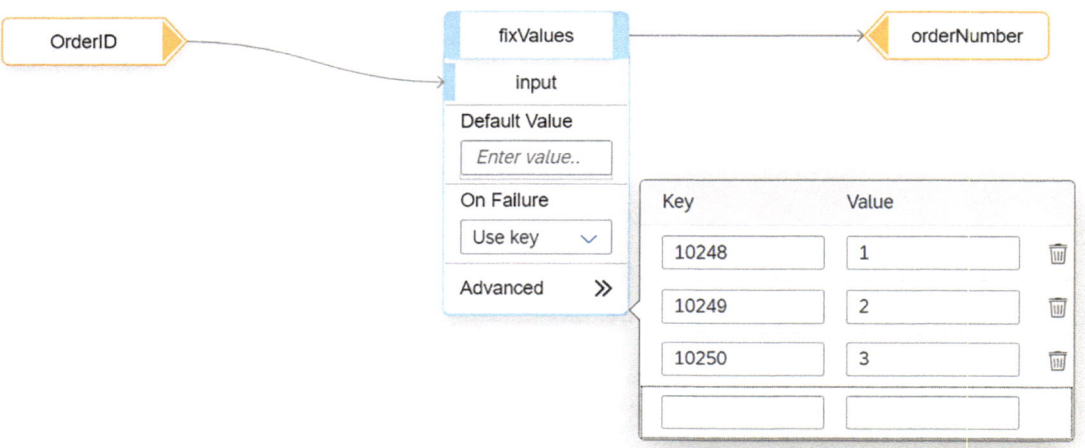

Figure 4-81. *Fix Values in Message Mapping*

Constant/Concat

Different functions can be used simultaneously in Message Mapping to build business requirements. Figure 4-82 uses a combination of Constant and Concat functions to concatenate ShipName and the constant text Abusiness with a delimiter string called supplierForename.

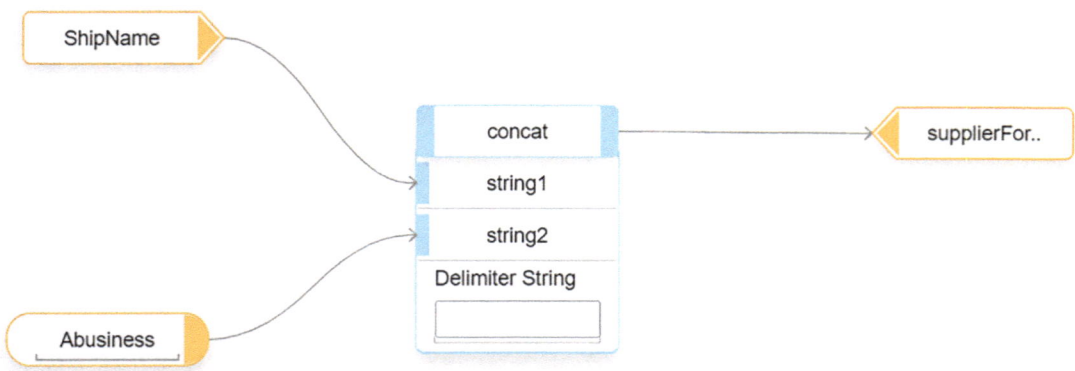

Figure 4-82. *Constant/Concat*

Constant/Add

Similarly, arithmetic functions (such as add) can be used to add numbers from the source structure and then map to the target structure, as shown in Figure 4-83.

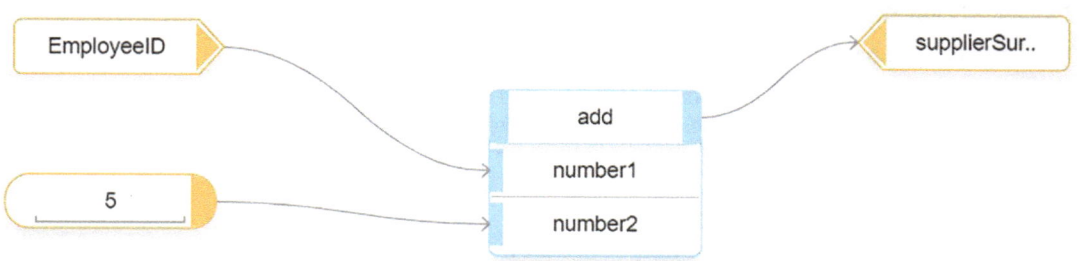

Figure 4-83. Constant/Add

You have now learned about the Message Mapping functions. The next section goes through Message Mapping testing.

Message Mapping Testing

It takes time to model and install a whole integration pipeline to test Message Mapping. Message Mapping cannot be tested independently and requires more time to model. The results of the mapping must therefore be displayed immediately in the test environment.

The Run Simulation button is located at the top of the Message Mapping page when you are in Edit mode, as shown in Figure 4-84. Simulating a mapping will build a virtual bundle, carry out your mapping, and return the results. You will have the choice to upload your input payload after clicking the Upload Input button.

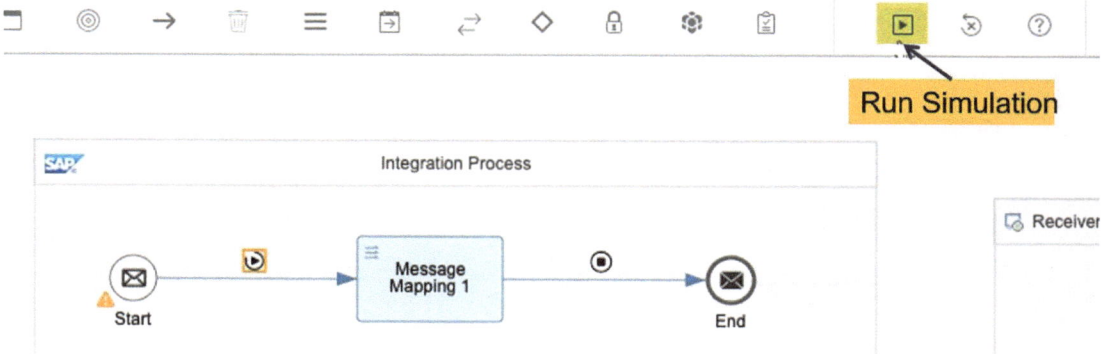

Figure 4-84. The Run Simulation button

You have learned all about the Message Mapping functions. To better understand these Message Mappings, you will go through an example in the next section.

Message Mapping: Practical Example

Message Mapping in the SAP Cloud Integration Flow Editor allows you to transform and manipulate data between different message formats or structures within integration scenarios.

To create Message Mapping in the SAP Cloud Integration Flow Editor, follow these steps:

1. Create an integration flow and open it in Edit mode.

2. Create a similar iflow, which you have used in request replies. Add Message Mapping to the flow. The iflow should look like Figure 4-85.

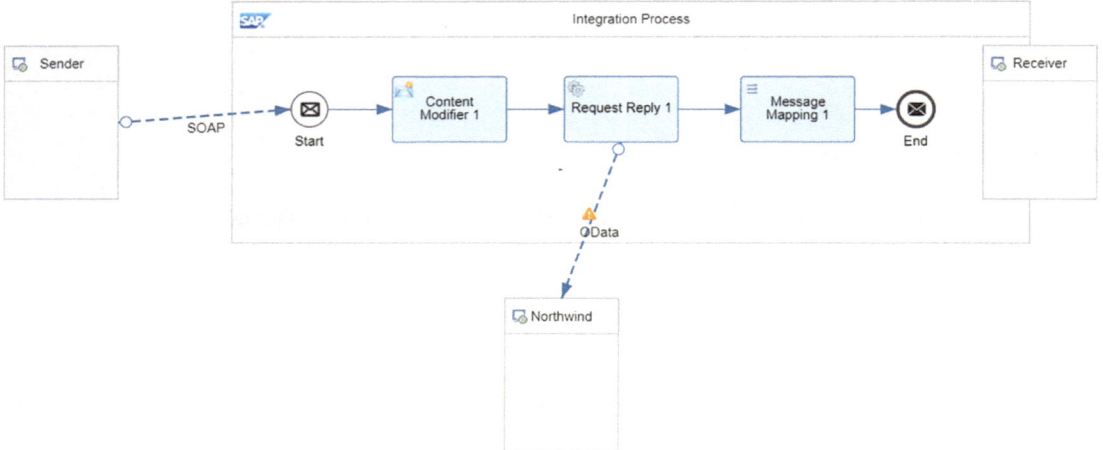

Figure 4-85. *iflow for Message Mapping*

3. You can create Message Mapping by clicking the symbol beside the Message Mapping step, as shown in Figure 4-86.

Figure 4-86. *Create Message Mapping*

4. Add your source and target files (which you intend to map with) to the Integration Flow ➤ Resources section.

5. In this example, EDMX is the source file, and WSDL is the target file. You can use the EDMX file, for example, from Webhook or Northwind.

6. You can now see the files in the Resources section of the integration flow.

7. When you create the Message Mapping file, you will see the source and the target field after selecting the source and target file.

8. You can map any field from the source and the target, as shown in Figure 4-87.

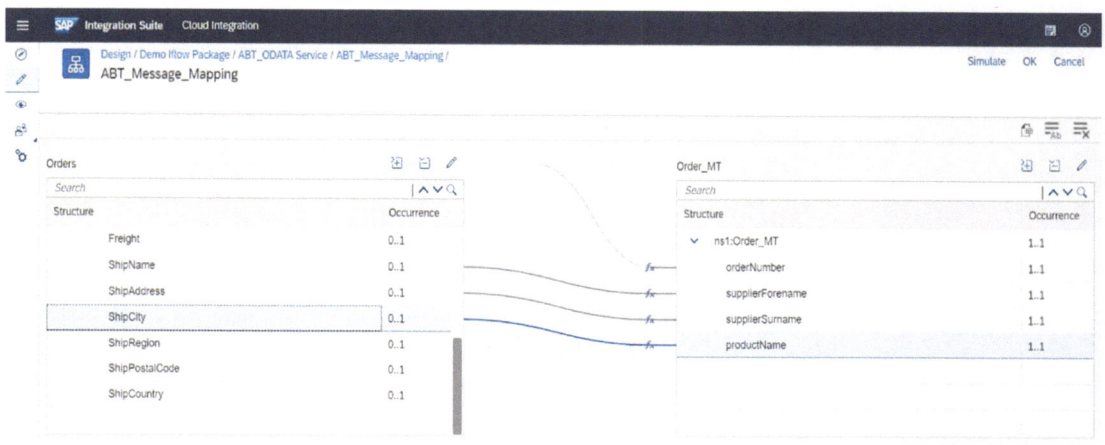

Figure 4-87. Mapping a source with target fields

4.3 Summary

This chapter discussed practical examples of packaging integration content, including creating, importing, editing, and exporting integration packages. It also cover predelivered integration packages from SAP partners and provided specific examples, such as the integration between SAP SuccessFactors and JIRA and SAP S/4HANA and the ServiceNow API.

The chapter explained in depth the participants, processes, events, and message transformations found in iFlow design object elements. You have become familiar with a variety of message-transformation methods, including the EDI Extractor, the Content Modifier, the XML Modifier, and converters, decoders, and encoders. Using scripting and mapping for integration was also covered in this chapter. It also covered the basic steps for the development of integration flow (iFlow) using the integration flow design object elements in the web-based user interface.

The next chapter looks deep into the advanced functions available for development in the SAP Cloud Integration. The SAP Cloud Integration Development Part II chapter focuses on functions that are required for more complex scenarios, such as routers, external system calls, multicasts, gathers, encryptors/encoders, and so on.

CHAPTER 5

■ ■ ■

SAP Cloud Integration: Development Part II

In Chapter 4, you learned about the development functions available in SAP Cloud Integration. This chapter is a continuation of Chapter 4, which goes through the advanced functions available in SAP Cloud integration for development. It examines the numerous design object features and elements necessary for creating integration projects using SAP Cloud Integration.

The chapter examines the call design object element, including local and remote calls, and provides useful illustrations of content enrichers and polling enrichers. You explore the routing design object element with a real-world example of router configuration, including aggregator, gather and join, multicast, splitter, and router. With practical examples of PGP encryption and decryption, you also dive deeply into security and message-level security use case setups, including signers, decryptors, and encryptors.

5.1 Iflow Design Object Elements

The integrated development environment (IDE) called Web IDE is a web-based IDE that SAP offers for creating and testing integration processes that link different applications, systems, and services. It offers a set of tools and templates for developing, deploying, and maintaining integrations in a cloud environment and enables developers to create, edit, and test integration flows using a visual design interface. Within a team, it also enables collaboration and version control for integration flows.

The following are the predelivered SAP palettes/functions/features that can be used in SAP Cloud Integration Development. You can also use Groovy/JavaScript if one of the predelivered functions does not meet your requirements.

■ **Note**   WebIDE is still used in the SAP Integration Suite even though SAP created Business Application Studio, which is available in the SAP BTP.

The following sections explain the different sections of the elements that are available in SAP Integration Flow. Let's first look at the participants.

© Jaspreet Bagga 2023
J. Bagga, *A Practical Guide to SAP Integration Suite*, https://doi.org/10.1007/978-1-4842-9337-9_5

5.1.1 Call

There are numerous steps that can be created in the integration flow that call remote (external) components or subprocesses.

The Call element has subelements like these:

- External calls

- Local calls

You now learn about these calls, one by one.

External Calls

An external call is a step in a flow that allows the integration flow to call an external service or system. A number of protocols—including HTTP, SOAP, and OData—can be used to accomplish this. The External Call step can be used to call web services, RESTful APIs, or other external systems that are needed to complete the integration flow, as shown in Figure 5-1.

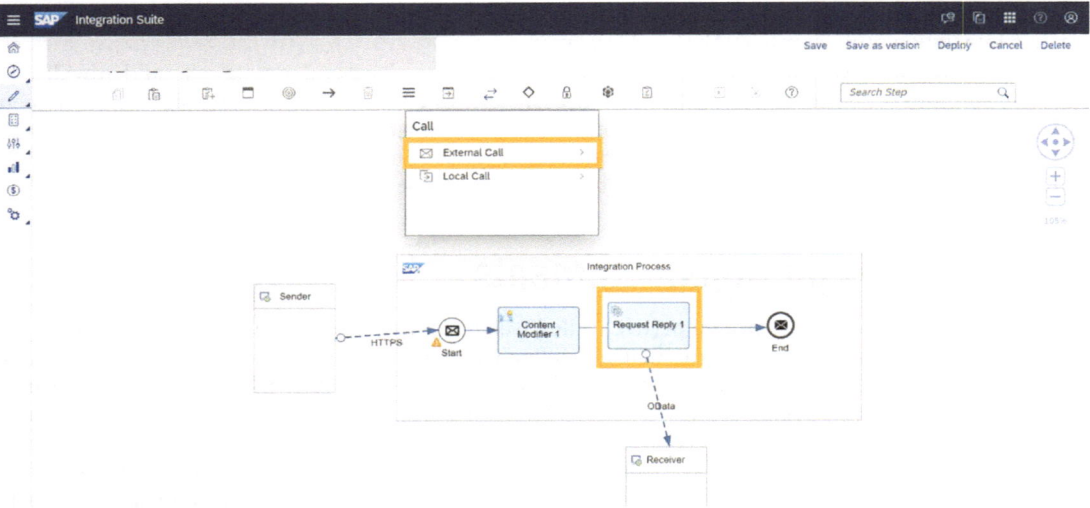

Figure 5-1. *External calls*

This allows you to configure request details such as the target URL, headers, and payload. In addition, response handling includes status codes, headers, and payloads. To add the external call to the integration flow, navigate to the call in the palette and select the call element, as shown in Figure 5-2.

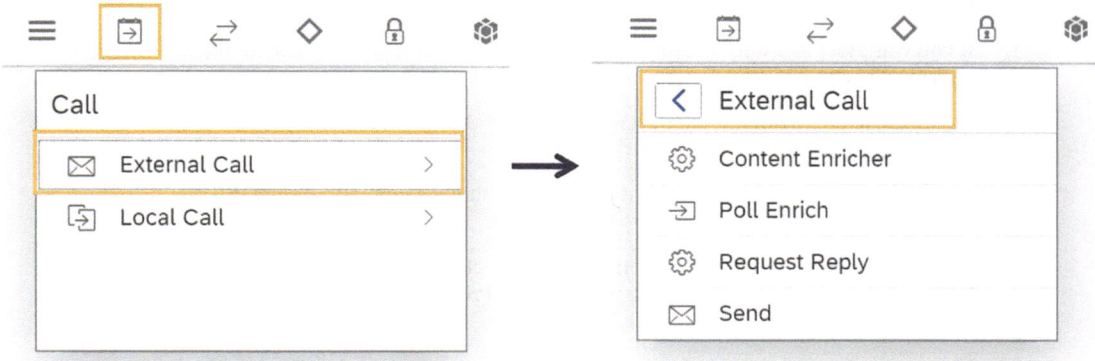

Figure 5-2. *External call in the SAP Cloud Integration Flow Editor*

Request Reply

Request Reply calls an external system (receiver) and gets a response. The integration flow is expected to communicate with an external service to retrieve, post, delete, and update the data and further process it synchronously. Figure 5-3 shows Request Reply used in the integration flow.

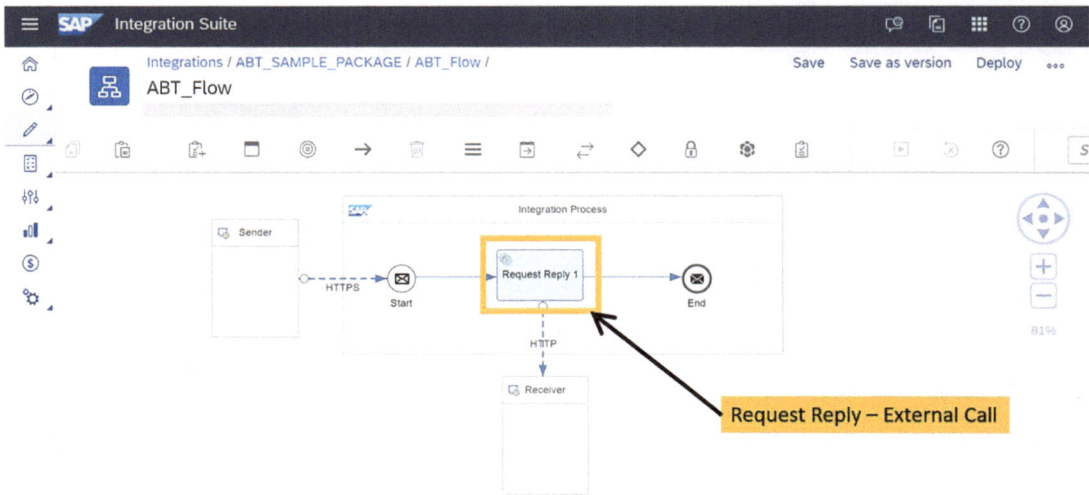

Figure 5-3. *Request Reply*

A sender can send a message to a recipient and then wait for a response using a Request Reply integration flow in SAP Cloud Integration. As a result, the sender and receiver can engage in real time using synchronous communication. A request message is sent, and the recipient must wait for a response message. The integration flow moves on to the following steps after receiving the response message.

Procedure:

1. If you want to add a Request Reply step to the integration process, the following substeps must be finished. Any integration flow is acceptable.

    - Choose Call from the palette (External Call). Access the submenu, and then choose Request Reply.

    - In the integration process, add the Request Reply shape and define the message path.

2. To add this shape to the integration flow model, click anywhere within the integration process box. There are no additional properties required for this step type.

3. Choose the Receiver option by dragging it outside the Integration Process pool after selecting the Participants (icon) button in the palette.

4. Create a channel between the Request Reply step and the receiver by connecting to it and setting up a connection, as shown in Figure 5-4.

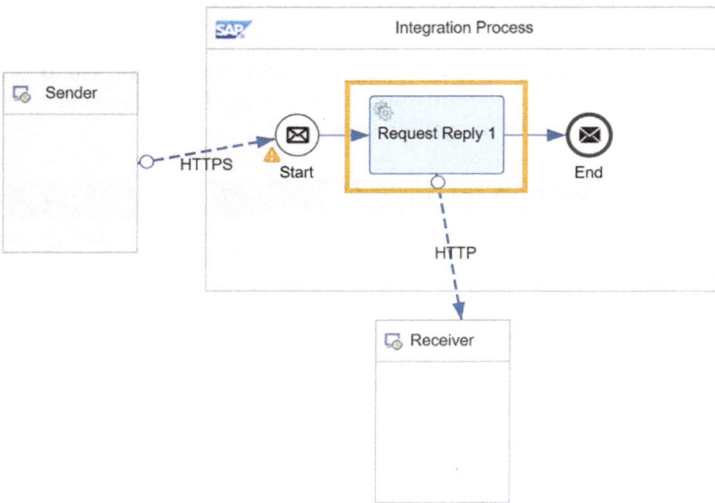

Figure 5-4. *Request Reply*

The next section explains the Content Enricher.

Content Enricher: Practical Example

Content Enricher is a tool that SAP Cloud Integration provides to enrich the tools when interacting with external resources.

Request Reply and Content Enricher are similar. The only difference is that you must combine the input values—which you obtain via the SOAP UI tool—and the OData responses—which you obtain from the Northwind OData service.

1.   Create an integration flow and open it in Edit mode.

2.   Connect the Sender with Start using the SOAP Protocol. Add the Content
      Modifier (choose Message Transformers ➤ Content Modifier), as shown in
      Figure 5-5.

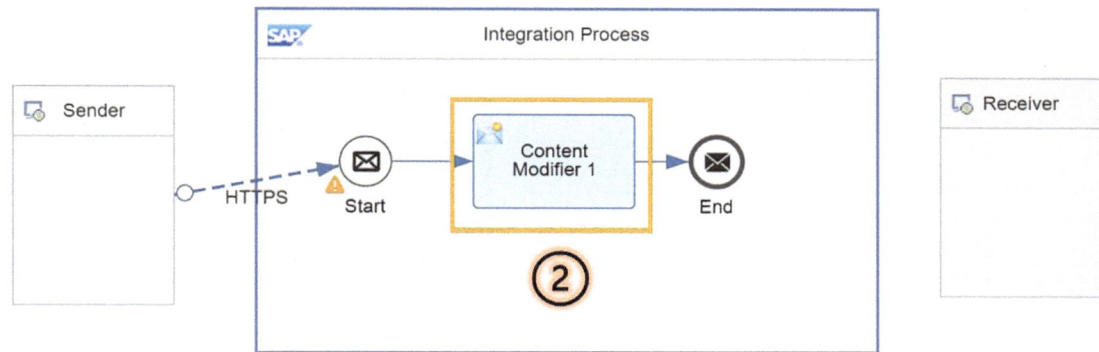

Figure 5-5. *Adding the Content Modifier to the integration flow*

3.   The message header of the Content Modifier provides the following details, as
      shown in Figure 5-6:

      • Action—Create

      • Name—OrderNo

      • Source Type—XPath

      • Source Value—//in

      • Data Type—java.lang. String

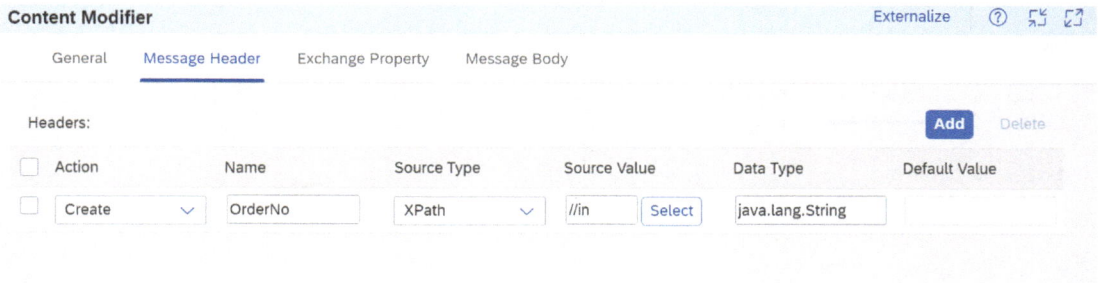

Figure 5-6. *Message header*

4. Add a Request Reply (Choose Call ➤ Enrich Call ➤ Content Enricher) to the flow, as shown in Figure 5-7.

5. Connect the Receiver and Request Reply to the OData V2 communication, as shown in Figure 5-7.

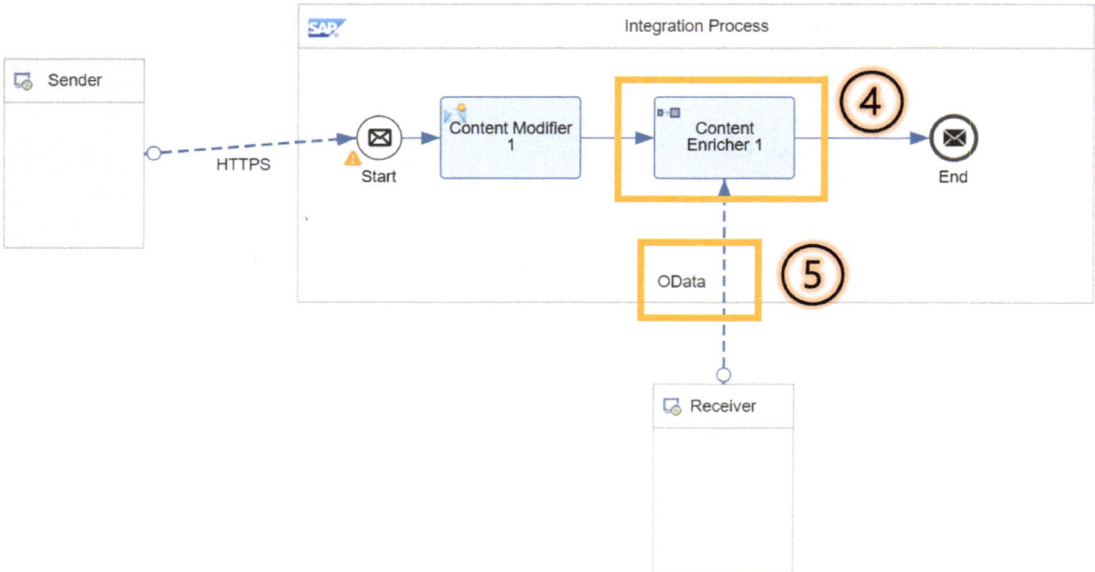

Figure 5-7. *Configuring the Request Reply and outbound communication*

6. In the Connection tab of OData, provide the following address: `http://services.odata.org/V2/Northwind/Northwind.svc.`, as shown in Figure 5-8.

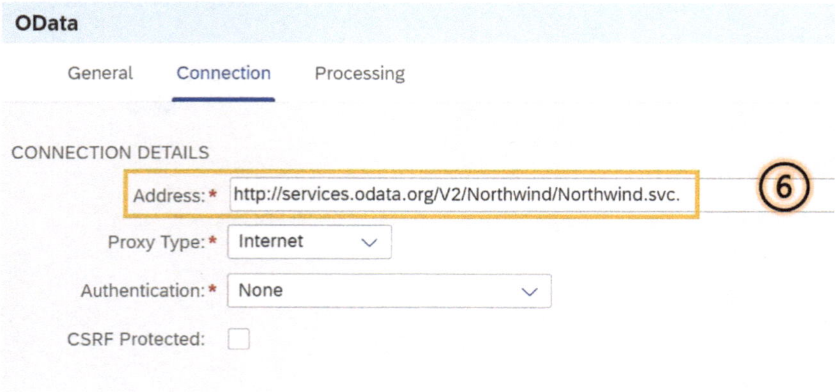

OData

General   Connection   Processing

CONNECTION DETAILS

Address: * `http://services.odata.org/V2/Northwind/Northwind.svc.`   ⑥

Proxy Type: *   Internet   ⌄

Authentication: *   None   ⌄

CSRF Protected: ☐

Figure 5-8. *Connection tab of OData*

7. In the Processing tab of OData, take these steps:

    ●   In the Resource path, select Order for the entity.

    ●   Connect to the system, recheck the entries, and move to Step 2, as shown in Figure 5-9.

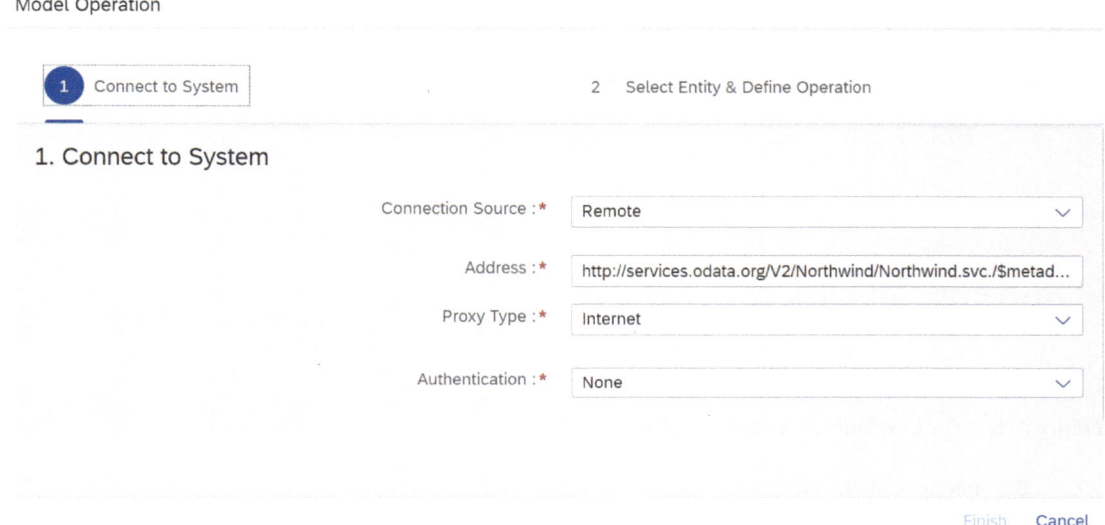

Figure 5-9. Connecting to the system

    ●   Step 2 is Select Entity & Define Operation. Select the field that you want to extract, as shown in Figure 5-10.

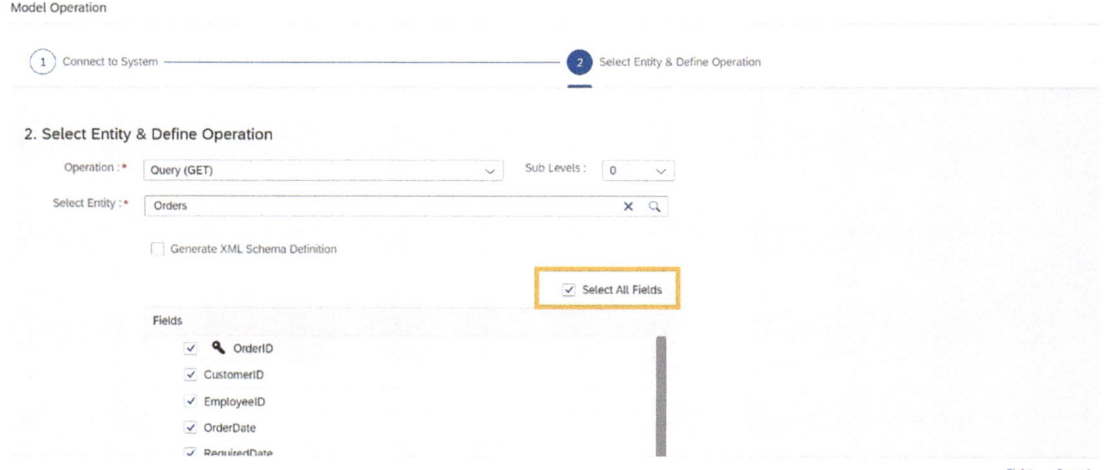

Figure 5-10. The Select Entity & Define Operation step

- In the third step, you can filter and sort the upcoming data. Filter the data by OrderID, which is equal to ${header. OrderNo}, as shown in Figure 5-11.

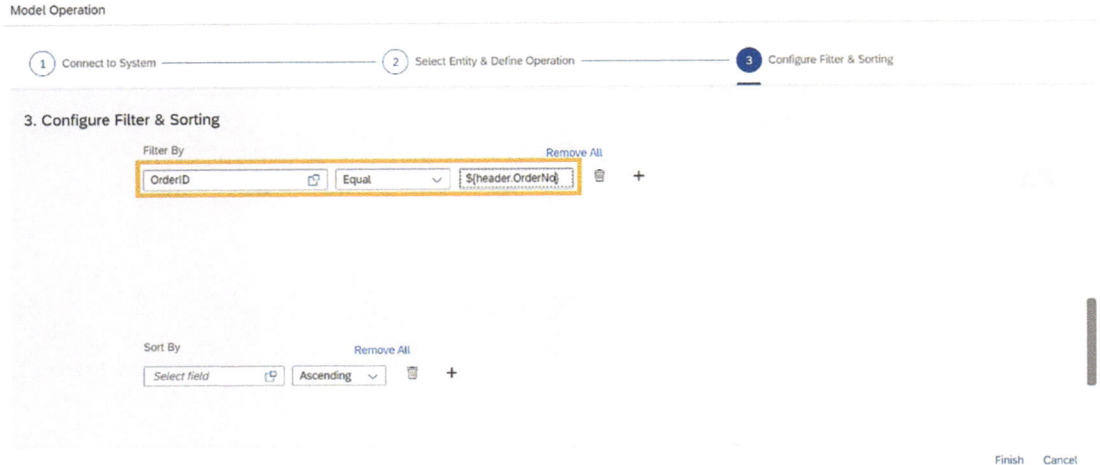

Figure 5-11. *The Configure Filter & Sorting step*

8. Recheck all the entries and deploy the integration flow after saving it.

9. After deploying, copy the endpoint and open SOAPUI 5.7.0. Create a new project and paste the endpoint. You can also run the endpoint in Postman.

10. Try to retrieve the OrderNo (10250), as shown in Figure 5-12.

11. The Content Enricher combines the contents of the external message that was returned with the original message. This merges two distinct messages into one payload, as shown in Figure 5-12.

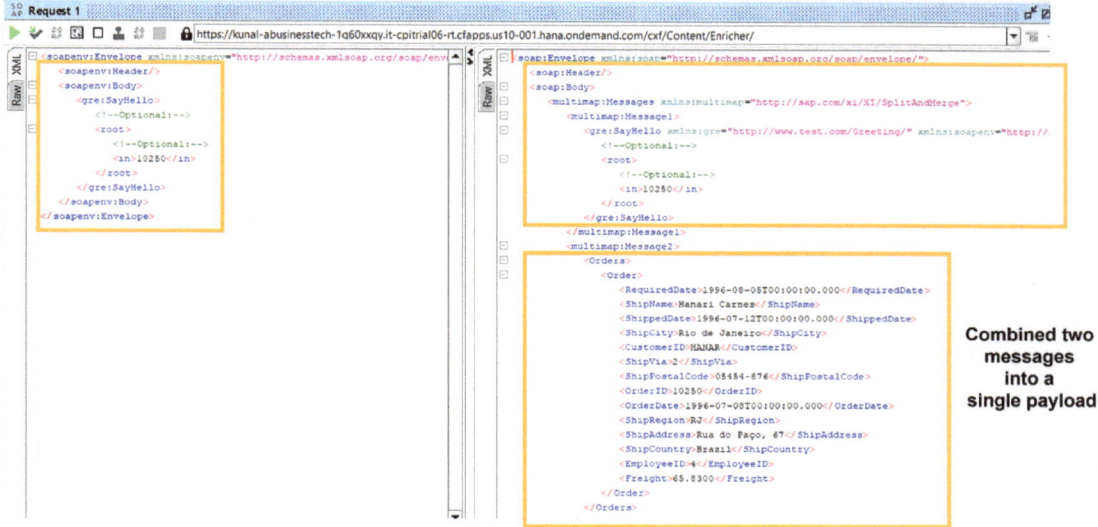

Figure 5-12. *Combining two messages into a single payload using the Content Enricher*

You have learned about the Content Enricher using this practical example. The next section goes through the Poll Enricher.

Poll Enricher

An HTTP inbound adapter initiates message processing at the beginning of the flow. Certain attributes are defined in a later Content Modification process and used dynamically by the SFTP adapter when polling the file from the SFTP server. The message is finally transmitted to the recipient system. Figure 5-13 shows the Poll Enricher used in the integration flow.

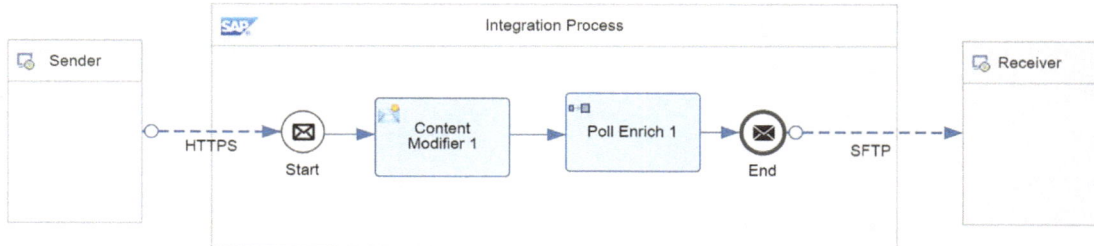

Figure 5-13. *Poll Enricher in integration flow*

Gather information from an outside source and add it to the original message. By reading messages from an external component and adding the content to the first message, you can use the Poll Enricher to read messages in the middle of the message processing sequence.

Procedure:

1. Add the Poll Enricher step to the integration flow of your choice.

2. By dragging and dropping, connect the Sender form to the Poll Enricher shape.

   - When combined with a polling sender adapter, such as the SFTP sender adapter, the Poll Enricher step actively polls messages from an external component. The Sender shape is hence linked to the Poll Enricher form.

3. Choose SFTP for the adapter type.

4. Click the connection line between the Sender shape and the Poll Enricher step to enter the SFTP sender adapter settings.

5. Choose the integration flow model's step for poll enrichment.

6. Set the parameters in the Processing tab, then save the flow.

You have now gone through the Poll Enricher thoroughly. In the next section, you deeply explore the Send element.

Send

A service call to a receiver system is configured using a Send step type for situations and adapters when no response is anticipated. Figure 5-14 shows the Send used in the integration flow.

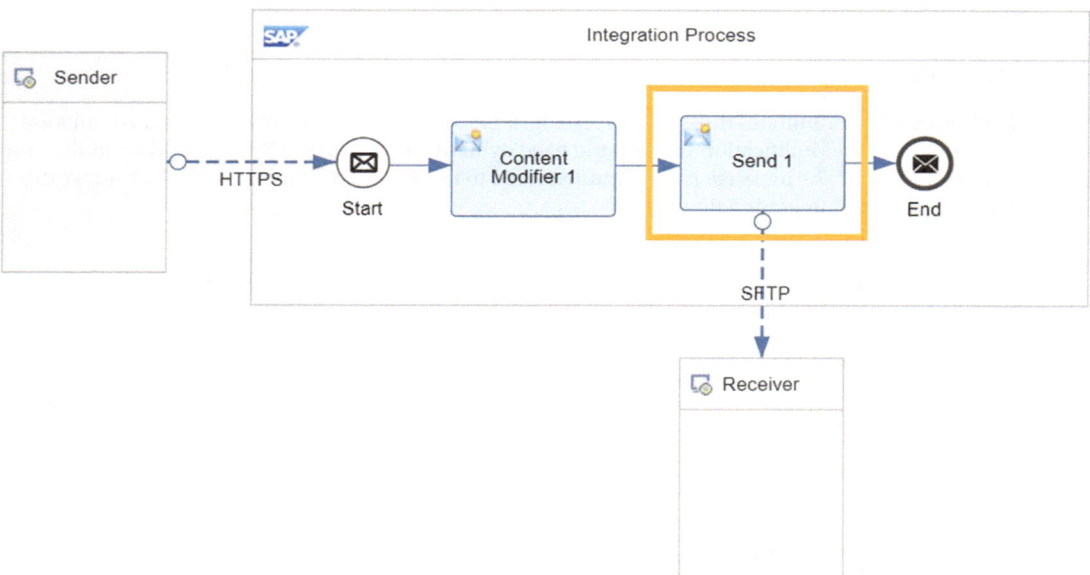

Figure 5-14. *Sending in an integration flow*

For this step to make sense, only one of the following adapter types should be used on the channel between the Send step and the receiver:

- AS2 adapter

- FTP interface

- JMS converter

- Postal adapter

- SAP RM adapter for SOAP

- SFTP interface

- (Quality of Service "Exactly Once") XI adapter

With the Request Reply, Content Enricher, Poll Enricher, and Send features, you have learned about the External Call. The next section goes through the Local Call.

Local Call

Using a Local Call, you can start a local integration process from the main integration process. To use a Local Call, choose Call ➤ Local Call and select the calls you need from the palette, as shown in Figure 5-15.

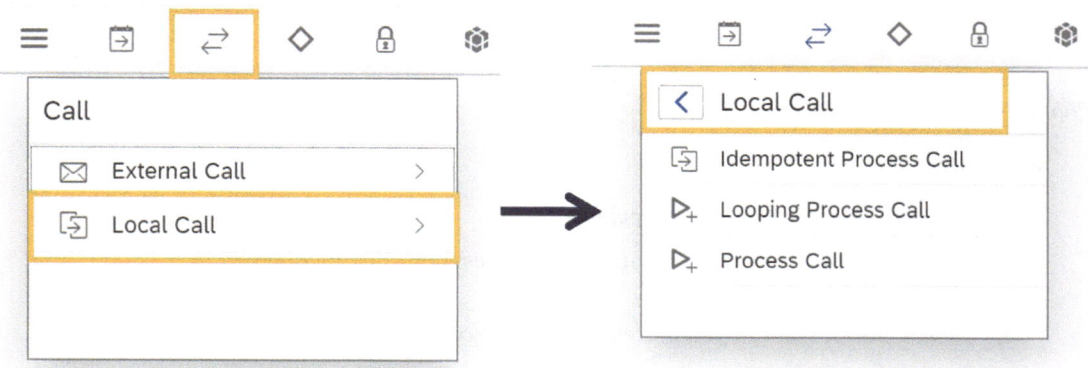

Figure 5-15. *Local Call in the SAP Cloud Integration Flow Editor*

This Local Call is further subcategorized into different elements, which the following sections go through one by one. Let's start with the Process Call.

Process Call

With this step type, you can start a local integration process from the main integration process.

Local integration strategies are employed to limit the size of a process model to a manageable amount. You can then divide the main integration process into smaller chunks in this manner (represented by local integration processes). To create the whole message processing design of your integration flow, you merge these parts, as shown in Figure 5-16.

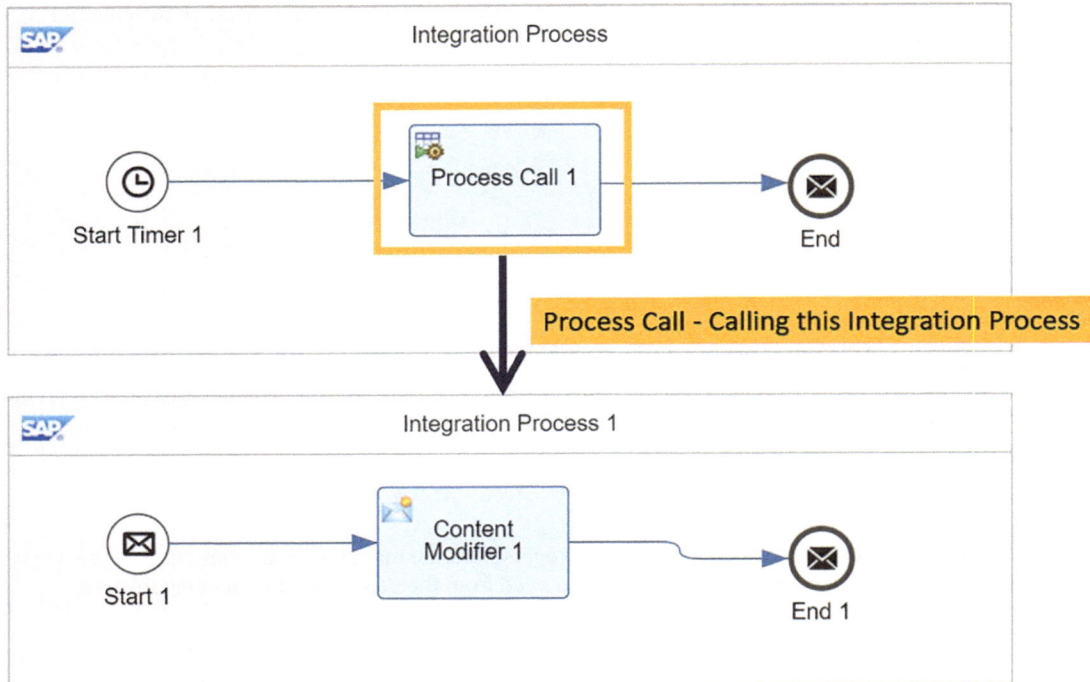

Figure 5-16. *Process Call*

Procedure:

1. Select the Process Call from the palette.

2. In the integration flow model's Integration Process box, click anywhere in any integration flow you chose.

3. To find the local integration process you want to connect to, choose Select.

4. Deploy and save the integration flow.

The next element of Local Call, Looping Process Call, is covered in the next section.

Looping Process Call

You can run a looping local integration procedure. To do this, choose any integration flow that you created earlier, for example, the one created in Section 4.1.7, or you can create a new integration flow.

Procedure:

1. Select the Looping Process Call from the palette, as shown in Figure 5-17.

2. Click anywhere in the integration flow model's integration process box.

3. Specify a name in the General tab.

4. Enter the following details in the Processing tab, as shown in Figure 5-17:

- Local integration process

- Expression type

- Condition expression

- Max. no. iterations

- Action when the max iteration is reached

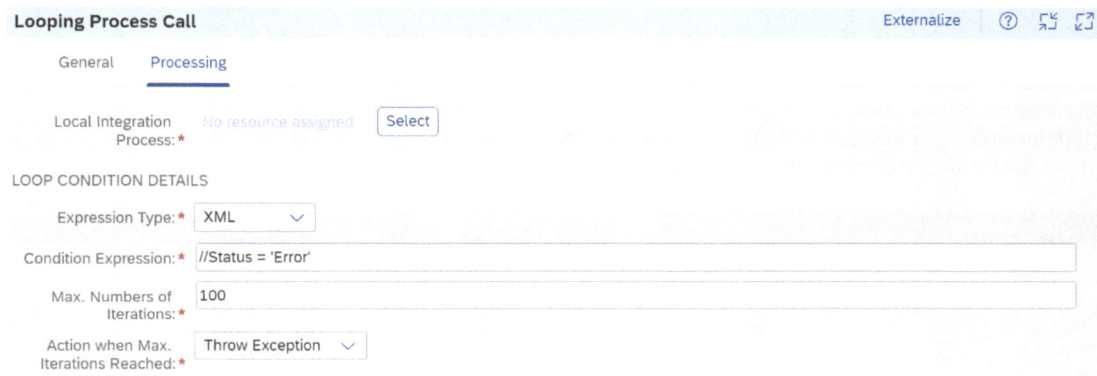

Figure 5-17. *Looping Process Call*

5. Save and deploy the integration flow.

The next section explains the next element of Local Call, the Idempotent Process Call.

Idempotent Process Call

The Idempotent Process Call recognizes when a message ID has already been successfully processed and logs the successful process's status in the idempotent repository. If a message is executed twice with the same message ID, the called subprocess can be skipped or the message can be marked as a duplicate (for instance, if the sender system retries). Afterward, you can choose how to handle the duplication in the subprocess.

The At-Least-Once or Exactly Once handling in the integration flow can be modeled with the help of the Idempotent Process Call. For instance, you can call a receiver system from within an Idempotent Process Call if the receiver system (e.g., a third-party legacy system) is unable to handle duplicate messages.

Procedure:

1. You can choose any integration flow, say the one created in Section 4.1.7, or you can create a new integration flow.

2. Select Idempotent Process Call from the palette.

3. Click anywhere in the integration flow model's integration process box.

4. Enter a name in the General tab.

5. Enter the following details in the Processing tab:

- Local integration process

- Message ID

- Skip process calls for duplicates

6. Save and deploy the integration flow.

You have now learned about the call elements and determined the different elements subcategorized in the call. In the next section, you learn about another element of the iflow design artifact, Routing.

5.1.2 Routing

Message routers can be used to specify a message path. The message can also be divided based on predetermined parameters, and the split messages can be sent through different message paths. Figure 5-18 shows the routing process used in the integration flow.

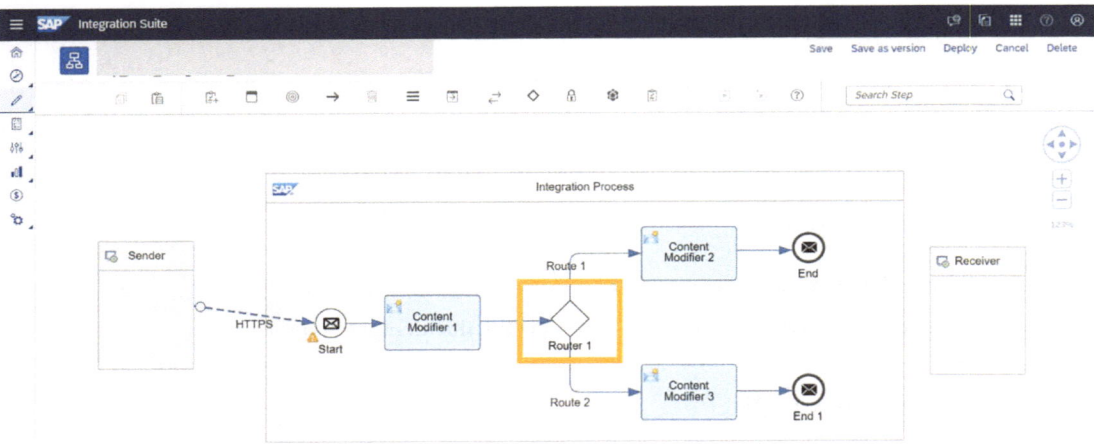

Figure 5-18. *Routing process*

From the palette, choose Routing ➤ Router, as shown in Figure 5-19.

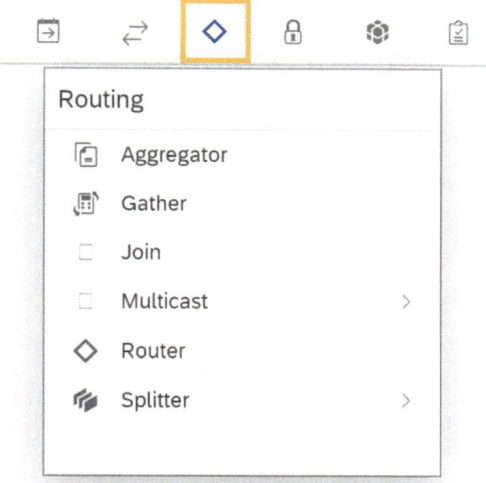

Figure 5-19. Routing in the SAP Cloud Integration Flow Editor

Aggregator

In SAP Cloud Integration, the aggregator is a message-processing step that enables you to combine many communications into one message depending on a set of parameters. It aggregates many input messages into a single output message as part of the Message Mapping process. Figure 5-20 shows the aggregator used in the SAP Cloud Integration Flow.

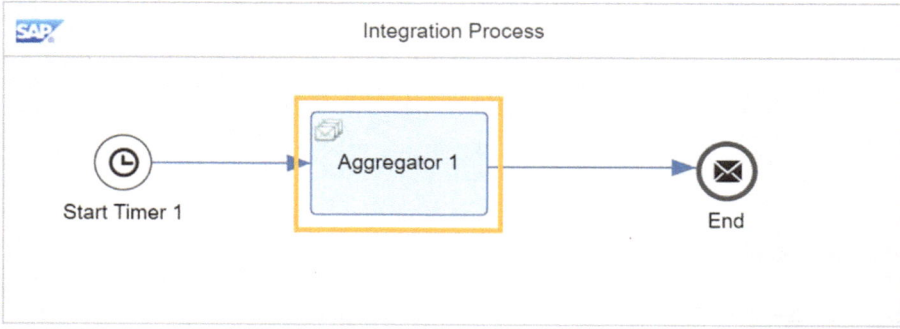

Figure 5-20. Aggregator in an integration flow

You can group similar individual messages together so that they can be handled in bulk. You can gather and store individual messages using an Aggregator pattern until a full set of connected messages is received. The real receiver is then informed of the aggregated message. Every five minutes, the message is resent until it is successfully delivered. You can handle any errors in the Exception Subprocess, ending in an End Event to stop the retry procedure.

Procedure:

1. Select Aggregator from the palette after selecting Aggregator.

2. Establish the message path and include the aggregator in the integration process in the integration flow you created in Section 4.1.7.

3. Name the aggregator's characteristics:

   - Correlation Expression (XPath)

   - Incoming Format

   - Aggregation Algorithm

   - Message Sequence Expression (XPath)

   - Last Message Condition (XPath)

   - Completion Timeout

   - Data Stored Name

4. Save and deploy the iflow.

The next section explains the Gather and Join message-processing phases.

Gather and Join

With SAP Cloud Integration, there are two message-processing phases—called Gather and Join—that let you merge several messages into a single one. Although the goals of the two processes are similar, they are accomplished in distinct ways. Figure 5-21 shows the Gather and Join phases in the integration flow.

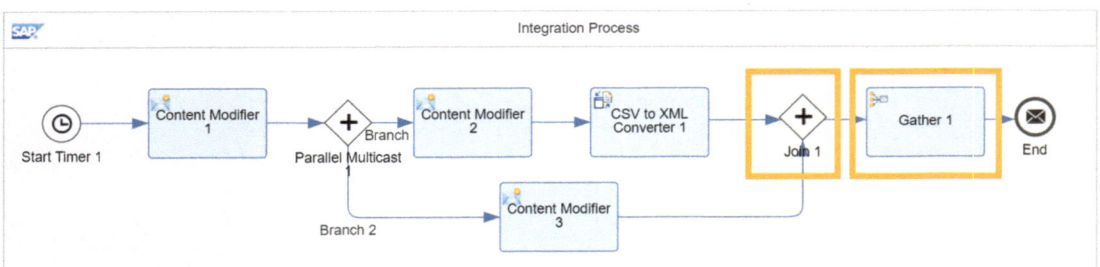

Figure 5-21. *Gather and Join phases*

The collect phase combines messages from several routes, with the option to set unique methods for how to combine the initial messages (into a single message). The Join and Gather steps are combined. By maintaining the messages' original substance, it merges the messages from many channels.

You can specify conditions based on the type of merged messages in the Gather step. You have the option of gathering:

- XML messages in various formats

- XML messages with the same structure

- Simple text messages

- Ally format

Before integrating them into one message, messages can be gathered from various routes using the Join element. You combine the Gather element with this step. Join does not change the substance of messages; it merely brings together messages from several routes.

You select the method to mix the two messages based on your needs:

1. XML messages with the same format can be combined.

2. The only thing you can do for XML messages in various forms is combine the messages.

3. You can only choose concatenation as the combination mechanism for plain text messages.

You can also select the ZIP approach to combine the input from all three incoming formats into a single archive (.ZIP) file. In that situation, a filename pattern must be specified. Provide a valid XPath expression with namespace prefixes if the inbound payload contains namespace declarations, including default namespace declarations.

If the inbound payload includes namespace declarations, including the default namespace, be sure to specify the XPath with namespace prefixes. Do not forget to define the namespace prefix mapping in the integration flow's runtime setting as well (in the Namespace Mapping field). If any element in any branch of the incoming XML is not referenced by the XPath, the scenario fails with an exception.

The next section explains the Multicast step.

Multicast

The Multicast step can be used to deliver duplicate messages to various routes. Using Parallel Multicast or Sequential Multicast, copies can be sent to all routes at once or in a specific order, respectively. Figure 5-22 shows the Multicast used in the integration flow.

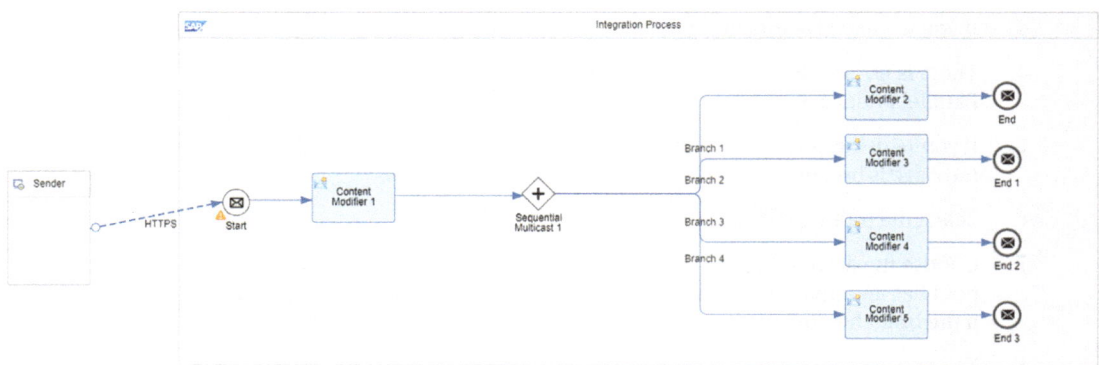

Figure 5-22. Sequential Multicast

As a result, several operations on the same message can be carried out within a single integration process. To do this work without Multicast, you would need multiple integration procedures. To add the Multicast from the palette, choose Routing ➤ Multicast and select the multicast option according to your requirements, as shown in Figure 5-23.

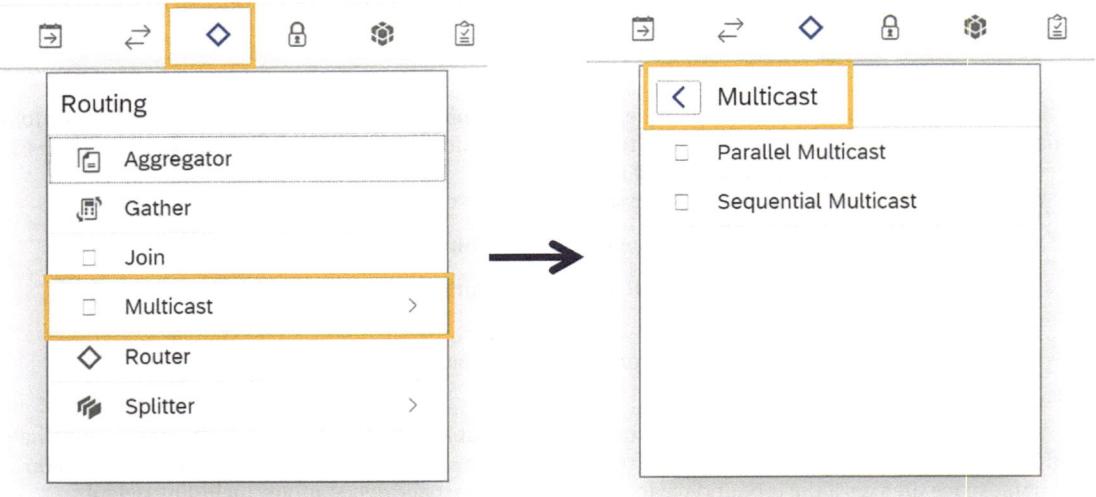

Figure 5-23. *Multicast in the SAP Cloud Integration Flow Editor*

Procedure:

1. Select Message Routing from the palette, then multicast to the integration flow of your choice. You can create a new one.

2. Depending on your needs, decide whether to include the step inside the integration process as a Parallel or Sequential Multicast. Create the message pathways in accordance with the scenario.

   - The order in which you establish the message path determines how the system recognizes each message route branch, such as Branch 1 and Branch 2.

3. If you want to give this step a name, type it in the Name area.

4. There is no additional settings needed for the multicast stage if you're using Parallel Multicast.

5. If you're using Sequential Multicast, it is possible to alter the order in which the message is broadcast to the Sequential Multicast branches.

6. Select the routing sequence.

7. Choose the desired branch. The sequence in which you defined the message path has an impact on the branch name. The branch name will not change even if the branches are rearranged.

8. You can deploy or save the settings.

The next section covers the Splitter step.

Splitter

A single message can be split into many messages depending on a set of parameters using the Splitter message processing step in SAP Cloud Integration (SCI). It separates a single input message into several output messages as part of the Message Mapping process. Figure 5-24 shows the splitter used in the integration flow.

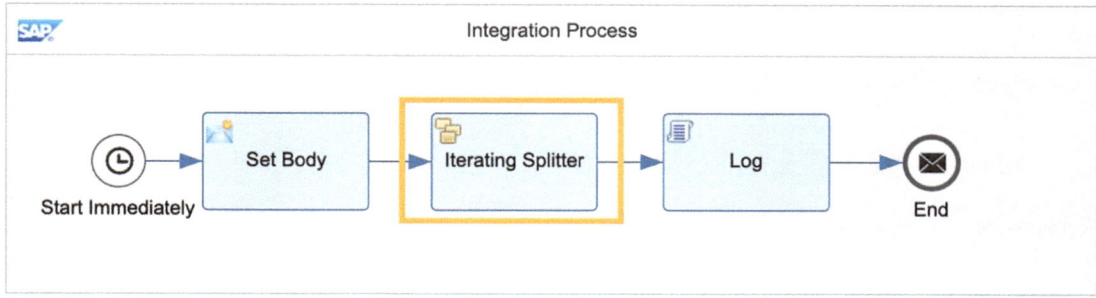

Figure 5-24. Splitter

You can divide a message into smaller chunks that can each be processed separately, thanks to splitters. There are four types of splitters available:

1. **Iterating Splitter**—Creates individual IDoc from a set of IDocs. The transmitter or receiver channel must be an IDoc channel for it to function. In this situation, there is no way to carry on processing in the event of an exception, as shown in Figure 5-25.

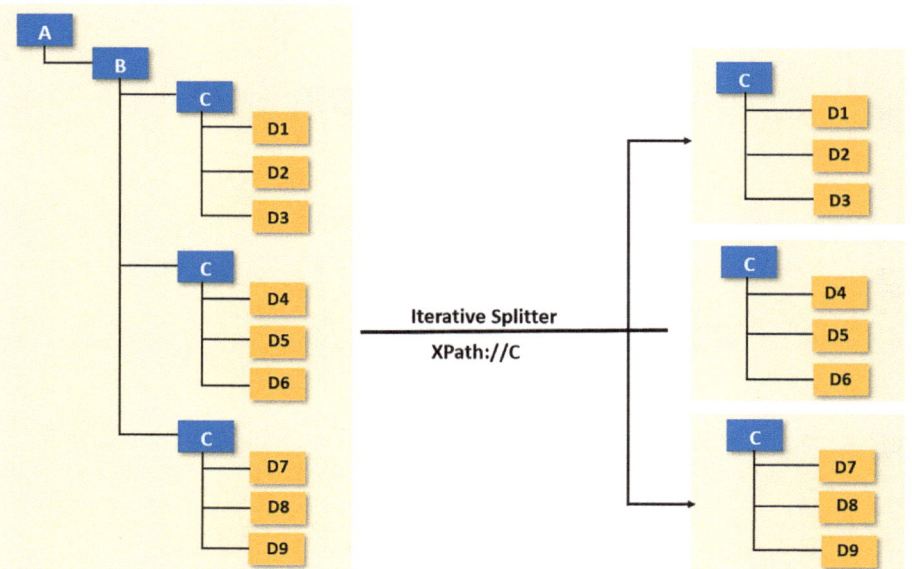

Figure 5-25. Iterating Splitter

2. **IDoc Splitter**—Creates individual IDocs from a set of IDocs. The transmitter or receiver channel must be an IDoc channel for it to function. In the event of an exception, there is no chance to continue processing here, as shown in Figure 5-26.

- IDoc Splitter – Divides a group of IDocs into individual IDocs. It only works if either the sender or the receiver channel is an IDoc channel. Here there is no chance to continue processing in case of an exception.

Define a splitter to break down composite message into series of messages

Name:* IDoc Splitter

o **Name** – can be any name for the Splitter

Figure 5-26. *IDoc Splitter, source (SE)*

3. **General Splitter**—A splitter that divides a message into multiple messages from a message that comprises many messages. This splitter retains the root nodes' context for each message it separates, as shown in Figure 5-27.

The General Splitter divides a composite message made up of N messages into N separate messages, each of which contains one message together with the composite message's enclosing parts. The split point is included in the details, which you refer to as the enveloping characteristics. It should be noted that components that come before the one designated as a split point in the first message (but on the same level) are not counted as enveloping elements. These will not be included in the messages that are produced.

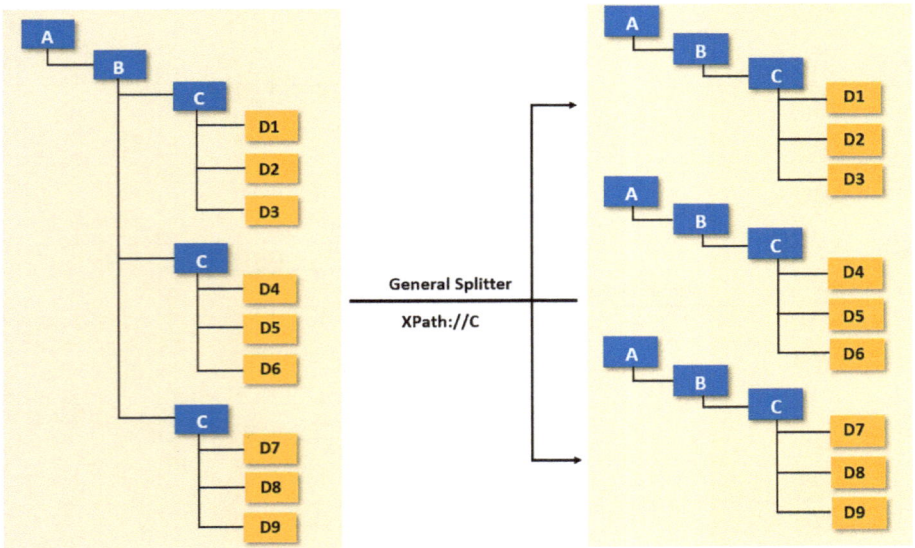

Figure 5-27. *General Splitter*

4. **PKCS (Public Key Cryptography Standard) Splitter**—This is employed to sign and encrypt messages, separating the payload from the signature and presenting them as split messages, as shown in Figure 5-28.

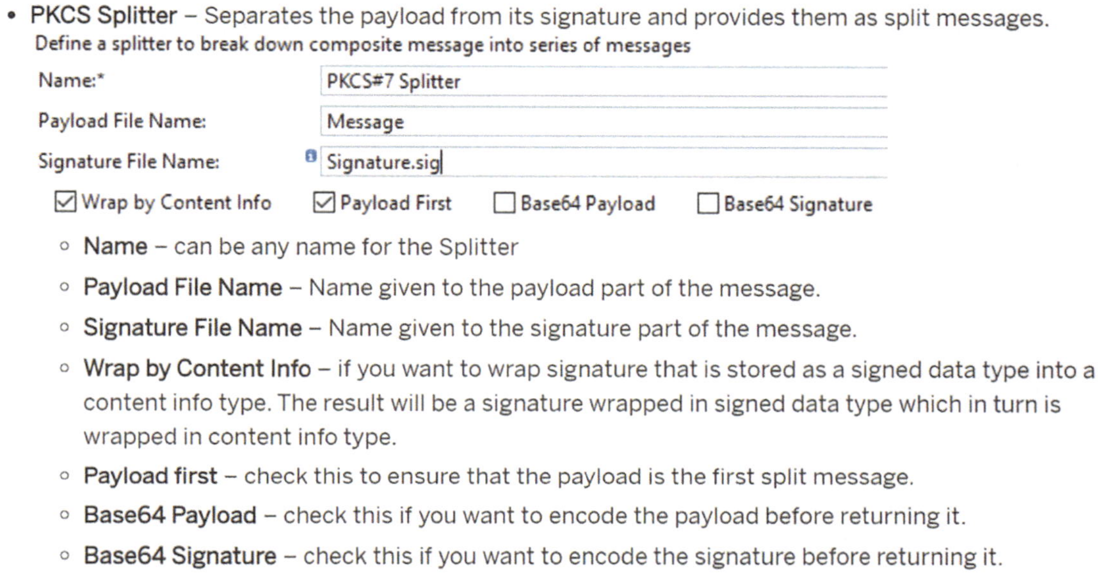

- PKCS Splitter – Separates the payload from its signature and provides them as split messages.

Define a splitter to break down composite message into series of messages

Name:*	PKCS#7 Splitter
Payload File Name:	Message
Signature File Name:	Signature.sig

☑ Wrap by Content Info   ☑ Payload First   ☐ Base64 Payload   ☐ Base64 Signature

- ○ **Name** – can be any name for the Splitter

- ○ **Payload File Name** – Name given to the payload part of the message.

- ○ **Signature File Name** – Name given to the signature part of the message.

- ○ **Wrap by Content Info** – if you want to wrap signature that is stored as a signed data type into a content info type. The result will be a signature wrapped in signed data type which in turn is wrapped in content info type.

- ○ **Payload first** – check this to ensure that the payload is the first split message.

- ○ **Base64 Payload** – check this if you want to encode the payload before returning it.

- ○ **Base64 Signature** – check this if you want to encode the signature before returning it.

Figure 5-28. *PKCS Splitter*

The next section explains the Routing element called Router. You learn about Router through a practical example.

Router: Practical Example

The router is a filter that routes the message based on a condition. In SAP Cloud Integration, the routing conditions can be written using an XPath expression or Apache Camel's Simple Language using the Router component.

In some integration scenarios, messages can be forwarded to different recipients depending on certain forwarding conditions. You can use the Router tool to perform such an operation in SAP Cloud Integration.

Take a look at the router steps in the following sample integration flow. Create your new integration flow for this example. Then, define your condition. Define the XML path as follows:

1. If the condition is Alphabet, return "learn SAP CI with ABT":

```
<root>
<condition>Alphabet</condition>
</root>
```

2. If the condition is Number, return: "ABT_021":

```
<root>
<condition>Number</condition>
</root>
```

3. If the condition is not Alphabet or Number, then return: "Invalid Condition".

Steps:

1. Open the integration flow created in Section 4.19 in Edit mode.

2. Connect the Sender to the Start using the HTTPS adapter. Provide the address path in the Connection tab.

3. From the palette, choose the router and place it between the Start and the End options.

4. From the Events in the palette, select the End Message and place it under the already present (End). Route the path from the Router to End 1.

5. To describe the condition of Route 1, select the Route 1 path and, in the processing tab of the router, choose XML for the expression type. Define the condition as /root/condition = 'Alphabet', as shown in Figure 5-29.

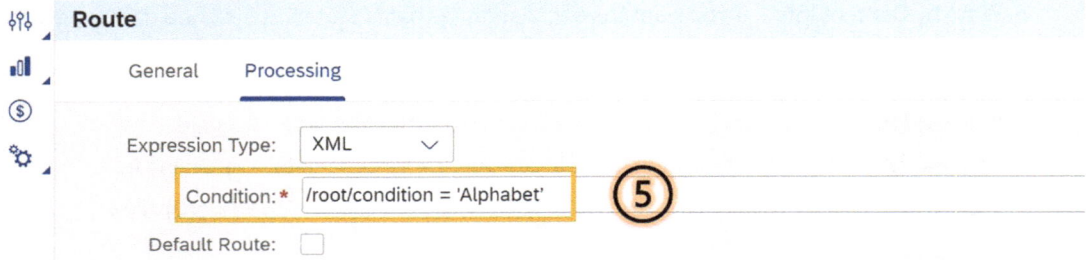

Figure 5-29. *Route conditions*

6. Select the Content Modifier from the palette and place it on Route 1 so you can define the body of the message, as shown in Figure 5-30.

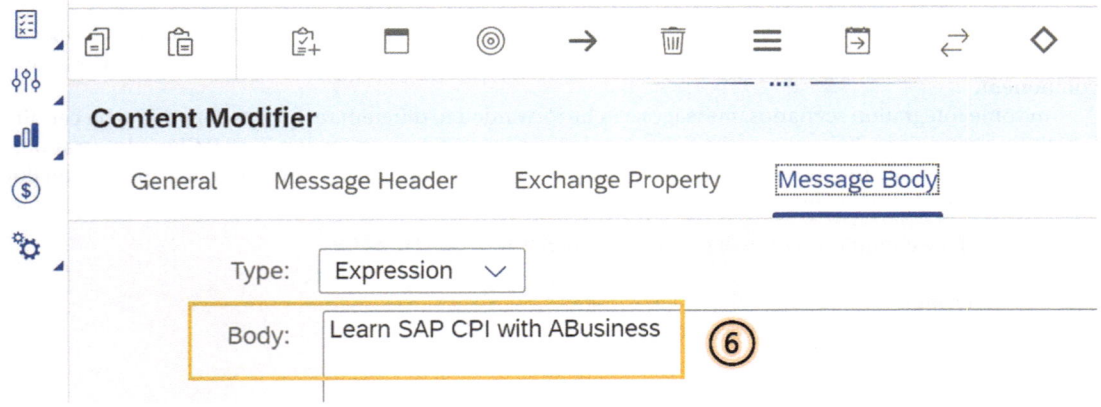

Figure 5-30. *Message body of the Content Modifier*

7. Do the same for Route 2 and change the condition for Route 2 to /root/
   condition = 'Number'.

8. Add the Content Modifier to display the message body.

9. From the palette, select another end message and route it through the router.
   Make this route the default one in the condition of the route. To display the
   message body, place a Content Modifier in Route 3.

10. The final integration flow with all the values is shown in Figure 5-31.

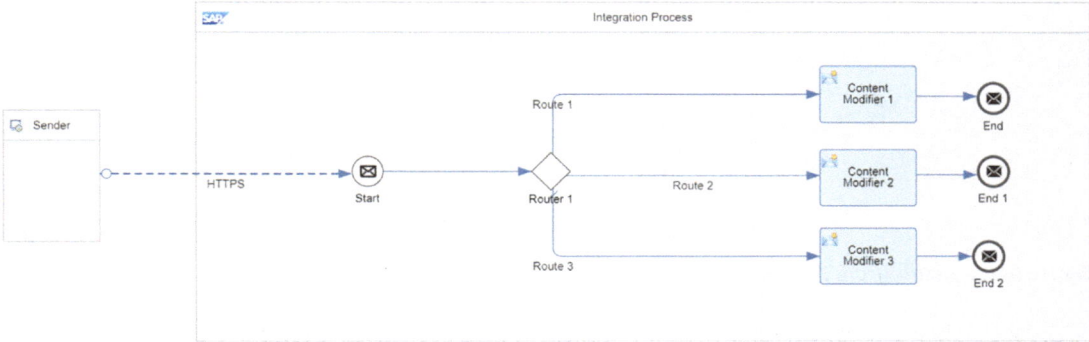

Figure 5-31. *Final integration flow*

Result

After successfully saving and deploying the integration flow, copy the endpoints from the Monitor tab
and run the endpoint in Postman.

1. In Postman, you'll see the following XML body:

   ```
   <root>
       <condition>(Alphabet)/(Number)/(Invalid Content)</condition>
   </root>
   ```

   When you give the input as Alphabet in the body of Postman, you see the result
   as "Learn SAP CPI with ABT", as shown in Figure 5-32. This is the same as what
   you gave in the message body of the Content Modifier.

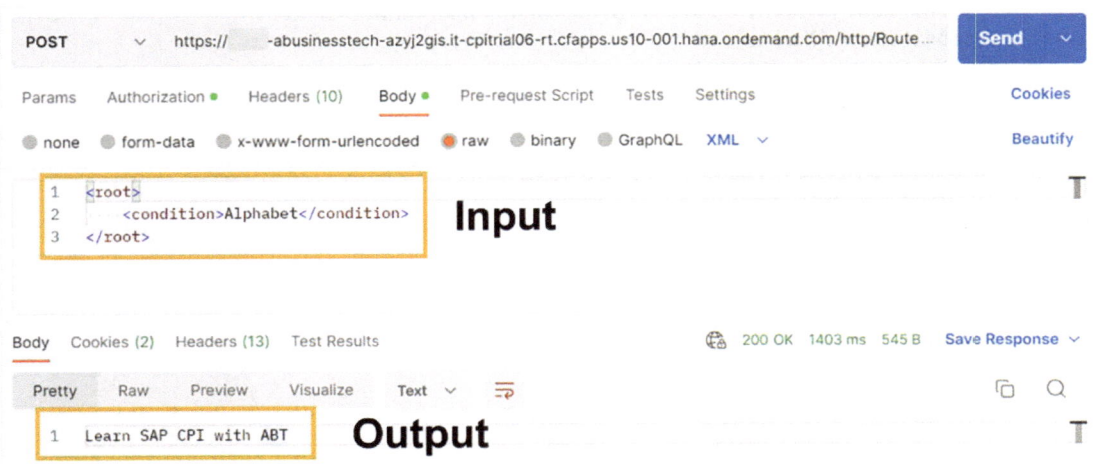

Figure 5-32. *Results from Postman for Splitter*

2. When you give the input as Number in the body of Postman, you see the result as `"ABT_CI_021"`. This is the same as what you have in the message body of the Content Modifier.

3. When you give the input as xxxx in the body of Postman, you see the result as `"INVALID CHOICE"`. This is the same as what you gave in the message body of the Content Modifier, as shown in Figure 5-33.

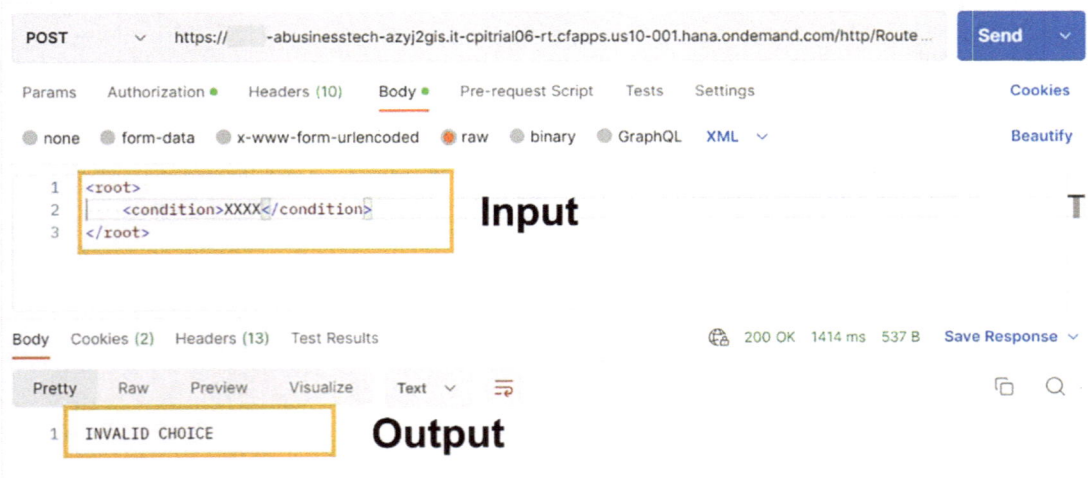

Figure 5-33. *Postman results for Splitter*

You have now learned about the routing integration flow design object elements and the various elements associated with routing. The next section covers security and message-level security.

5.1.3 Security and Message-Level Security Use Case Configuration

Message exchange can be made even more secure by digitally signing and encrypting it in addition to using a secure transport channel that is based on HTTPS or SFTP.

You can utilize many security standards to do that.

- Inbound: Message-Level Security Using XML Digital Signature and PKCS#7

You can add message-level security features on top of a secure transport channel (based, for instance, on HTTPS). In this manner, communication can be secured via digital signature or encryption. The tenant serves as a receiver in the incoming scenario and either decrypts or validates a message.

To implement message-level security for the PKCS#7, WS-Security, and XML Digital Signature standards, you use X.509 certificates (the same type of certificates used for HTTPS-based transport-level security). However, SSL transport-level security and message-level security often use different keys. The XML digital signature only supports the use cases of message signing and message verification.

Configure Sender

The sender keystore should be set up as follows:

- Make a key pair (and get it signed by a Certificate Authority).

- Enter the sender keystore with the tenant public key imported.

- Verify communications delivered to the tenant by giving the tenant administrator the public key.

Configure Integration Flow Steps:

- Set up the XML Signature Verifier or PKCS7 Verifier step. To choose the appropriate keys from the tenant keystore, enter the public key aliases.

- PKCS7 decoder setup. If you selected Enveloped, Signed, and Enveloped Data, or Signed and Enveloped Data for signatures in the PKCS7 message, be sure to include the public key aliases of all anticipated senders.

- Inbound: Message-level security with OpenPGP.

You have the option to supplement a secure transport channel based, for example, on HTTPS with message-level security measures. This allows for the encryption or digital signature of the transmission to be safeguarded. The tenant serves as a receiver in the incoming scenario and either decrypts or validates a message. PGP keys are used to implement message-level security for OpenPGP.

Configure Sender

Open the Monitor section of Cloud Integration and choose Manage Security ➤ PGP Keys.

1. Create and set up the PGP secret and public keyrings as well as the sender system's key storage locations.

2. Finish setting up the sender system by importing the relevant public keys from the tenant into its public PGP keyring.

Provide the public key to the tenant administrator (it's used to verify messages sent to the tenant). Don't worry, as you learn about uploading and PGP keys in the next chapter.

Configure Integration Flow Steps

Open your Integration Flow Editor and create the integration flow where you are using the PGP Verifier and Decryptor steps. Figure 5-34 shows the use of Security Encryptor and Decryptor in the integration Flow.

Integration Process

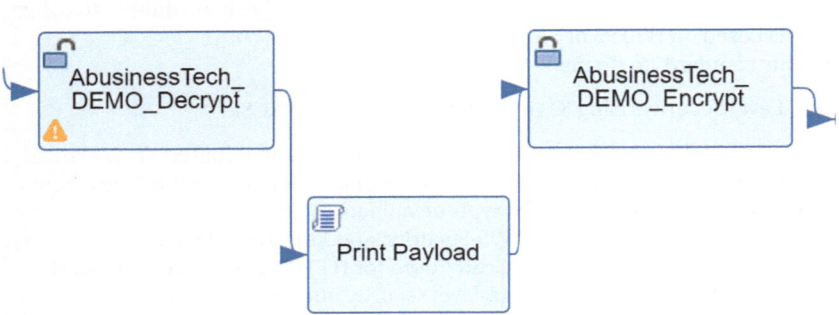

Figure 5-34. *Encryptor and Decryptor*

1. Set up the integration flow stages for security.

2. Set up the PGP Verifier and Decryptor steps.

If signatures are needed, make sure to include the signer's user ID of the public key(s) for each intended sender.

Based on the signer's user ID of the key(s) sections, the public key (for message verification) is looked for in the PGP public keyring. By using the signer's user ID of the key(s) key parts provided at this stage as an authorization check, the list of intended senders is constrained.

- Outbound: Message-Level Security Using XML Digital Signature and PKCS#7

You have the choice to add message-level security features on top of a secure transport channel (based, for instance, on HTTPS). In this manner, communication can be secured via digital signature or encryption. The tenant serves as the sender in the outbound scenario and either encrypts or signs a communication.

The same type of certificates used for HTTPS-based transport-level security are used for message-level security for the PKCS#7, WS-Security, and XML Digital Signature standards. These certificates are called X.509 certificates. However, SSL transport-level security and message-level security often use different keys. Only message signing and verification use cases are covered by the XML Digital Signature.

Configure the Receiver

Open the Monitor section of Cloud Integration and choose Manage Security ➤ PGP Keys. The receiver keystore should be set up as follows:

- Make a key pair (and get it signed by a CA).

- Import the receiver keystore with the tenant public key.

- Provide the public key to the tenant administrator (used to encrypt messages sent to the receiver).

Configure the Integration Flow Steps

Open your Integration Flow Editor and create the integration flow. Configure the security-related integration flow stages based on the preferred choice.

- Configure the PKCS7 Encryptor phase.

  Enter the public key aliases to select the correct key from the tenant keystore. In the event that you selected Signed and Enveloped Data, you must also specify the Private Key Alias to select the proper private key for signing (as signatures).

- Set up the XML Digital Signer or PKCS7 Signer phase.

  To choose the required keys from the keystore, be careful to mention the private key aliases.

An alias is often a reference to a keystore entry. One or more public keys can be present in a keystore. You can choose a certain key from a keystore by using an alias as a shortcut.

- Outbound: Message-level security with OpenPGP

You can add message-level security features on top of a secure transport channel (based, for instance, on HTTPS). In this manner, you can secure the message using digital signatures or encryption. The tenant serves as the sender in the outbound scenario and either encrypts or signs a communication. PGP keys are used to implement message-level security for OpenPGP.

Configure the Receiver
Open the Monitor section of Cloud Integration and choose Manage Security ➤ PGP Keys.

1. Create a PGP secret and the public keyrings, as well as the PGP keys, for the receiver system.

2. Finish configuring the receiver system by importing the pertinent public keys from the tenant into the receiver's public PGP keyring.

3. Provide the public key to the tenant administrator (used to encrypt messages sent to the receiver).

Configure the Integration Flow Steps
Open the Integration Flow Editor and create the integration flow. Configure the security-related integration flow stages based on your preferred choices.
Set up the PGP encryption phase:

1. Provide the user ID of the key(s) from the public keyring to select the relevant public receiver keys from the PGP public keyring.

2. Provide the signer user ID of the key(s) from the secret keyring if you want to sign the payload.

3. Next, select the appropriate private key from the PGP secret keyring.

The integration flow's security components let you encrypt, decrypt, sign, and verify messages. By doing this, you can be sure that only the intended receiver can access the message. During message exchange, encryption also stops nonrepudiation of messages. Open the Integration Flow Editor and choose the Security element from the palette, as shown in Figure 5-35.

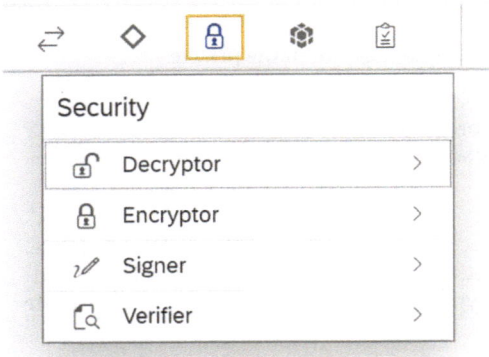

Figure 5-35. *Security-related steps in the SAP Cloud Integration Flow Editor*

The following sections further discuss the security elements in the SAP Integration Flow Editor, starting with the signer, covered in the next section.

Signer

In SAP Cloud Integration, a signer is a component that is used to digitally sign data. A digital signature is a mathematical method that proves that a communication or document is authentic. It allows the recipient of the message or document to confirm the identity of the sender and the integrity of the data.

A signer in SAP Cloud Integration can be used to sign data before it is sent to an external system or to verify the signature received from an external system. The signer component can be used in integration flows to sign and verify digital signatures using various algorithms, such as RSA and DSA.

The signer component uses a private key to sign the data and a public key to verify the signature. This ensures that the data has not been tampered with in transit and that it came from the expected sender. From the palette, choose the Security element and select Signer. Under Signer, you select additional elements, as shown in Figure 5-36.

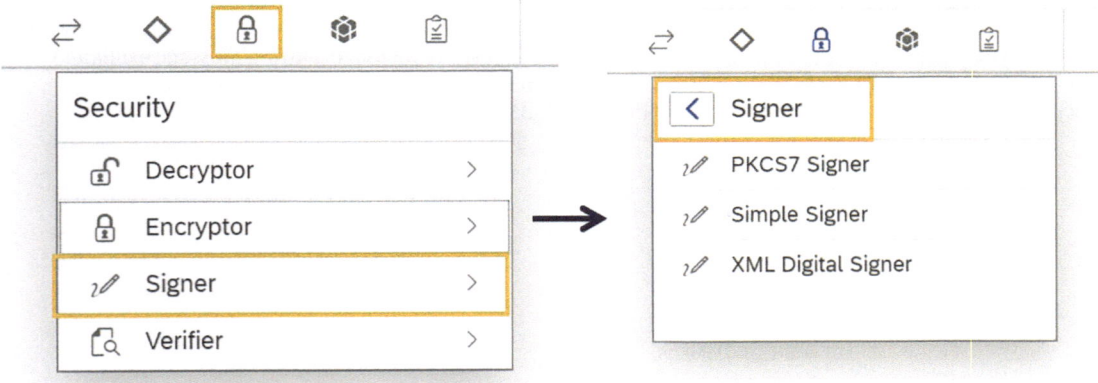

Figure 5-36. *Signer in the SAP Cloud Integration Flow Editor*

Signer, a subcategorized element of signer, is further divided into various elements, which you will learn about one by one in the following sections. The next section covers the PKCS7 Signer.

PKCS7 Signer

You collaborate with the PKCS#7/CMS signer to let the participants know who you are and thereby guarantee the validity of the cloud messages you deliver. This assignment secures your identity by utilizing a signature algorithm to sign the messages with one or more private keys.

Simple Signer

Easy Signer makes it easy to sign communications to ensure authenticity and data integrity while sending data to cloud participants.

You collaborate with the Simple Signer to reveal the sender's identity to the recipient or recipients and therefore guarantee the communication's veracity. By employing a private key and a signature algorithm to sign the communications, this task ensures the sender's identity.

XML Digital Signature

An XML digital signature is used to ensure message authenticity and data integrity while sending an XML resource to cloud participants.

The next section explains what a decryptor is.

Decryptor

A decryptor is a phase in an integration flow in SAP Cloud Integration that is used to decrypt data that has been encrypted previously in the flow. Data that has been encoded or encrypted must be decrypted to be restored to its original, unencrypted state. A particular algorithm and a secret key are often used for decryption. Figure 5-37 shows the decryptor used in the integration flow.

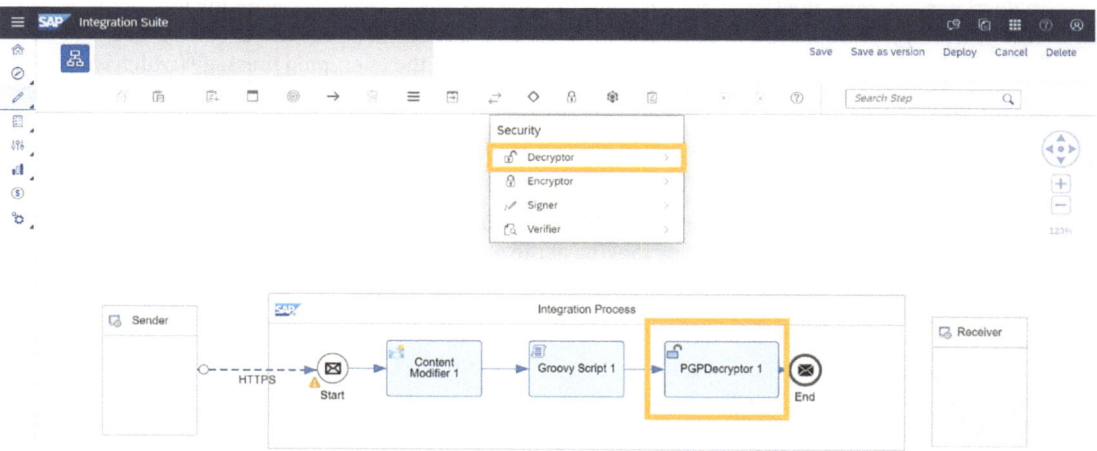

Figure 5-37. *Decryptor*

The decryptor step can be used to decrypt data that has been encrypted using a variety of encryption algorithms, such as AES or RSA. It allows you to configure the algorithm, the key, and the initialization vector.

This step is typically used in integration flows where sensitive data, such as passwords and personal information, need to be protected while in transit. It allows data to be protected while it is being transferred between different systems, ensuring that only authorized parties can access the data. From the palette, choose Security ➤ Decryptor and then choose the decryptor according to your requirements, as shown in Figure 5-38.

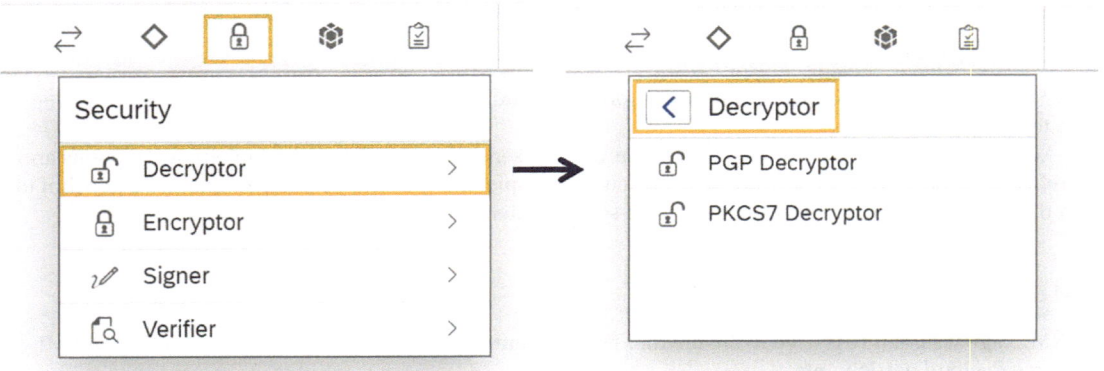

Figure 5-38. *Decryptor in the SAP Cloud Integration Flow Editor*

The decryptor is further divided into two elements, which are covered in the following sections.

PGP Decryptor

You can decrypt OpenPGP-encrypted mail by using the PGP Decryptor.

To decrypt a message, the Decryptor step requires you to deploy a private key on the tenant (as part of a PGP Secret Keyring). The PGP Secret Keyring allows you to store several private keys.

To guarantee that the right private key is used for decryption, the encrypted message (which was processed by the decryptor) contains a reference that allows the system to explicitly identify the proper key.

PKCS7 Decryptor

You decrypt communications from a participant in the cloud using the PKCS#7 decryptor. A signed message's validity can also be confirmed by looking at the object's SignedAndEnvelopedData signature. Figure 5-39 shows the PKCS7 decryptor used in the integration flow.

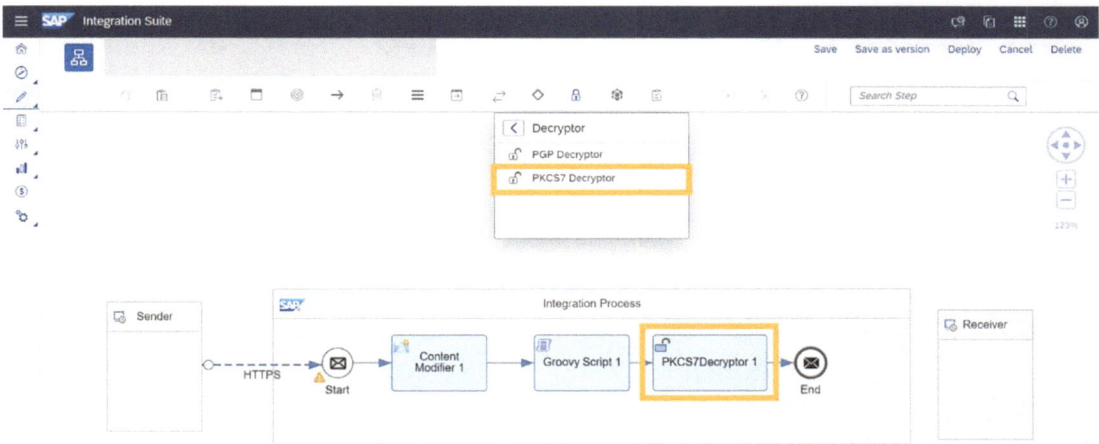

Figure 5-39. *PKCS7 Decryptor*

To decrypt a message, the decryptor requires a private key that must be set up on the tenant (as part of a key pair). Remember that the tenant keystore can include several key pairs. To guarantee that the right private key is used for decryption, the encrypted message that the decryptor processes contains a reference that allows the system to explicitly identify the proper private key. The message particularly includes the certificate for the PKCS#7/CMS decryptor's accompanying public key's issuer-differentiated name and issuer-specific serial number.

You have briefly learned about the decryptor available in SAP Cloud Integration. The next section covers the encryptor.

Encryptor

A component called an encryptor is used in SAP Cloud Integration to encrypt sensitive data, including passwords and other private information. When plain text is encrypted, it is turned into a coded version that only those with the right decryption key can decipher. This helps protect sensitive data from unauthorized access and ensures that it is only visible to authorized users or systems.

Encryptors in SAP Cloud Integration can be used to encrypt data before it is sent to an external system or to decrypt data that has been received from an external system. The encryptor component can be used in integration flows to encrypt and decrypt data using various encryption algorithms, such as AES and RSA.

Overall, the encryptor component ensures that the sensitive data is protected and that only authorized parties can access it. To access the encryptor, select Security from the palette and then select Encryptor, as shown in Figure 5-40.

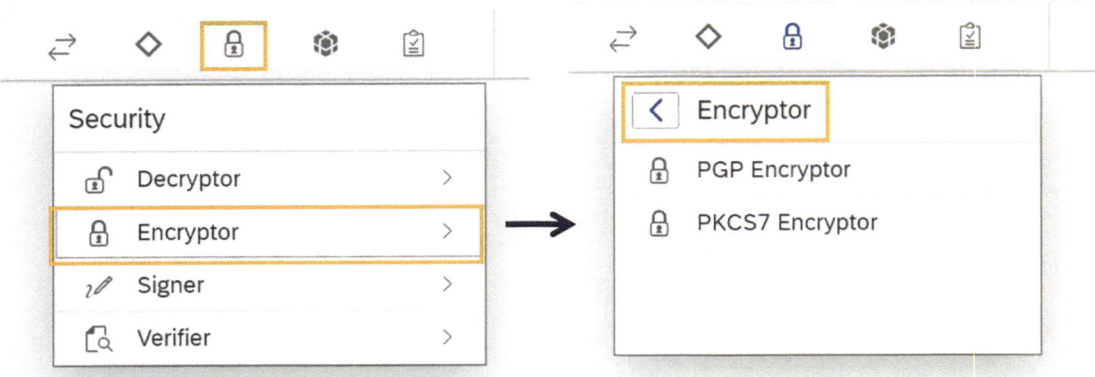

Figure 5-40. *Encryptor in SAP Integration Flow Editor*

The encryptor is further subcategorized into two elements that you will briefly learn about in the following sections.

PGP Encryptor

The payload is encrypted or signed and encrypted using the OpenPGP standard in the PGP Encryptor. Figure 5-41 shows the PGP Encryptor used in the integration flow.

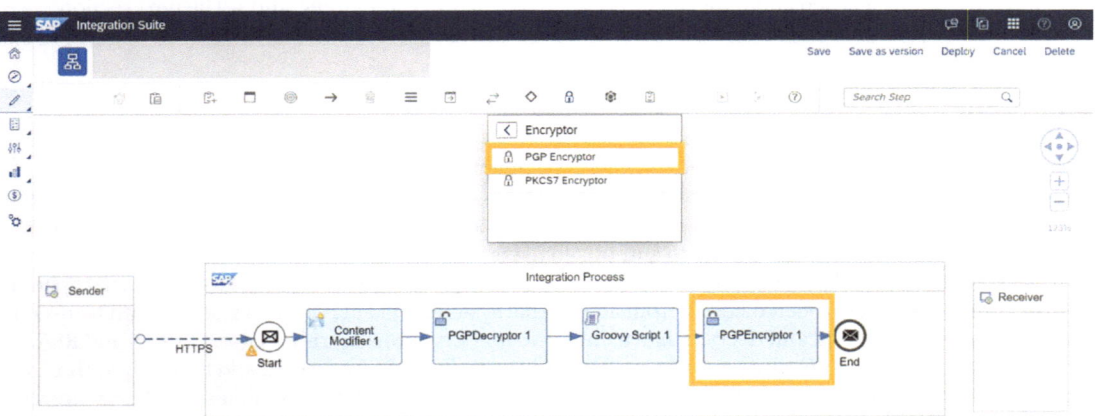

Figure 5-41. *The PGP Encryptor*

PKCS7 Encryptor

You perform this task by encrypting the message content. You indicate the public key alias, content encryption technique, and secret key length while configuring the encryptor in the integration flow model. The encryptor uses one or more receiver public key aliases to find the public key in the keystore. The encryption procedure encrypts data using a symmetric key with a predetermined length. The symmetric key is encrypted using the cypher by the public recipient key. The encryption is impacted by the kind of content encryption algorithm you use. Together with the encrypted content, the receiver information also contains the symmetric encryption key. Figure 5-42 shows the PKCS7 Encryptor used in the integration flow.

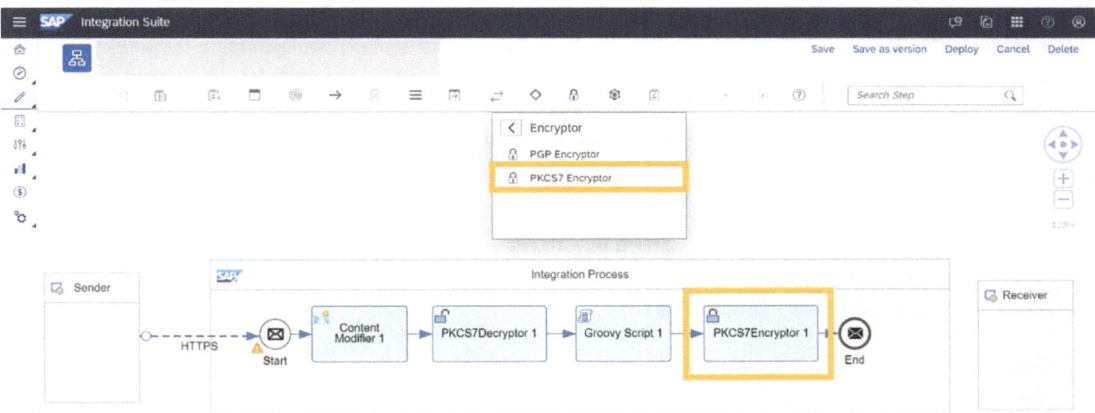

Figure 5-42. *PKCS7 Encryptor*

You can sign the message content in addition to encrypting it to let the other participants know who you are and to guarantee the validity of the messages you are delivering. This assignment secures your identity by utilizing a signature algorithm to sign the messages with one or more private keys.

The next section covers these encryptors and decryptors using a practical example.

PGP Encryption and Decryption: Practical Example

SAP Cloud Integration uses PGP (Pretty Good Privacy) encryption and decryption to secure data. PGP is a well-known encryption technology that enables users to send and receive data securely over unreliable networks, including the Internet.

Follow these steps to create PGP encrypted messages:

1. Open the integration flow in Edit mode. You can create a new integration flow.

2. Connect the Sender to Start using the HTTPS adapter. Provide the address path in the Connection tab.

3. From the palette, in the Security tab, choose the PGP Encryptor. Then choose Security ➤ Encryptor ➤ PGP Encryptor and place it between the Start and End, as shown in Figure 5-43.

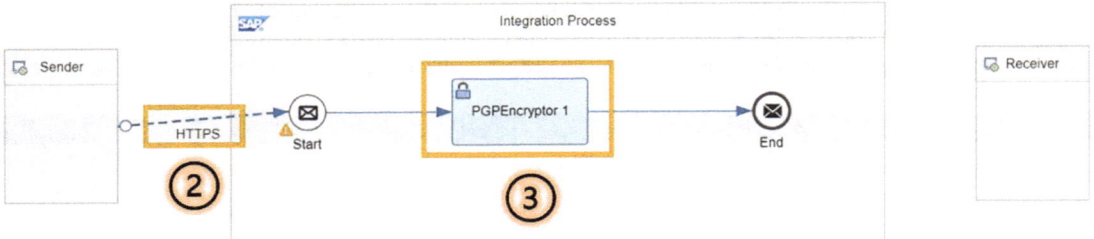

Figure 5-43. *Integration process*

4. Open the Processing tab of the PGP Encryptor. Choosing Including for the signature. The security key length is 256, and all the details are kept as their defaults, as shown in Figure 5-44.

Figure 5-44. *Configuring the PGP Encryptor*

5. In the Encryption Keys User ID, provide the Encryption User ID from public keyring as the open PGP key.

6. In the Signer Key User ID from Secret Keyring, provide the open PGP key.

7. Save the iflow and deploy it. Copy the endpoints and run it through Postman. You should see the encrypted message in Postman, as shown in Figure 5-45.

```
┌-----BEGIN PGP MESSAGE-----
Version: BCPG v1.47

hI4DSXmHo6/DeFOQAf484GdF/3WhW/2lU6WT1VMQOyQACx4rsGrx!
naYUGj6i1pXAzO9yfu+tieYtV4YW2zAMZaXENVAHAgCHsijA/9i7'
JpDmzf4Zeit4pNDz43PZzzk5Ygh2OxY+jFLkhGClzWZJ+wk/gdPbl
OnYBZtzJ/PaH1oSAOpAhJvXdil/OfDT5fgMIqOKm1uCY/yCafe18:
jfqkUooP+pSrqcti4le8JSlPRrqOVl/akNb6cda9smDt5SJXUGUT(
hy2hRZeX6NrHt3HEemGwHp/VL9omMDX4
=jXYY
-----END PGP MESSAGE-----
```

Figure 5-45. *Encrypted message*

Follow these steps to create PGP decrypted messages:

1. Open the integration flow in which you have created the decrypted message in Edit mode.

2. Connect the Sender to the Start using the HTTPS adapter. Provide the address path in the Connection tab.

3. From the palette, in the Security tab, choose the PGP Decryptor. Then choose Security ➤ Decryptor ➤ PGP Decryptor and place it between the Start and End, as shown in Figure 5-46.

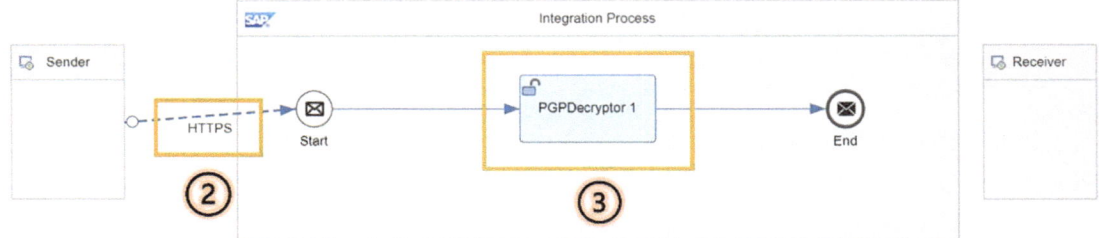

Figure 5-46. *Integration process for the PGP Decryptor*

4. Open the Processing tab of the decryptor to see signature options. From the dropdown list, you can specify the signature key while decrypting the message. After specifying the signature type, specify the signer key user IDs, as shown in Figure 5-47.

Figure 5-47. Configuring the PGP Decryptor

5. Save the iflow and deploy it. Copy the endpoints and run it through Postman. You can find the decrypted message in Postman.

Here are optional steps for decrypting a PGP message using the Kleopatra tool:

1. Open the Kleopatra app.

2. In the Dashboard, go to the Notepad and open it, as shown in Figure 5-48.

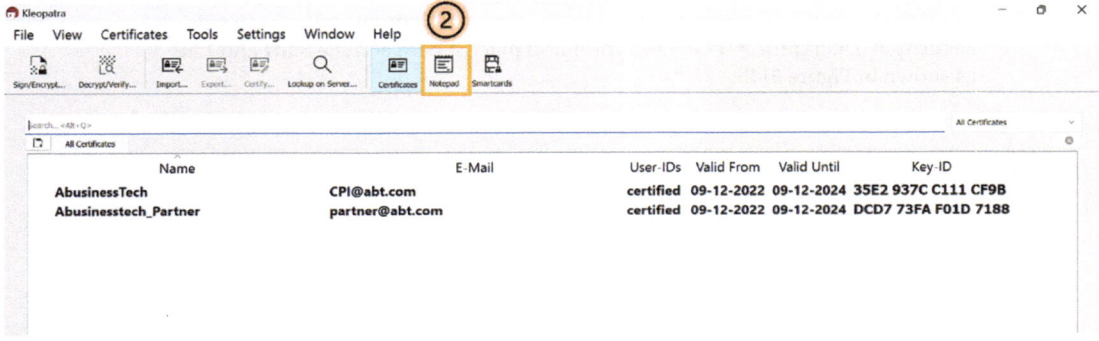

Figure 5-48. Open Notepad in Kleopatra

3. Copy the encrypted message from Postman and paste it in the Kleopatra Notepad. After pasting the message, click Decrypt/Verify Notepad.

4. After providing the passphrase, you will be able to see the encrypted message, as shown in Figure 5-49.

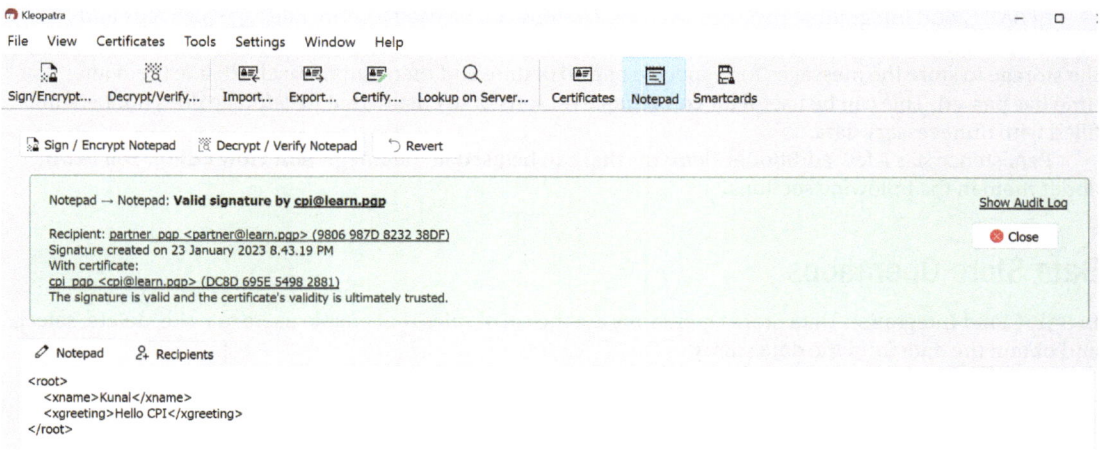

Figure 5-49. Crypted message that you sent from Postman

You have now learned about PGP encryption and decryption in depth with a practical example. The next section thoroughly covers the next element of the iflow design, Persistence.

5.1.4 Persistence

In SAP Cloud Integration, message persistence steps are used to store message payloads and metadata in persistent storage, such as a database or a file system. This allows the messages to be stored for a certain period of time and retrieved later for further processing or auditing purposes.

The message persistence step can be used in integration flows to store messages at different stages of the flow. For example, it can be used to keep the message's original form before any processing is done or the message's altered form after processing. Figure 5-50 shows the persistence step used in the integration flow.

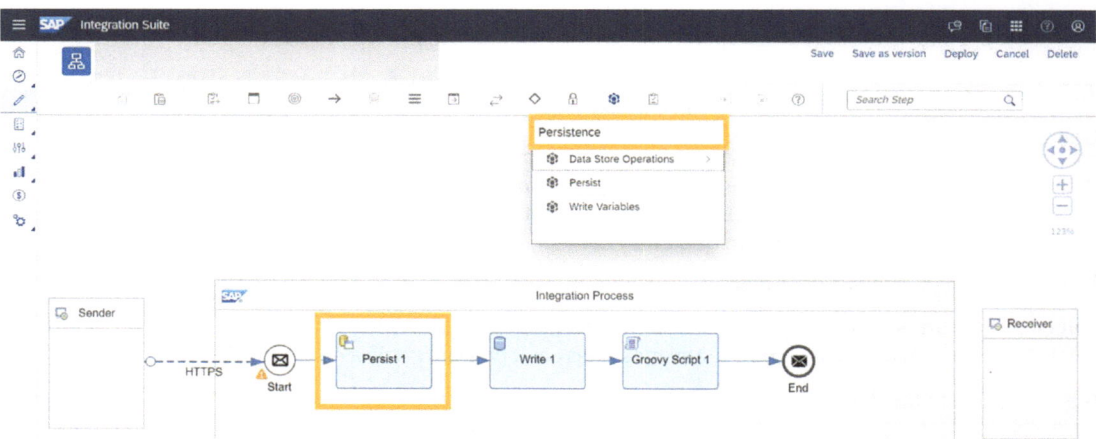

Figure 5-50. Persistence

In SAP Cloud Integration, message persistence steps can be used to store message payloads and metadata in different types of storage, such as files, SFTPs, SCPs, SCPs, and databases. You can also configure the storage to store the messages for a specific period of time and then automatically delete them after that time has passed. This can be useful for compliance reasons or to ensure that the storage does not become filled with unnecessary data.

Persistence has a few additional elements that can be used in the Integration Flow Editor. You learn about them in the following sections.

Data Store Operations

In SAP, Cloud Integration Data Store Operations are the set of rules that enable users to write, delete, select, and obtain the data from the data stores.

The data store can be used to temporarily save messages. Figure 5-51 shows how to select Data Store Operations from the palette in the SAP Cloud Integration Flow.

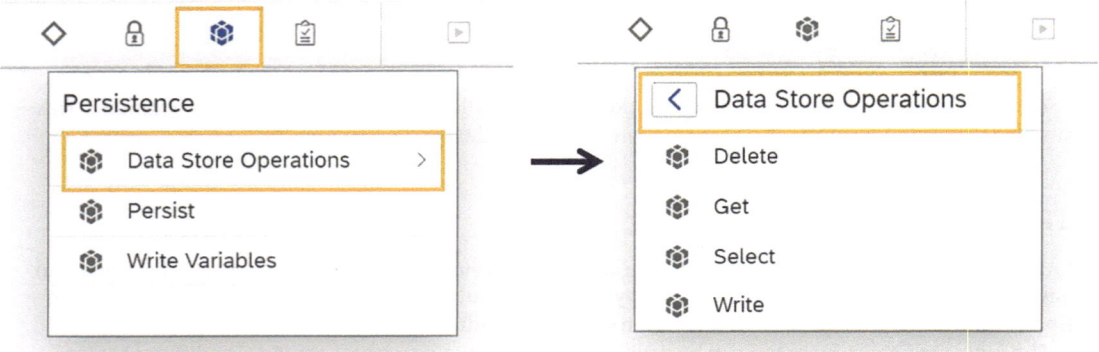

Figure 5-51. *Data store operations*

Table 5-1 shows the data store operations that are available in SAP Cloud Integration.

Table 5-1. *Data Store Operations*

Operation	Used To ...
Write	Keep the messages in the data store for the time being. You can save the messages in the data store if you perform a Write operation by specifying the data store name and a unique Entry ID.
Delete	Start the data store's message-deletion process.
Select	Download messages in bulk from the data repository. You can also decide how many messages each poll can retrieve at one time.
Get	Go to the data storage and retrieve a specific message.

The next section explains another element of the iflow object element, Persist.

Persist

The reason to save a message is to make it accessible for analysis at a later time. A Persist step can be added to an integration flow to save a message at a specific point in the workflow. The message storage feature is especially useful for auditing.

Consider the following case as an example: Your integration flow receives product orders from a sender system, processes those orders, and then updates a product catalog in line with the order details. Your IT department must be able to prove that it meets certain compliance requirements.

The logical storage location that the Persist step uses is the message store. In terms of physical consumption, the data store and message store share the same tenant database. This component stores information about your renter. The maximum amount of disc space that can be used is 32 GB.

The data store is the only way to access the content of the message store via an interface. To access its content, you must use the Cloud Integration OData API. Additionally, a message store entry cannot be viewed while the integration flow that generates it is in progress. Only the message store content that an integration flow has written is available for viewing after it has been processed.

In the next section, you learn more about the Write variables in an example.

Write Variables: Practical Example

To transmit data between several integration flows, you construct variables (deployed on the same tenant). If you want to use a variable across several integration processes that are installed on the same tenant, you need to make it a global variable.

Use the write variables step type if you want to create a variable at a certain point in the message processing sequence. The variable might be consumed using a Content Modifier in either a different stage of the same integration flow or in a different integration flow. This step enables you to modify an exchange property or header based on a variable.

1. Open the Integration Flow Editor for any integration flow of your choice. Or you can create a new one.

2. Choose Persistence ➤ Write Variables from the palette, as shown in Figure 5-52.

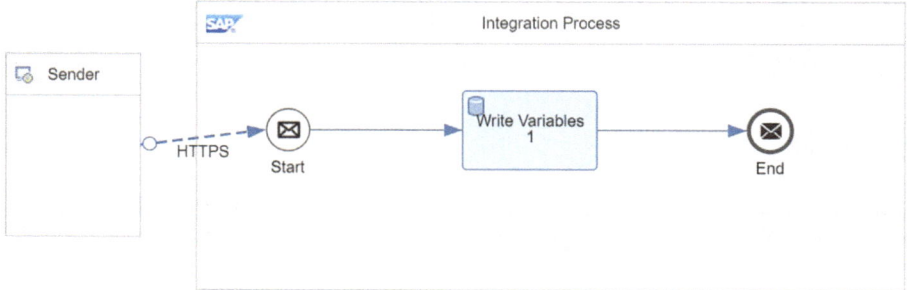

Figure 5-52. Write variables

3. Connect the Sender to the Start using HTTPS.

4. In the Processing tab of the Write Variable, configure the following variables as per the needs of the organization. This can be any value; for example Test124. This will give the result of Test 124, as shown in Figure 5-53.

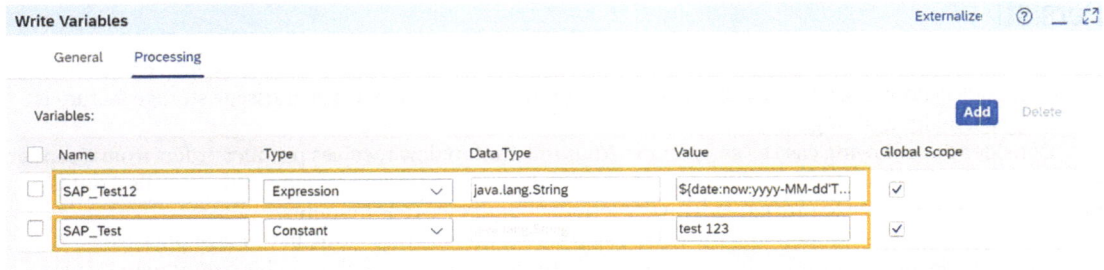

Figure 5-53. *Processing tab of Write variables*

5. Save the iflow and deploy it.

6. In the Message Payload, you can see in the header that you obtained two variables, as shown in Figure 5-54.

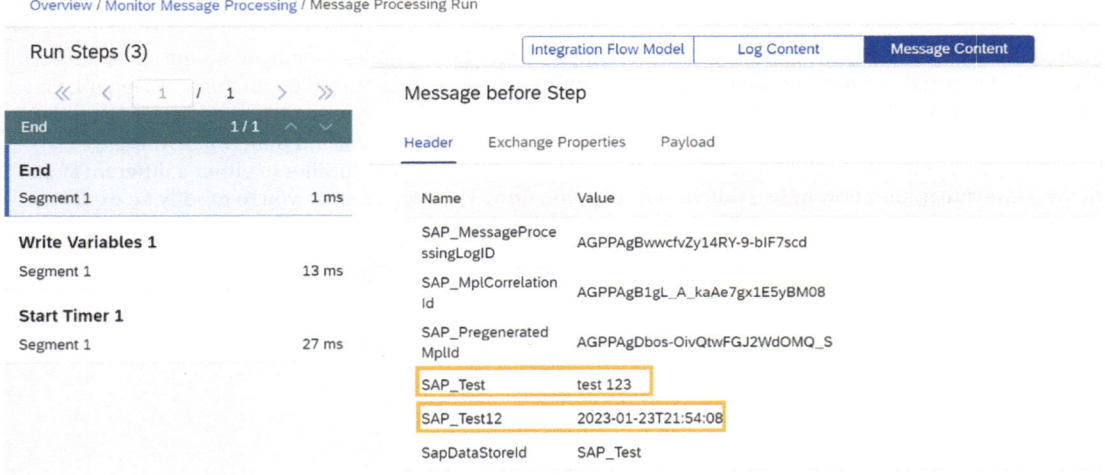

Figure 5-54. *Message Payload*

You now have briefly learned about the persistence elements and about their subcategorized elements. The next section goes through the Validator, which is another element of the iflow design object.

5.1.5 Validator

A defined schema is checked by valuators against the message's content. Figure 5-55 shows the Validator used in the integration flow.

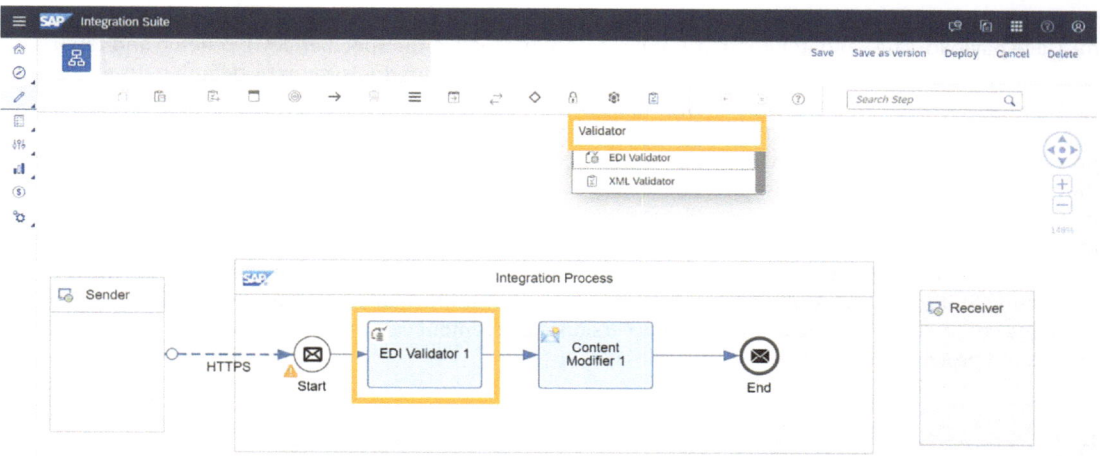

Figure 5-55. *Validator*

From the palette, choose the Validator and select the EDI or XML Validator, as shown in Figure 5-56.

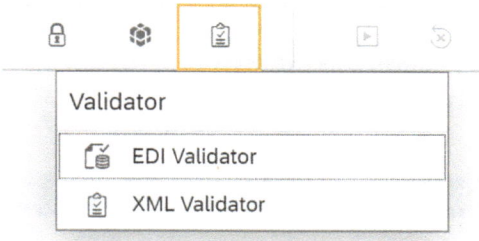

Figure 5-56. *Validator in SAP Integration Flow Editor*

The next section covers the EDI Validator. The Validator is also subcategorized into different elements.

EDI Validator

The defined XSD schema is checked by the EDI Validator against the message payload in the EDI flat file format. Document types for UN-EDIFACT, ODETTE, and ASC-X12 are supported by EDI validators. This technique is used to allocate XSD files for message payload validation in a process step. When there is a difference between the message payload and the defined XSD schema, the validator throws an exception.

XML Validator

The defined XML schema is checked by the XML validator against the message payload in XML format. Your integration flow project's src.main.resources.xsd location now contains the XML schema (XSD files). If the required location does not exist in your project, you must first create it before adding the XSD files.

With the help of these steps, you can assign an XML schema (XSD files) to a process step to validate the message payload. The validator compares the message payload to the specified XML schema and reports any discrepancies. If the validation is unsuccessful, the Cloud Integration system by default halts all message processing.

You have learned that designing iflow objects is a crucial step in building effective integration processes in the SAP Integration Suite. However, once these objects are in place, how can you ensure that they continue to perform reliably and consistently over time? That is where version management comes in. By keeping track of changes to your iflow objects and providing a clear audit trail of revisions, version management can help you maintain the integrity and reliability of your integration processes, even as they evolve and grow more complex. The next section covers the version management capabilities of the SAP Integration Suite, from creating and managing versions of your iflow objects to monitoring changes and resolving conflicts. You'll dive in and see how version management can help you build more robust, scalable integration processes.

5.2 Version Management: Practical Example

Version management can be used in Cloud Integration to distinguish production versions from active versions in development.

Whenever you modify an iflow, it is advised to create a new version and work on it to preserve the existing changes in an older version, which can be recalled later (if needed).

Use the following steps to create a new version:

1. Open the integration flow that you created in the integration package in Section 4.1. Open the integration flow created in Section 4.17 in Edit mode. Perform the required changes to the integration flow and save it as a new version, as shown in Figure 5-57.

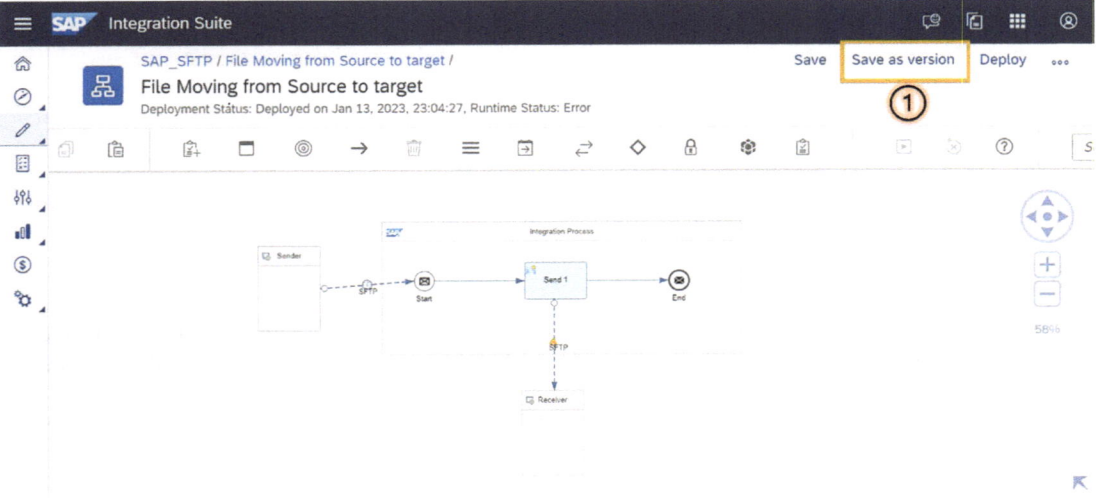

Figure 5-57. *The Save as Version option for an integration flow*

2. Provide a version number and add comments about it. Click the OK button, as shown in Figure 5-58.

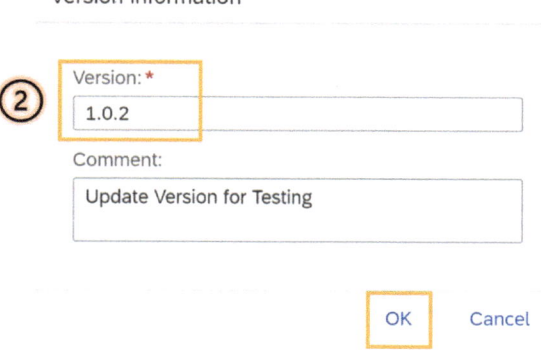

Figure 5-58. Version information

This section has briefly discussed version management. In the next section, you learn about sorting and reverting to different versions you created in the integration flow.

5.2.1 Version Management: Restore/Revert

1. Check your package in. A new version of the artifact will be added, as shown in Figure 5-59.

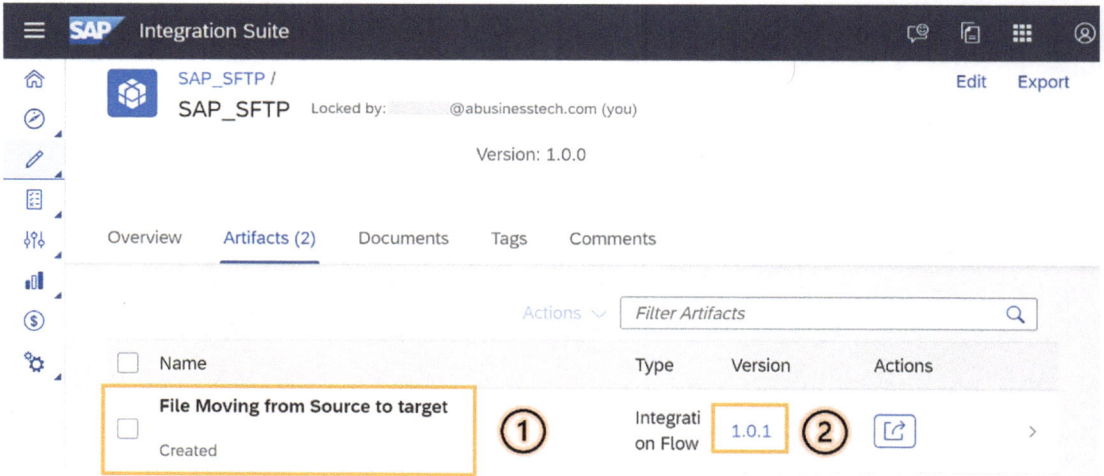

Figure 5-59. Click the version

2. If you click the new version link, you will see histories of the various versions, as shown in Figure 5-60.

Version history

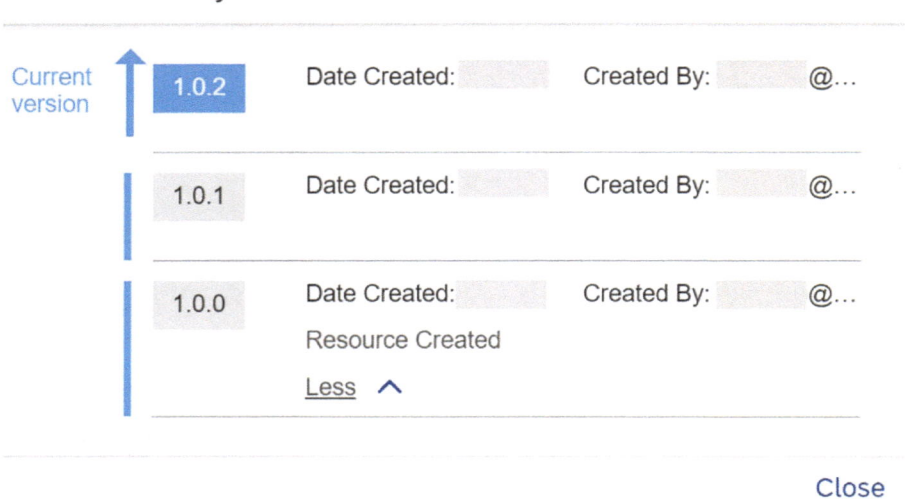

Figure 5-60. *Version history*

3. You often need to recall an older version when your current version is throwing errors (which you need to correct in the development system). If you do not recall the older version that you want to revert, you can select each version and deploy it to determine the successful iflow result.

4. For this purpose, you need to revert to an older version of the iflow to correct the production errors. You can work on the new version to fix the issues in the interim (in development). Once this version is fixed and thoroughly tested, it can be moved back to production.

5. To recall an older version, click the Revert button in the version history (besides the older version). This will revert to the older version, as shown in Figure 5-61.

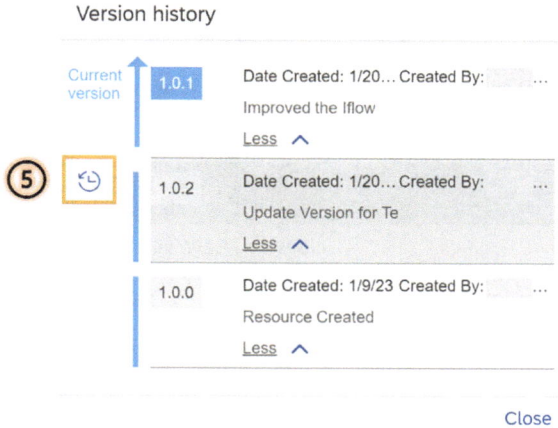

Figure 5-61. *Click the Revert button*

You have learned about version management and went through the prescribed steps to revert to older versions. In the next section, you learn about externalized parameters.

5.3 Externalized Parameters: Practical Example

Integration developers use SAP Cloud Integration tools to create their integration flows, which they then transfer to test and production systems once the development phase is complete. During this development phase, they become aware that the configuration of adapters or flow steps can be altered for the same integration flow to function as intended across several systems. To overcome this situation, the integration developers use the externalization feature offered by SAP Cloud Integration tools.

An integration developer can specify parameters for certain adapter configurations or flow steps of an integration flow using the externalization function without changing the integration flow itself. The parameters' values can then be provided later.

The following definitions apply to an integration flow's externalization parameter value field:

- **Default value**—The parameter's default value is the one you choose by modifying it in the integration flow. It is a predefined integration value that the integration developer can change.

- **Configured value**—The configured value of a parameter is the value that is specified in the configuration view.

There are few steps for externalization that you will see as you move further in this section.

5.3.1 Externalization Editor

Users can externalize integration flow elements, including endpoint URLs, connection information, and credentials, using the SAP Cloud Integration functionality known as the Externalization Editor. It is simpler to manage and modify these parameters without having to modify the integration flow code, thanks to externalization, which refers to the storage of these parameters outside of the code. To externalize the parameters, the first step is to open the Externalization Editor following these steps:

1. Open an integration flow that you created in Section 4.1.7, or any integration flow in which you created the practical examples in Section 4.2.4.3.1. Select the adapters or the flow steps to load them in the bottom pane. The Externalize button will be visible if it is supported for that component. Click Externalize, as shown in Figure 5-62.

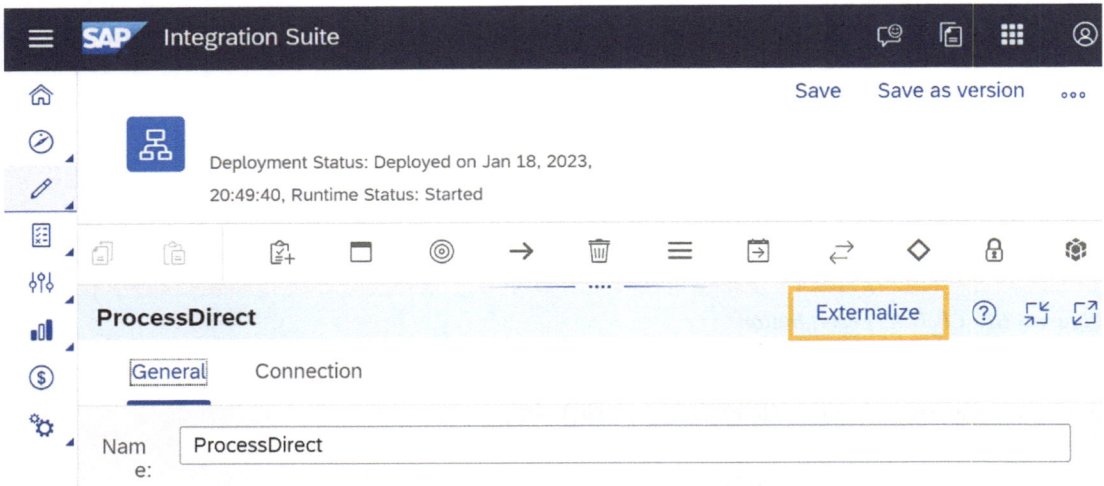

Figure 5-62. *Externalize Editor*

2. The Externalization Editor is launched, and it contains all of the configurable settings for the chosen component that can be externalized.

5.3.2 Create a New Parameter

In SAP Cloud Integration, creating a new parameter for externalization parameters involves adding a new parameter to an existing configuration, which can then be used to reference dynamic values at runtime. This process allows you to define parameters that can be set externally, rather than being hardcoded into your integration flows. To create a new parameter, follow these steps:

1. To externalize a field, you can create a new parameter in the specified column using the {{<parameter_name}} format. A token for the new parameter is created in the value column once you tab out of the parameter column, as shown in Figure 5-63.

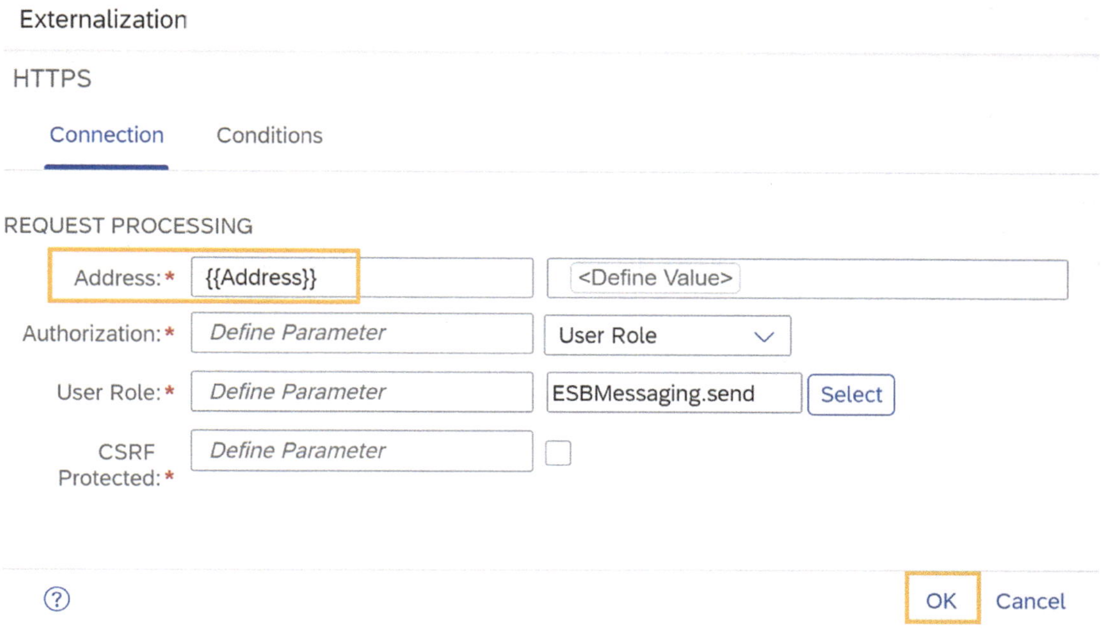

Externalization

HTTPS

Connection   Conditions

REQUEST PROCESSING

Address: *	{{Address}}	\<Define Value\>
Authorization: *	*Define Parameter*	User Role ∨
User Role: *	*Define Parameter*	ESBMessaging.send Select
CSRF Protected: *	*Define Parameter*	☐

⑦   OK │ Cancel

Figure 5-63. *Externalization*

> 2. You can specify the configured value if the default value is already provided in the parameterized column, as shown in Figure 5-64.

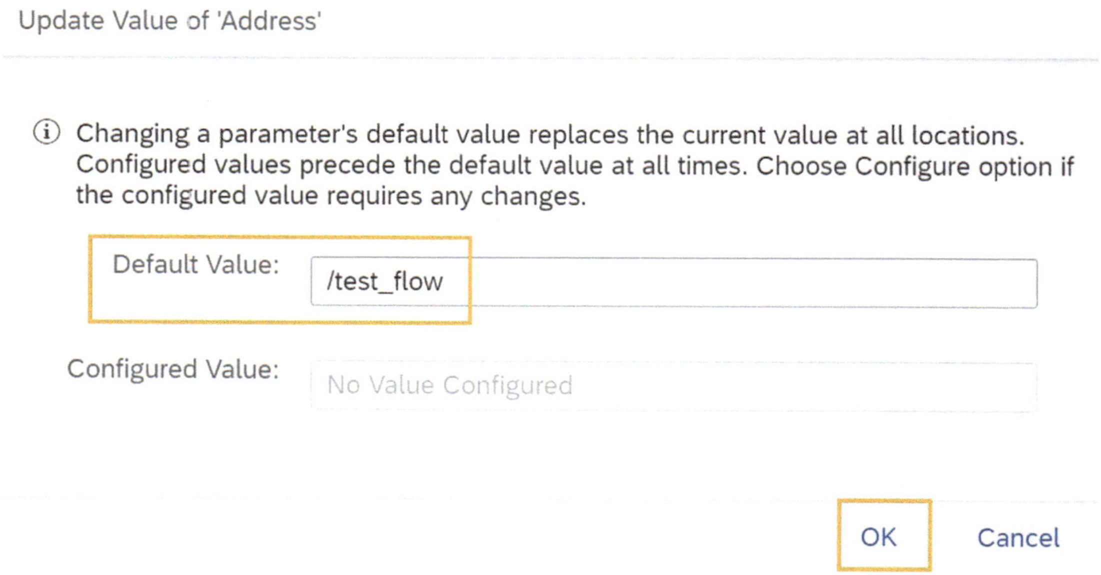

Update Value of 'Address'

ⓘ Changing a parameter's default value replaces the current value at all locations. Configured values precede the default value at all times. Choose Configure option if the configured value requires any changes.

Default Value:	/test_flow
Configured Value:	No Value Configured

OK   Cancel

Figure 5-64. *Update value of the Parametrized field*

5.3.3 Reusing Existing Parameters

Reusing existing parameters in externalization refers to leveraging existing parameter values across multiple integrations in SAP Cloud Integration. By externalizing commonly used parameters, such as endpoint URLs or API keys, you can avoid the need to re-enter these values for each integration, saving time and reducing the risk of errors.

When you externalize a parameter in SAP Cloud Integration, you can assign it a unique name and value. These parameters can then be referenced in your integration flows using the ${parameter_name} syntax. This means that if you need to update a parameter value (for example, if an API endpoint changes), you can do so in one place and have the change automatically propagate to all integrations that reference that parameter.

1. Within the same integration flow, it is also feasible to reuse current parameters between adapters or flow phases. The autosuggest function activates as soon as you enter two curly braces ("") in the parameter column. It offers you a list of all currently used parameters and their associated values that can be reused, as shown in Figure 5-65.

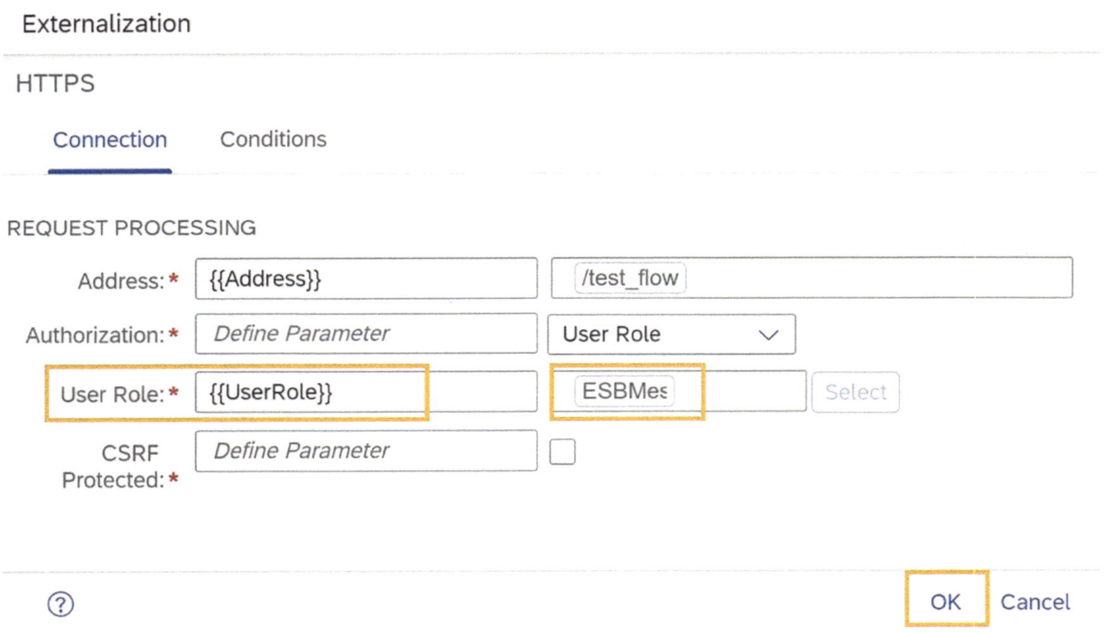

Externalization

HTTPS

Connection   Conditions

REQUEST PROCESSING

Address: *	{{Address}}	/test_flow
Authorization: *	Define Parameter	User Role ⌄
User Role: *	{{UserRole}}	ESBMes Select
CSRF Protected: *	Define Parameter	☐

⑦   OK   Cancel

Figure 5-65. *Reusing an existing parameter*

2. By clicking the token in the value column, you can alter the value of an existing parameter. Be careful, because this will also change all the other locations used in the integration flow.

5.3.4 Removing Parameters

Externalization of a field can also be undone by removing the parameter from the Parameter column. The parameter's value is kept in the Value column on the tab, but the field will not be externalized because there is no parameter to use.

The parameter is not deleted when a field's externalization is removed, since it might still be used elsewhere in the integration flow.

5.3.5 Managing Externalized Parameters

The Externalized Parameters view was added as part of the integration flow's configuration, and from there, you can manage all the externalized parameters. This view offers a list of all the parameters used in the integration flow. It enables you to enter new values for parameters or modify those that are already in use from a single location.

However, it would be beneficial if you utilized caution when making parameter changes from this point on, because the modified value will be applied to all instances of the same parameter across the integration pipeline.

Additionally, unnecessary parameters can be deleted by selecting Remove Unused from the Externalized Parameters view.

To view the externalized parameters of your integration flow, follow these steps:

1. Open the integration flow in which you made the parameters externalized.

2. All the parameters are visible when the Externalized Parameters tab is opened, as shown in Figure 5-66.

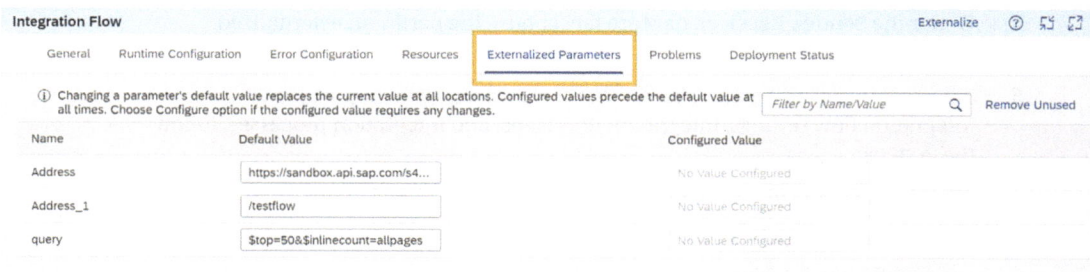

Figure 5-66. Managing externalized parameters

5.3.6 Configure Externalized Parameters

You can configure a single integration flow or many integration flows at once using Cloud Integration. You can specify the integration flow's description and the runtime and error configurations.

For the integration flows on the Monitor and Discover tabs, the quick configure option is not available. The integration flows can be set up only from the Design tab of the customer workspace.

1. Decide which integration package holds the configuration of the integration flow.

2. Choose Content for the package. You see a list of every artifact present in the chosen integration package.

3. Select configure in the Actions tab of the integration flow artifact as shown in Figure 5-67.

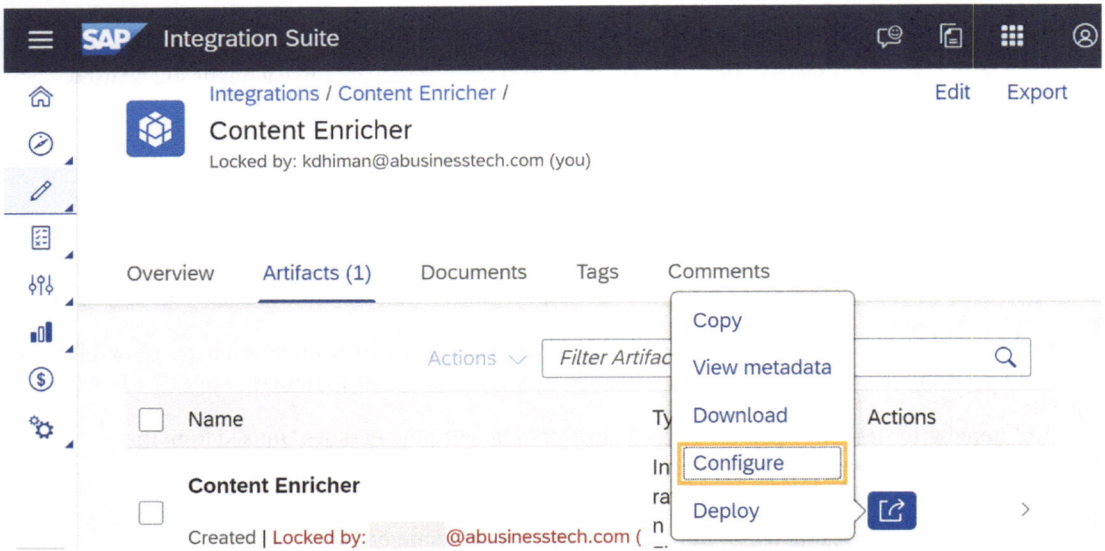

Figure 5-67. *Configuration parameters*

4. Under the Sender, Receiver, or More tab, choose the pertinent externalized component data.

5. To examine and edit externalized values for local integration processes, integration flow settings, integration flow steps, and integration processes, follow these steps:

    • Choose the component type.

    • Choose the Component Name (ID) to configure the integration flow. You cannot choose a name here.

6. Choose Save as Version and enter the parameters for the version you want to save this customized integration flow as. Deploying the integration flow is another option.

7. Select the appropriate options to save or deploy the integration flow to avoid errors.

5.3.7 Error-Handling Strategies

When runtime message processing fails, errors can be handled using the error configuration process. Figure 5-68 shows the Error Configuration tab in the integration flow.

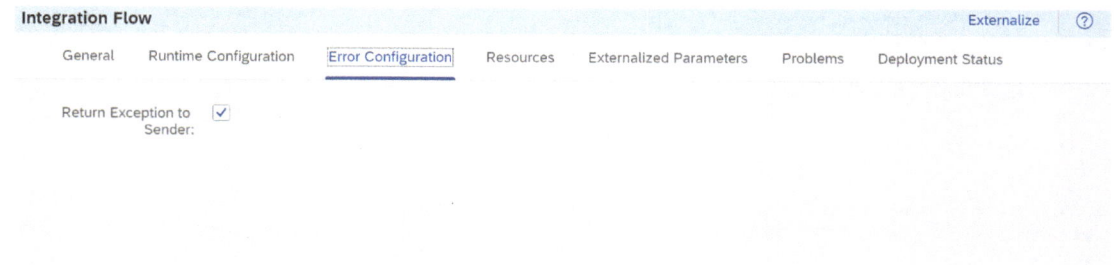

Figure 5-68. *The Error Configuration tab*

Choosing an error-handling technique is based on the following descriptions:

- **None (Default Setting)**—If an error occurs and a message exchange can be performed, no error management technique is employed.

- **Raise Exception (Deprecated)**—If an IDoc or SOAP channel is present in the integration flow but a message exchange cannot be handled, an exception is raised to the sender.

- **Raise Exception**—If a message exchange cannot be handled, an exception is reported to the sender.

You have briefly learned about the externalized parameters and all its steps with a practical example. The next section covers the different artifacts that can be created in the integration package. You see how to build API-based artifacts in the integration package.

5.4 Develop API-Based Integration Artifacts

You can create integration scenarios using REST, SOAP, and OData APIs, deploy them, and manage your artifacts all in one location.

REST, SOAP, and OData API integration artifacts are all possible. A default integration design is generated by the API-based artifacts, which you can then modify to suit your needs. When your integration design is finished, you can deploy and save the artifacts. Finally, to view your deployed API-based artifacts, you can design your own tiles in the monitoring section.

You now see the different artifacts that can be created in the integration package and try to build the SOAP, REST, and ODATA artifacts in the following section.

5.4.1 SOAP, REST, and OData API Artifact: Practical Example

In this section, you create the different artifacts in the integration package. You create the SOAP API, REST API, and OData API in the integration package. You also learn about each step needed to create the artifacts.

In the integration package, you have the following options, as shown in Figure 5-69—REST API, SOAP API, and the OData API—that can be created in the integration package.

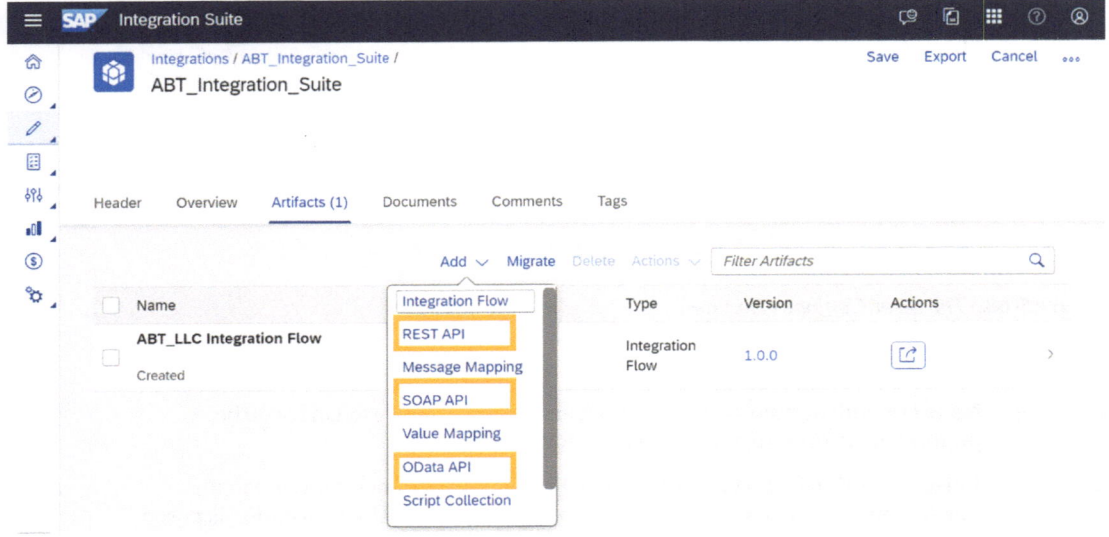

Figure 5-69. *Adding SOAP, REST, and OData API Artifacts*

Let's start by creating the SOAP API Artifacts.

SOAP API Artifacts Example

A SOAP API Artifact in SAP Cloud Integration is a configuration item that symbolizes a SOAP-based online service. It helps integration developers design integration flows that communicate with the web service by defining the structure and behavior of the web service.

1. Create an integration package. In the Artifact tab, click Add.

2. Select SOAP API from the dropdown list, as shown in Figure 5-70.

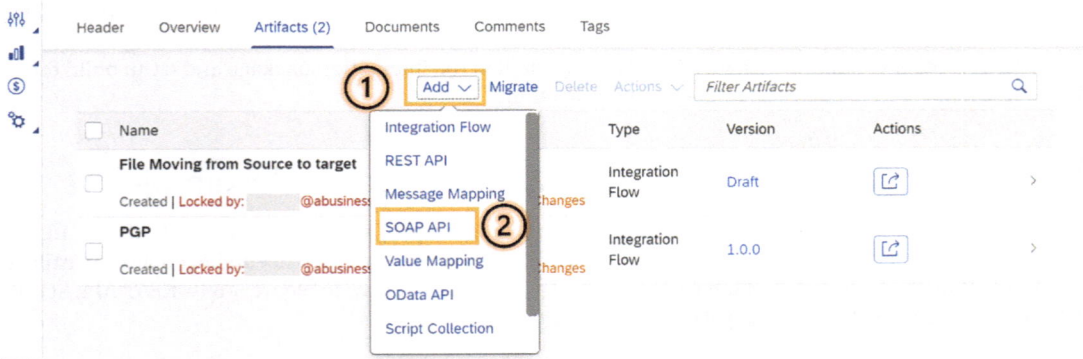

Figure 5-70. *Creating a SOAP API Artifact*

3. You can create and upload the SOAP API Artifact.

4. Specify the SOAP API details, as follows:

   - Name—Name the API

   - SOAP API—Upload the ZIP file (only when using Upload)

   - ID—Uniquely identify the API Artifact

   - Description—Describe the SOAP API Artifact

5. Select OK to create the artifact, as shown in Figure 5-71.

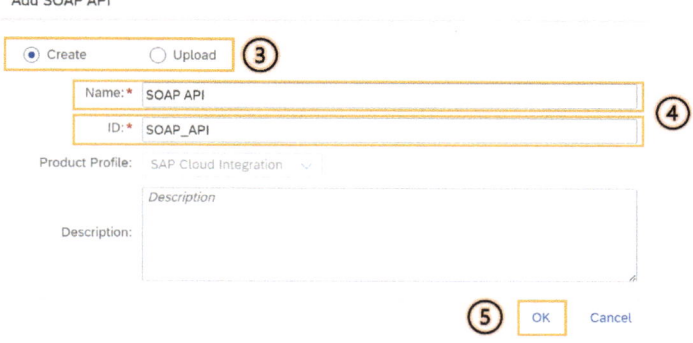

Figure 5-71. Adding the SOAP API

6. Select Deploy to deploy the artifact when it's ready for execution, as shown in Figure 5-72.

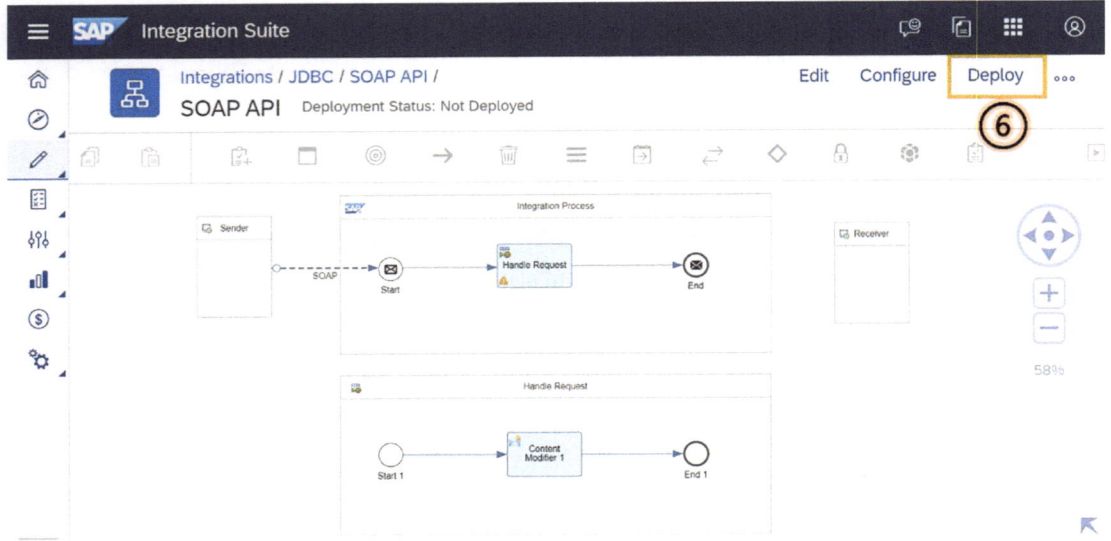

Figure 5-72. *Deploying the artifact*

You have successfully created the SOAP API Artifact. The next section covers the REST API Artifact.

REST API Artifact Example

A REST API Artifact in SAP Cloud Integration is a design-time entity that simulates a REST API. An HTTP protocol-based online service known as a REST API is intended to be straightforward, adaptable, and lightweight. REST APIs are widely used for building modern web applications, mobile apps, and IoT devices.

1. Create an integration package.

2. Select Add SOAP API from the Artifacts tab.

3. You can create and upload the REST API Artifact.

4. Specify the REST API details:

   - Name—Name the API

   - REST API—Upload the ZIP file (only when using Upload)

   - ID—Uniquely identify the API Artifact

   - Description—Describe the REST API Artifact

5. Select OK to create the artifact, as shown in Figure 5-73.

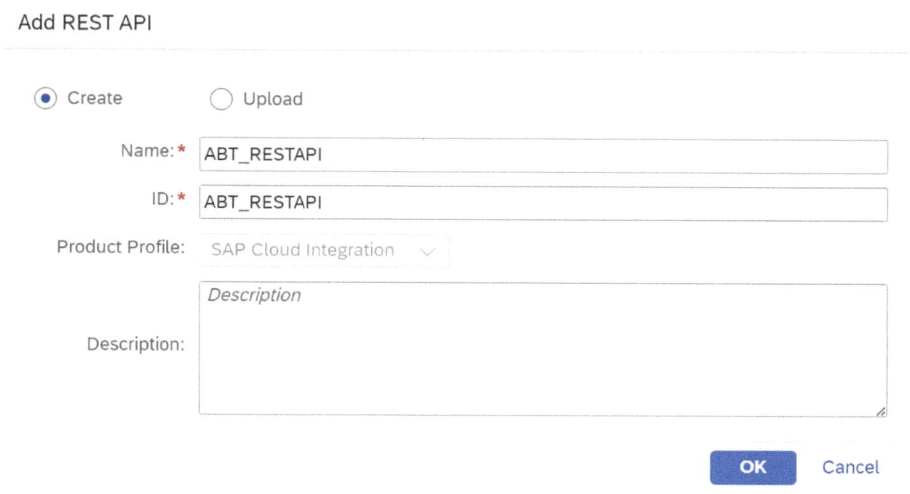

Add REST API

⦿ Create   ◯ Upload

Name: * | ABT_RESTAPI |

ID: * | ABT_RESTAPI |

Product Profile: | SAP Cloud Integration ⌄ |

Description: | Description |

OK   Cancel

Figure 5-73. *REST API configuration*

6.   Select Deploy to deploy the artifact when it's ready for execution.

7.   The steps are nearly identical for REST API, as shown for SOAP API.

You have successfully created REST API Artifacts in the SAP Integration package. The next section covers the OData API Artifact.

OData API Artifact Example

You can expose SAP Cloud Integration artifacts as OData services using the OData API Artifact component in SAP Cloud Integration. OData is a REST-based protocol for web-based data access and modification. OData allows you to make assets from SAP Cloud Integration such as integrations, interfaces, and mappings available as OData services that can be used to create web apps or be consumed by other applications.

1.   Create an integration package.

2.   Select Add OData API from the Artifacts tab.

3.   You can create and upload the OData API Artifact.

4.   Select Create Using Template in the dialog box to add an OData API.

5.   Enter the OData API's name and description. The ID field and product profile cannot be changed, as shown in Figure 5-74.

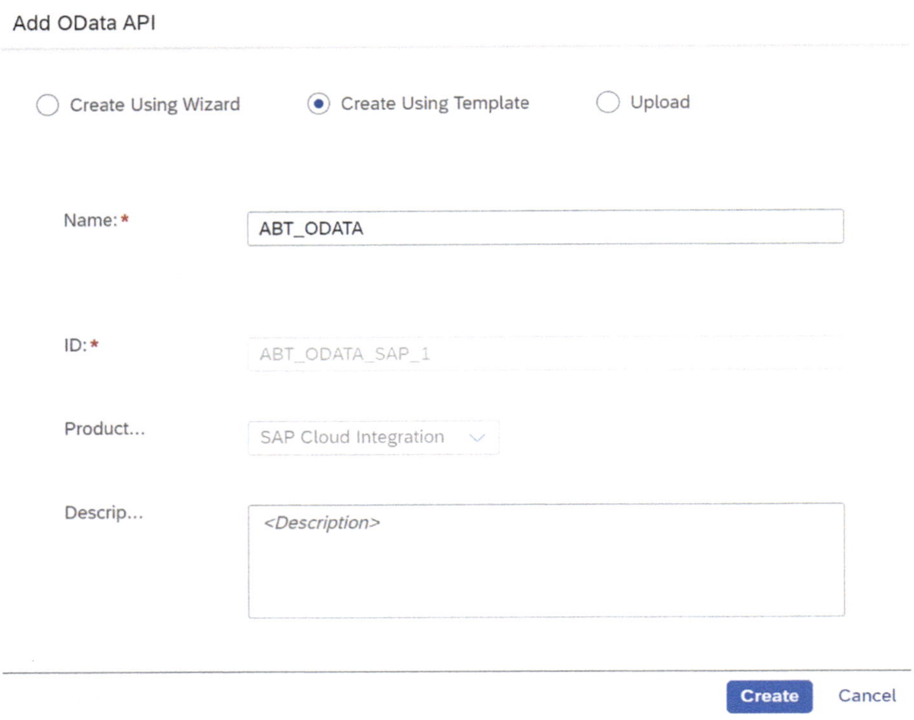

Add OData API

○ Create Using Wizard   ◉ Create Using Template   ○ Upload

Name: *

    ABT_ODATA

ID: *

    ABT_ODATA_SAP_1

Product...

    SAP Cloud Integration   ∨

Descrip...

    <Description>

Create   Cancel

Figure 5-74. OData API Artifact

6. To create the artifact, select OK. The integration package contains the artifact.

7. Integrate package should be saved.

8. To open an artifact, select it. The default integration flow model is displayed in the Integration Flow Editor.

9. Select Edit. A palette with access to all integration flow steps that you can include in the model appears at the top of the integration flow model.

10. Select Save. When your artifact is prepared for use, select Deploy.

11. The steps are nearly identical for the ODAT API, as shown for the SOAP API.

You have learned about the three artifacts—REST API, SOAP API, and OData API. The next section shows an example of creating an OData API and exposing it using SAP Cloud Integration.

5.4.2 OData API Project in SAP Cloud Integration: Practical Example

You can develop OData APIs that expose preexisting data sources, such as SOAP, as OData endpoints. Any custom app, including SAP Fiori apps and SAP BTP Mobile Services, can utilize these OData APIs to develop user-centric scenarios.

You can create and deploy an OData API as a service developer without having to spend time writing code. Creating an OData API artifact in an integration package is the first step in the procedure. Creating an OData API has more details. The following steps give you the opportunity to develop the OData API using cloud integration.

Simple modeling steps are used to import an OData model. If you like writing code, Cloud Integration also enables you to create an OData model by hand or modify an existing one:

1. Utilize integration flows to link OData objects to a data source.

2. The integration flows should be adjusted to improve business logic.

3. Set up and keep an eye on the OData API.

You can create the OData API Project by following these steps:

1. Open the integration package created in Section 4.1.1. in Edit mode.

2. Add the OData API to the Artifacts tab, as shown in Figure 5-75.

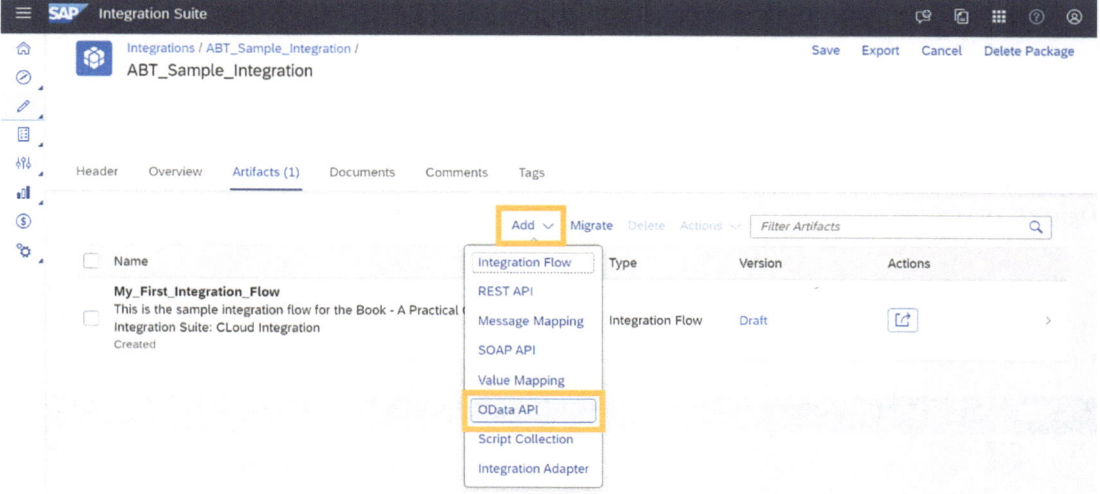

Figure 5-75. *Adding the OData API*

3. Provide the following details as shown in Figure 5-76:

    •   Name—ABT_OData_API

    •   Description—Provide a description of the OData API

    •   ID—This is automatically generated

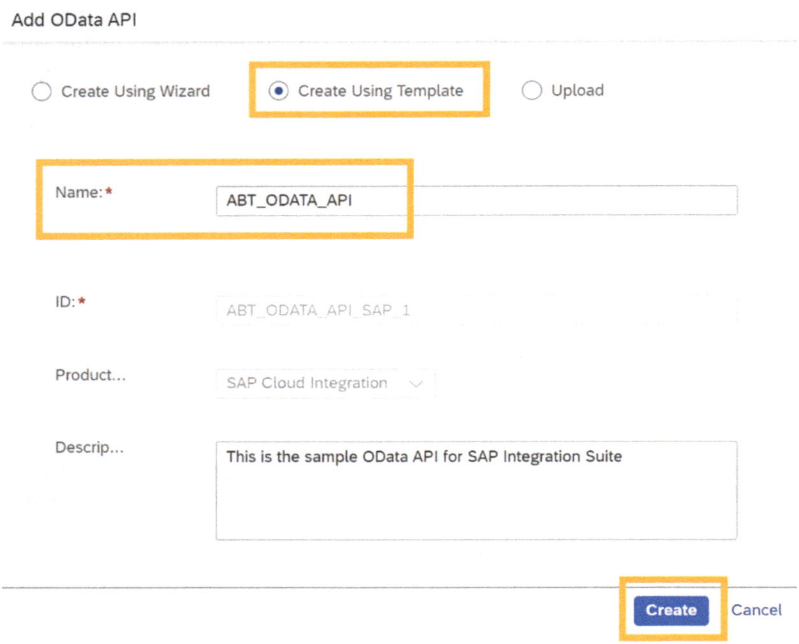

Add OData API

◯ Create Using Wizard   ⦿ Create Using Template   ◯ Upload

Name:*   ABT_ODATA_API

ID:*   ABT_ODATA_API_SAP_1

Product...   SAP Cloud Integration   ⌄

Descrip...   This is the sample OData API for SAP Integration Suite

Create   Cancel

Figure 5-76. *OData API*

4. Open the OData API created in the Artifacts of the integration package. You will
    be directed to the integration flow, where you can edit it by opening it in the Edit
    mode, as shown in Figure 5-77.

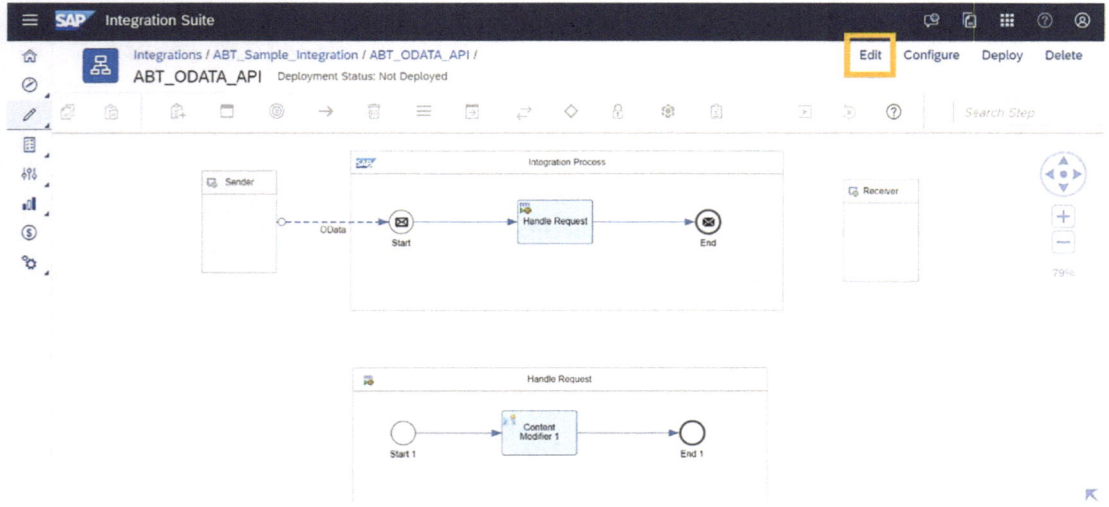

Figure 5-77. *Editing the integration flow*

5. The Content Modifier is added to the integration flow next to handle requests, and the header provides the following details, as shown in Figure 5-78:

    *   Action—Create

    *   Name—OrderNo

    *   Source—XPath

    *   Source value—//in

    *   Data type—java.lang. String

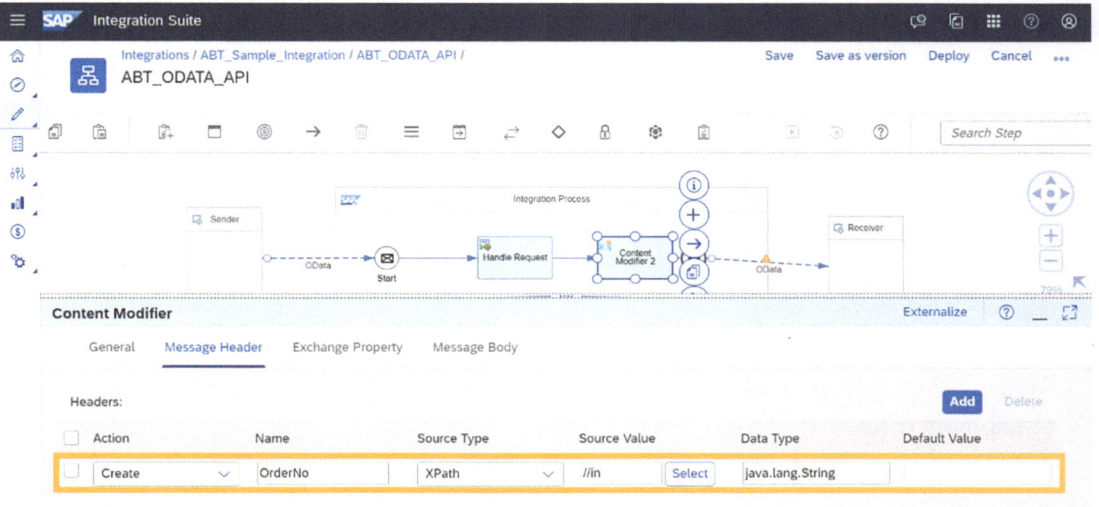

Figure 5-78. *The Content Modifier*

6. Connect the End Message to the receiver through the OData V2 adapter. In the Connection tab of OData, provide this address: http://services.odata.org/ V2/Northwind/Northwind.svc. This is the URL for Northwind, as shown in Figure 5-79.

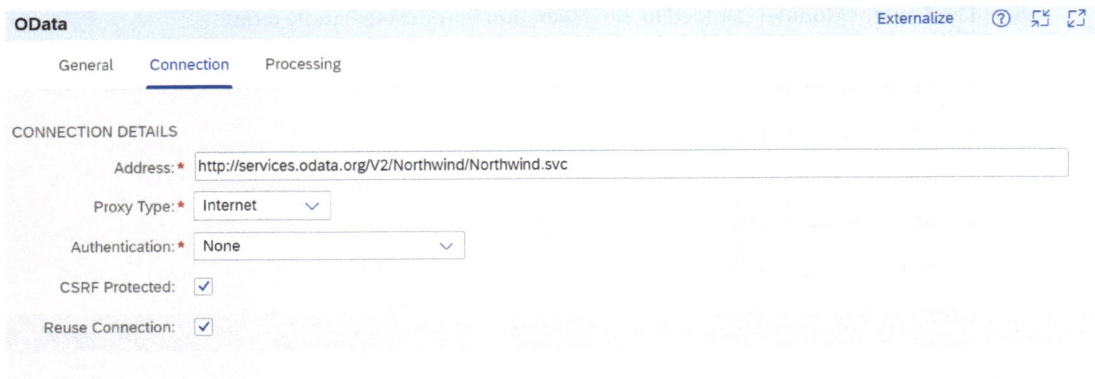

Figure 5-79. *Connection tab of OData*

7. Navigate to the Processing tab of OData and click Select in the Resource path, as shown in Figure 5-80.

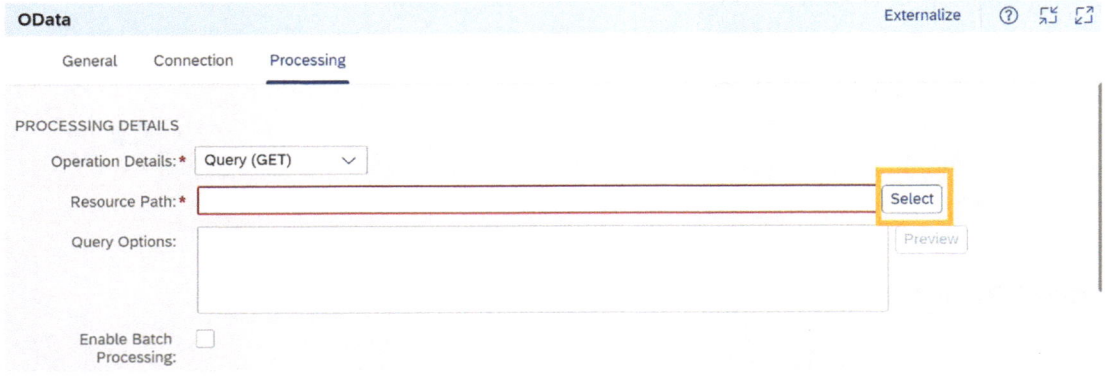

Figure 5-80. *Selecting the resource path*

8. A wizard will open with three steps. In Step 1 (Connect to System), check all the details entered and with the Authentication set to None. Then click the Step 2 button, as shown in Figure 5-81.

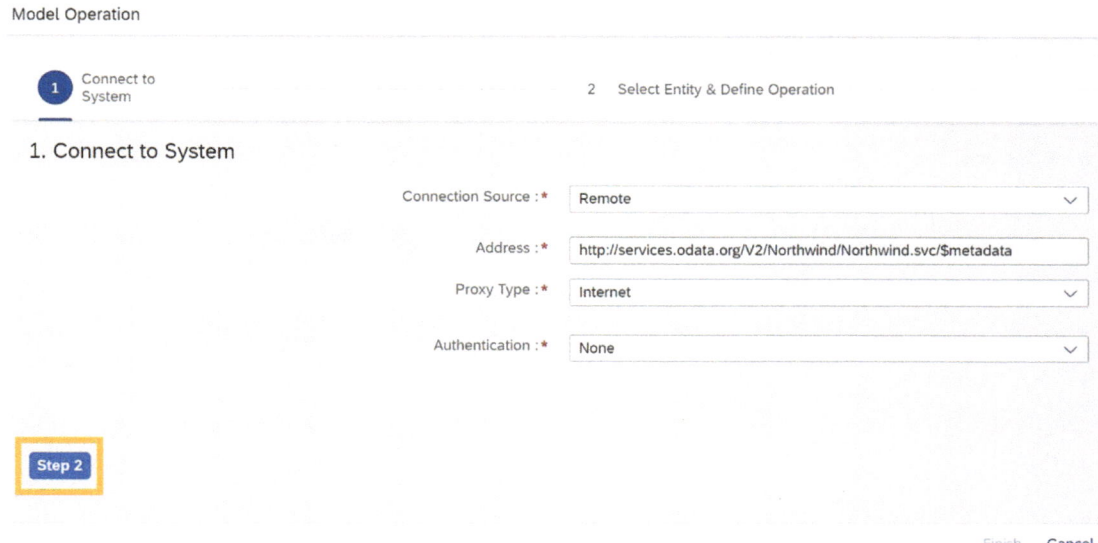

Figure 5-81. *Connecting to the system*

9. In Step 2, select Orders for the Entity and select All Fields for the fields, as shown in Figure 5-82.

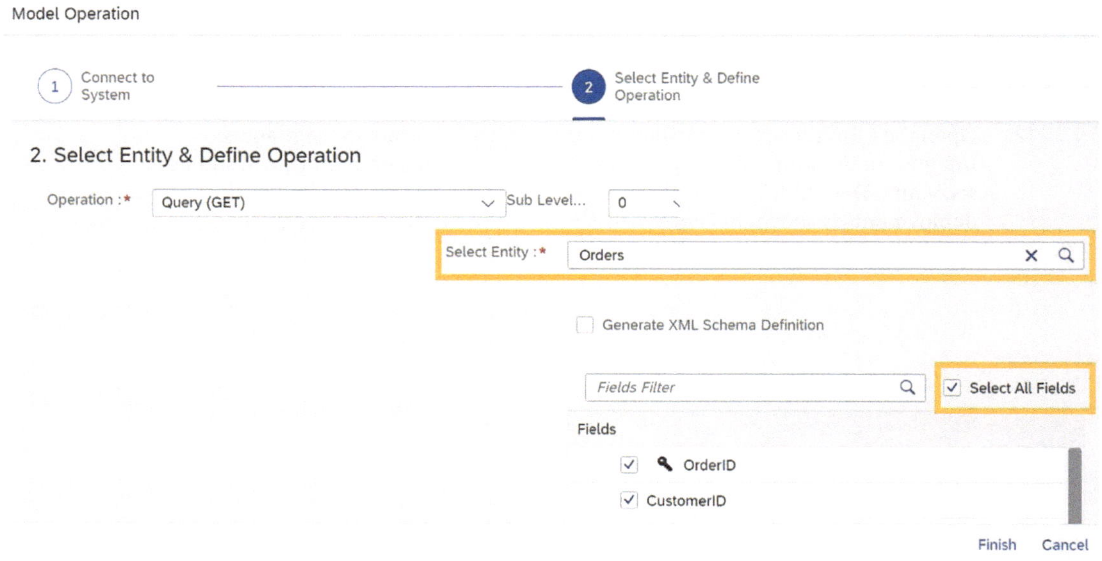

Model Operation

1 Connect to System ——————————— 2 Select Entity & Define Operation

2. Select Entity & Define Operation

Operation : * Query (GET)   Sub Level...   0

Select Entity : *   Orders   ✕ 🔍

☐ Generate XML Schema Definition

Fields Filter   🔍   ☑ Select All Fields

Fields

☑ 🔑 OrderID
☑ CustomerID

Finish   Cancel

Figure 5-82. *Selecting the entity and defining the operations*

10.   In Step 3, select OrderID for the unique identifier, which will be equal to the Header value of Order No. Provide the header code as `${header. OrderNo}`, the same code that you maintained in the Content Modifier header, as shown in Figure 5-83.

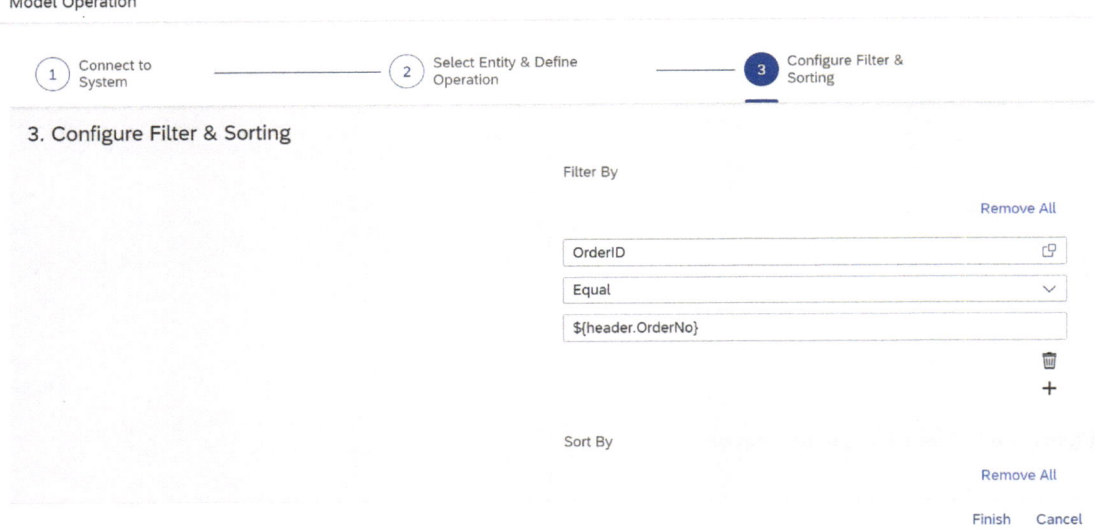

Figure 5-83. *Configuring the Filter and Sorting options*

11.   Save the integration flow and deploy it by clicking the Save and Deploy button in the top corner.

12.   Open the Monitor section of Cloud Integration and navigate to the Manage Integration Content. Choose your integration flow. In the endpoints, you should see your OData API. You can also find the EDMX file that was created during deployment, as shown in Figure 5-84.

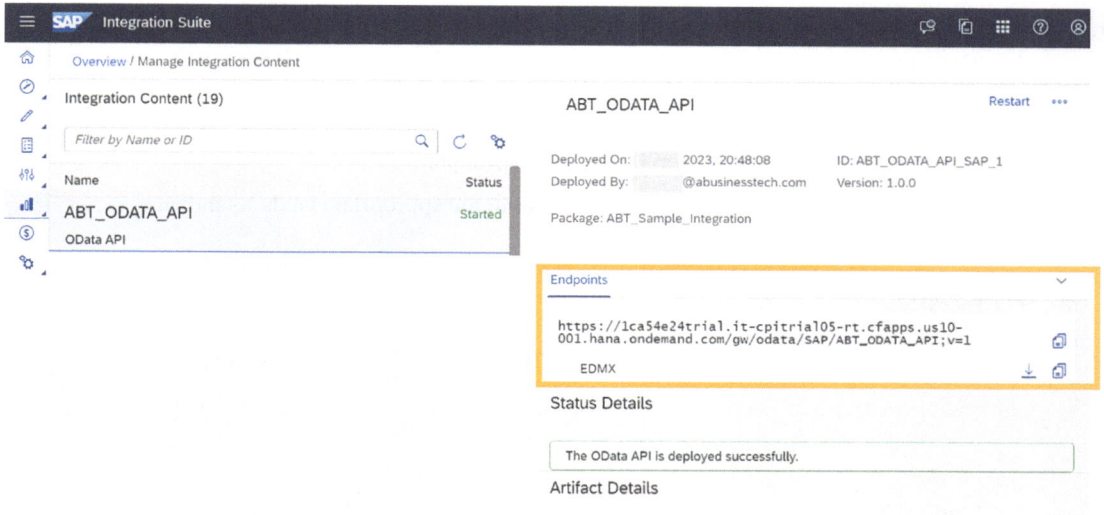

Figure 5-84. *Endpoints have been configured*

13. As you have obtained the EDMX file as well the OData API, you can use this EDMX file for further integration.

You have now learned about the three artifacts, looked into the OData API artifact, and exposed it through SAP Cloud Integration. The next section includes a practical example of the Process Direct Adapter.

5.5 Process Direct Adapter: Practical Example

A Process Direct Adapter (sender and receiver) is utilized to quickly establish direct communication between integration flows while lowering latency and network overhead.

There are two types of Process Direct Adapter:

- **Process Direct Sender Adapter**—Using the Process Direct Sender Adapter to consume data from other integration flows designates the integration flow as a consumer integration flow. In this case, the integration flow has a Process Direct Sender Adapter.

- **Process Direct Receiver Adapter**—If the Process Direct Adapter is used to send data to other integration flows, the integration flow is referred to as a producer integration flow. The integration flow in this case includes a Process Direct Receiver Adapter.

5.5.1 Configuring the Process Direct Sender Adapter

If both integration flows are available inside the same tenant, you can utilize the Process Direct Sender Adapter to establish a quick and direct connection while decreasing latency and network overhead.

To configure the Process Direct Sender Adapter, follow these steps:

1. Open the Integration Flow Editor for any integration flow. Connect Sender to Start using the Process Direct Adapter. Open the General tab and provide the following details.

2. Provide the name of the Process Direct Sender Adapter.

3. Choose the Connection tab, then enter values in the appropriate fields, as shown in Figure 5-85.

Figure 5-85. *The Process Direct Sender Adapter*

5.5.2 Configuring the Process Direct Receiver Adapter

If both integration flows are available inside the same tenant, you can utilize the Process Direct Receiver Adapter to establish a quick and direct connection while decreasing latency and network overhead.

The following attributes can be set once a receiver channel has been formed and the Process Direct Receiver Adapter has been chosen. See the Integration Flow Editor overview.

To configure the Process Direct Receiver Adapter, follow these steps:

1. Open the Integration Flow created in Section 4.1.7. Connect the receiver to the end with the Process Direct Receiver Adapter. Open the General tab and provide the following details.

2. Provide the name of the Process Direct Receiver Adapter.

3. Select the Connection tab and provide an URL of the system you are connecting to as your target. Consider the command /localiprequiresnew..

5.5.3 Basic Configuration for the Process Direct Adapter

The message body is sent from the producer to the consumer in this instance, where it is received via email at the consumer's integration flow. With the help of the Content Modifier, you are developing a producer and consumer integration flow.

Producer Integration Flow

For the producer integration flow, you employ an HTTPS adapter at the sender end and a Process Direct Adapter at the receiver end:

1. Go to the Connection tab and select the HTTPS adapter.

2. Provide the producer's address. Remember to uncheck the CSRF Protected checkbox if your project is not CSRF protected.

3.   After deciding on the Content Modifier, select the Message Body tab.

4.   Connect the End message to the receiver using the Process Direct Adapter, as shown in Figure 5-86.

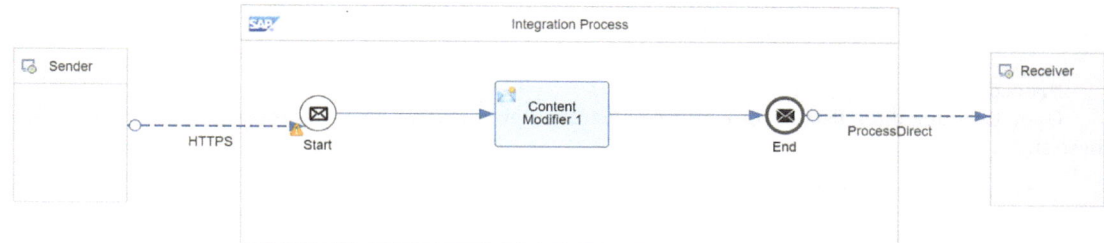

Figure 5-86. *Process Direct Adapter Producer Integration Flow*

Consumer Integration Flow

The consumer integration flow uses the Process Direct Adapter at the sender end and the SFTP adapter at the receiver end.

1.   Go to the Connection tab and choose the Process Direct Adapter.

2.   Use the same endpoint address you supplied in the producer integration flow at the receiver end in the address field.

3.   Choose the SFTP adapter and make the necessary adjustments, as shown in Figure 5-87.

Figure 5-87. *Process Direct Adapter Consumer integration flow*

You learned about the Process Direct Adapter using a practical example. In the next section, you learn about the SFTP adapter, including looking at a practical example.

5.6 Configuring the SFTP Adapter: Practical Example

To connect to a distant server and read files from it, an SAP Cloud Integration tenant must use the SSH File Transfer Protocol. The SSH File Transfer Protocol is also known as the Secure File Transfer Protocol (or SFTP).

How does SFTP sender work? Say you've set up a sender SFTP adapter. In that case, the following is how message processing is carried out at runtime: The data flow is in the opposite direction from the SFTP server to the tenant, even though the tenant sends a request to the SFTP server (consider this as the sender system). Figure 5-88 shows an explanation of the tenant and the file in between the SFTP server.

Figure 5-88. *Workings of the SFTP sender*

Procedure:

Open the integration flow of your choice or create a new integration flow for this example. Then follow these steps:

1. Click the Objects tab on the Display Configuration Scenario screen.

2. Double-click the Communication Channel in the Type column.

3. Click Edit to set the adapter's configuration.

4. Enter the channel parameters in the Edit Communication Channel screen.

5. Choose the necessary application user interface language in the display language field.

6. Select the Identifiers tab and enter the sender and receiver agencies' parameters.

7. Select the Module tab and enter the parameters that define the generic adapter modules.

8. Select the Sender radio button on the Parameters tab to instruct the SFTP adapter to process incoming messages.

9. Pick the necessary option in the Message Protocol field.

10. Select the Source tab and enter the values in the appropriate fields in accordance with the parameters' descriptions, as shown in Figure 5-89.

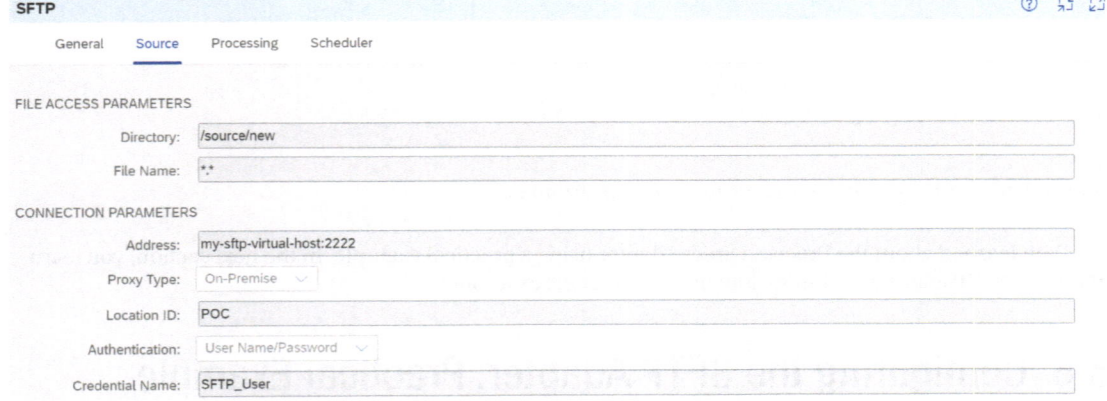

Figure 5-89. *Configuring the SFTP Sender Adapter*

11. Select the Processing tab and enter the values in the appropriate fields in accordance with the parameters' descriptions.

12. If you chose File Content Conversion as the message protocol, select the Content Conversion tab and fill in the relevant fields with the values specified in the parameter descriptions.

13. Select the Advanced tab and type the values in the appropriate fields in accordance with the parameters' descriptions.

14. The final integration flow is shown in Figure 5-90.

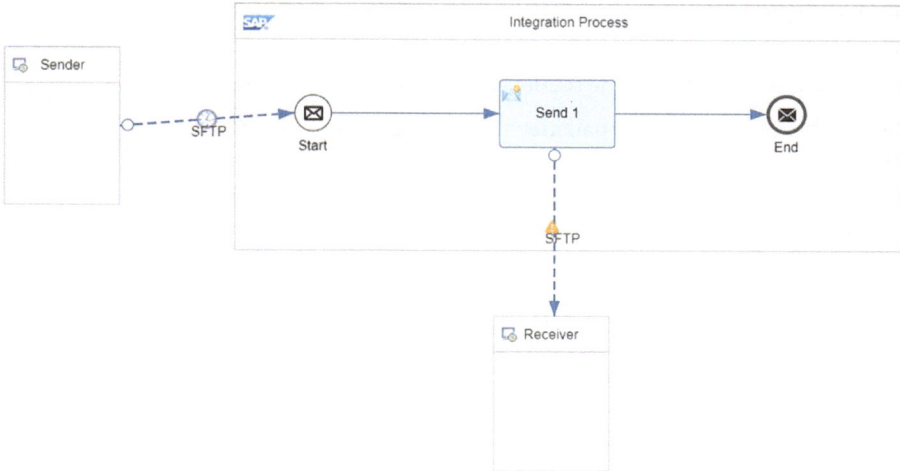

Figure 5-90. *Integration flow file-to-file*

5.6.1 Configure the SFTP Receiver Adapter

A connection is made using the SSH File Transfer Protocol between an SAP Cloud Integration tenant and a remote system so that files can be written to the system. The SSH File Transfer Protocol is also known as the Secure File Transfer Protocol (SFTP).

How does the SFTP receiver work? Say you've set up a receiver SFTP adapter. In that case, the following is how message processing is carried out at runtime: Consider an SFTP server as the receiver system that receives requests from tenants. Figure 5-91 shows the data flow from tenants to SFTP servers in the same direction. Alternatively, the tenant uploads files to the SFTP server (where the communication partner can read them).

Figure 5-91. *Workings of the SFTP Receiver*

Procedure:

Open the integration flow created in Section 4.17 or create a new integration flow for this example. Then follow these steps:

1. Click the Objects tab on the Display Configuration Scenario screen.

2. Double-click the Communication Channel in the Type column.

3. Click Edit to set the adapter's configuration.

4. Fill out the fields on the Edit Communication Channel screen with the channel parameters.

5. In the Display Language field, choose the required application user interface language.

6. Select the Identifiers tab and enter the sender and receiver agencies' parameters.

7. Select the Module tab and enter the parameters that define the generic adapter modules.

8. Select the Receiver radio button on the Parameters tab to enable the SFTP adapter to process outbound messages.

9. Select the Destination tab and enter the values in the appropriate fields in accordance with the parameters' descriptions.

10. Select the Processing tab and enter the values in the appropriate fields in accordance with the parameters' descriptions.

11. Select the Advanced tab and enter the values in the appropriate fields in accordance with the parameters' descriptions.

The Target tab of the SFTP receiver adapter is shown in Figure 5-92.

Figure 5-92. Configuring the SFTP Receiver

5.7 Summary

This chapter discussed the different types of calls, such as external and local calls. It then explored various call options, including Request Reply, Content Enricher, Poll Enricher, and Send. The chapter also covered routing options, such as Aggregator, Gather and Join, multicast, splitter, and router, with practical examples.

The section on security and message-level security use case configuration discussed the signer, decryptor, and encryptor options, including PKCS7 signer, simple signer, XML digital signature, PGP decryptor, PGP encryptor, PKCS7 encryptor, and decryptor. A practical example of PGP encryption and decryption was also included.

You also learned about Persistence, and the chapter delved into data store operations, persist, and write variables with a practical example. Finally, the chapter concluded with a discussion of the validator, covering EDI validator and XML validator. Overall, this chapter is a comprehensive guide to iflow design object elements and is a valuable resource for SAP Cloud Integration users.

Remember when you save and deploy the integration flow, you find the endpoints in the Monitor screen. You might wonder what this Monitor screen is and which actions are associated with it. You are now familiar with the phrase "integration flow" as well as the components and adapters it involves. The next chapter moves on to SAP Cloud Integration's Operations and Monitoring section to more deeply explore the world of Cloud Integration.

CHAPTER 6

■ ■ ■

SAP Cloud Integration: Monitoring and Operations

This chapter delves into the world of SAP Cloud Integration Monitoring and Operations. In the SAP Cloud Integration Monitor area, a number of tiles are accessible. Every function and property of each tile is displayed. The chapter examines every tile and covers the odd things you can observe, including the messages and the other runtime artifact logs. Every integration flow that you deploy has an identified endpoint, and you can control the integration content.

With regard to security, you will be at the forefront of managing security by examining the security content relating to user security, OAuth2 client credentials, OAuth2 SAML bearer assertion, and secure parameters. This chapter examines the keystore so you can learn about the many features and options for storing security keys. Additionally, there is a PGP tile that allows you to set and safeguard your PGP encryption and decryption by keeping the PGP keys on file. The various data stores and drivers are also covered in this chapter.

This chapter examines the many connectivity solutions that are available for successfully and securely connecting non-SAP apps, including TLS, SSH, FTP, SMTP, IMAP, AMQP, POP3, and Kafka. In the SAP Cloud integration, you will see the various access logs for Cloud Foundry and Neo. The various concepts of Cloud Integration monitoring and operations covered in this chapter can be made more effective, simple and automated by leveraging Cloud Foundry APIs as demonstrated in bespoke SAP App Store products like DOST Add-on®.

6.1 Monitoring Message Processing

To ensure that integration scenarios go smoothly, SAP Cloud Integration offers a number of monitoring options. The execution of integration flows must be tracked to spot any potential faults or problems. Real-time monitoring is a feature of the cloud integration platform that enables customers to keep track of integration situations and see statuses instantly. Additionally, this functionality offers thorough logs and warnings to inform users about any issues or faults that arise throughout the integration process. Moreover, SAP Cloud Integration provides a number of monitoring dashboards that provide customers with a thorough overview of the integration landscape and that help them quickly discover and resolve problems.

When running a SAP Cloud Integration, you can check the status of messages and integration artifacts for used tenants.

The following duties exist:

- Watching a cluster and the flow of messages through it. You can utilize the web-based monitoring program to accomplish that.

- Examine the state of the tenant cluster's messages and integration content artifacts.

Each of the following sections on the start page covers a different work area.

© Jaspreet Bagga 2023
J. Bagga, *A Practical Guide to SAP Integration Suite*, https://doi.org/10.1007/978-1-4842-9337-9_6

Table 6-1 describes the different sections of the Monitor screen of SAP Cloud Integration and explains what the users are allowed to do in that particular section.

Table 6-1. *Monitoring SAP Cloud Integration*

Container	Allows You To ...
Monitor Message Processing	Follow the tenant's communication processing. The amount and status of processed messages during a certain time window are displayed in this section's tiles.
Manage Integration Content	Control the tenant's integration content. The tiles in this section show the number and state of integration content objects (such as integration flows). The Integration Adapter can also be particularly monitored in the Cloud Foundry environment.
Manage Security	Manage the tenant's security artifacts. You can manage specific tasks associated with setting up secure connections between your tenant and distant systems using the tiles in this area. You can access and use security-related artifacts, such as user credentials artifacts, by using the Security Material tile. You can control the tenant keystore's content, as well as the lifetime of keys and certificates, by using the Keystore tile, which gives you access to its contents. You can view the PGP key details by using the PGP Keys tile. You can update access policies by using the Access Policies tile, which gives you a summary of the ones that are currently in place. You can control certificate-to-user mappings using the Certificate-to-User Mappings tile (relevant for the setup of inbound connections). The JDBC Material tile offers a summary of the JDBC drivers and the artifact connections that are utilized to interface with databases. You can manage user roles by seeing an overview of them in the User Roles tile. You can check the connection to a receiver system using the Connectivity Tests tile.
Manage Stores	Manage the tenant's temporary data storages. The Data Stores tile lists all of the tenant's storages that are currently being used to temporarily store different kinds of data during message processing. With the Variables tile, you can monitor the variables utilized in integration processes. The Message Queues tile gives you a list of a tenant's active queues and lets you control them. The number ranges used in business-to-business settings are described in the Number Ranges tile.
Access Logs	Access the System Log Files and examine any issues that happened while processing inbound HTTP (and documented in system log files). Obtain audit logs (resulting from system changes).
Manage Locks	Manage the entries for design artifact and message locks. You can manage the locks that are created in the in-progress repository using the Message Lock tile to stop the same message from being processed more than once concurrently. You can browse a list of the locked designtime artifacts and control them using the Designtime Artifact Locks tile.

SAP Cloud Integration allows you to monitor your integration flows and manage your different operations and stores. The Monitor screen contains different tiles, which you will learn about in brief. You will also learn about the Monitor message processing part, and you will look at the different tiles associated with it.

The message monitor shows a summary of the messages handled by a tenant and lets you inspect the specifics of each given message. You can open the message monitor by clicking the tile in the area labeled Monitor Message Processing.

Messages are shown in accordance with the tile's filter settings. Figure 6-1 shows the monitor message processing tiles.

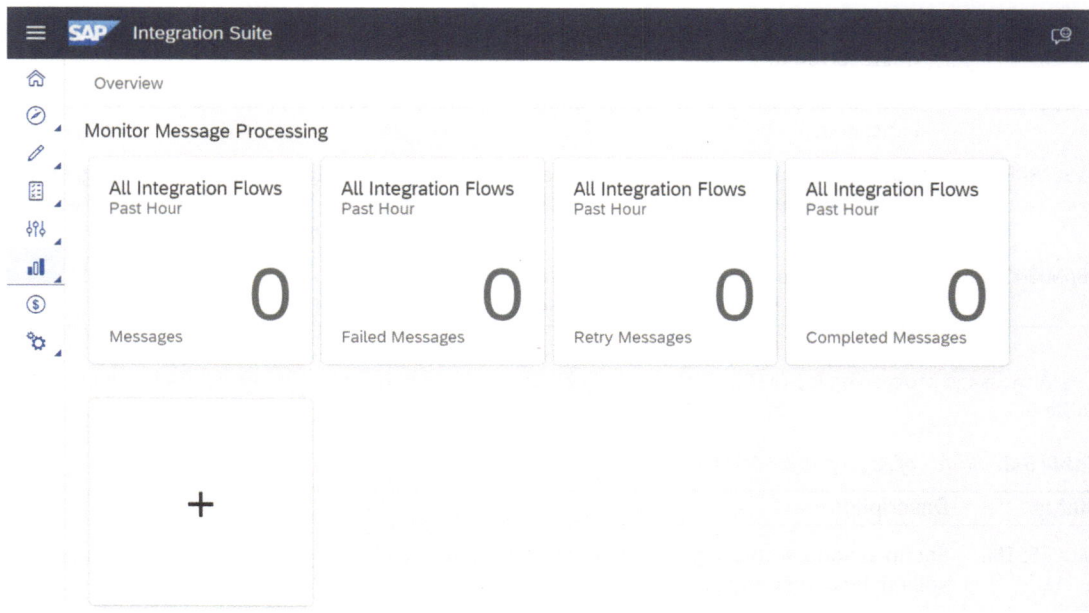

Figure 6-1. *Monitoring message processing tiles*

The following statuses can be seen in the message processing tiles, as explained in Table 6-2.

Table 6-2. *Status of Monitor Message Processing*

Status	Description
COMPLETED	Successful message delivery to the intended recipient.
PROCESSING	The message is currently processing.
RETRY	Status retry is set when an automatic retry was started as a result of a message processing error.
ESCALATED	A message processing error occurred, and a repeat attempt has not been made.
FAILED	There is no way to try again because the message processing failed and the recipient has not received it.
CANCELLED	JMS queue items that have been manually cancelled have their MPL status set to cancelled.
DISCARDED	The Message Processing Log is displayed on the worker node where scheduler-triggered integration flows first started message processing. On each subsequent start of message processing, the status of the message is changed to DISCARDED.
ABANDONDED	The processing of messages was halted, or the log was not updated for an unusually lengthy time. If processing is resumed, the status can change.

A message processing log (MPL) that has been aggregated can contain the status values listed in Table 6-3.

Table 6-3. *Status of Aggregated Messages*

Status	Description
PROCESSING	Set up as soon as the aggregation process starts and remains in place while the aggregate is still taking in incoming messages.
FAILED	Set when, after the aggregating process has been successfully finished, the aggregated message fails.
RETRY	When the aggregation process has been successfully finished and the message delivery attempt is made again because of an error, the value is set.

You learned about monitor message processing and the different attributes associated with it. Now, in the next section, you will learn about the different tiles associated with monitor message processing.

6.1.1 Message Logs

The message processing log shows the message's structural data.

A message can be chosen from the list. The header in the view section displays the name of the integration flow, the date and time it was last processed, and its processing status.

When you will navigate to the Monitor Message Processing tile. The number of runs is displayed in the logs section along with the following details listed in Table 6-4.

Table 6-4. *Runs*

Property	Description
#	Shows the run's sequential number.
Started At	Shows the time and date of the most recent run.
Duration	Shows the processing run's duration in milliseconds (ms).
Log Level	Displays the log level.
Process ID	Shows the worker node's ID that the run was executed on.
Status	Following this run, displays the message processing log's intermediate state.

You can view the message processing log by selecting the Monitor Processing log level. We will learn in deep about these log levels in the upcoming sections.

The Integrations Steps list is on the left in the detailed view. A red exclamation point in the run step list indicates whether there was a problem with one of the run steps. By clicking the exclamation point, you can view the error message.

To examine the log content, select Log Content after selecting a run step from the list. The content of the log shows the attributes and actions of the step. All of the steps that are associated with a flow element more than once are picked from the list. The element info bar's arrows allow you to go between steps and review the actions and characteristics as appropriate.

Only when the log level is set to trace is the message content section available. If the integration flow model is deployed, it is shown on the right side of the screen. The message's journey can be seen in the integration flow. A run step is highlighted in the integration flow model if it is chosen for the list.

The integration flow model uses envelope icons to represent the trace data availability. Table 6-5 shows the different envelope icons during the monitoring of the integration flow when the trace mode is activated.

Table 6-5. *Envelope Icons*

Icon	Description
	Success, Trace Data Available
	Error, Trace Data Available
	Success, No Trace Data Available
	Error, No Trace Data Available

In the absence of trace data, steps with accessible trace data and those with processing-only information are visually distinguished by inverted icons.

To print the message logs, you can use the Groovy script. If you create the integration flow and it deployed successfully and you did not get a message log in the payload, it will be hard to monitor your message. To print your message logs, you can use the Groovy Script step, which contains the specific code that prints your message log. You will see the practical example of the Groovy step and printing the log to your payload in the next section.

Using Groovy for Printing Logs: Practical Example

You learned a little about this Groovy step in Chapter 3, Section 3.3.

With a Groovy script step, the integration can be configured in such a way that the payload of the message is written to the message processing log.

1. Add the Groovy script in the iflow, as shown in Figure 6-2.

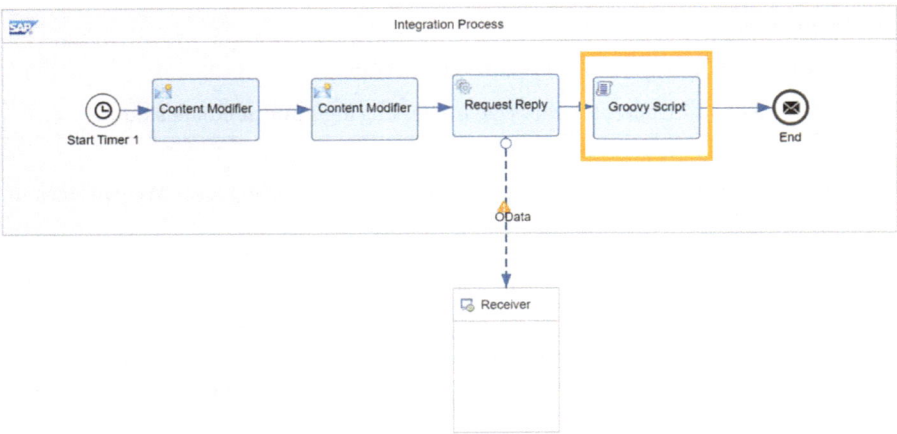

Figure 6-2. *Adding a Groovy script in integration flow*

2. Add the script code to the Groovy script step, as shown in Figure 6-3.

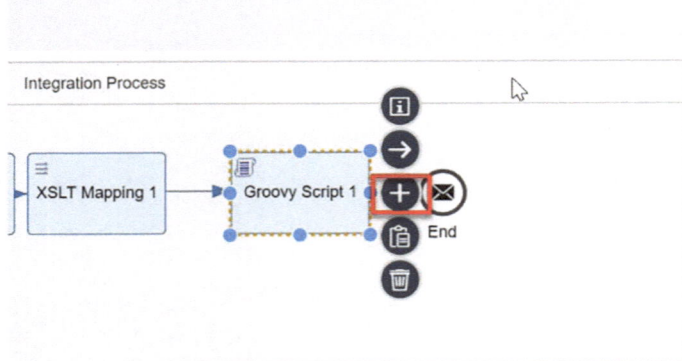

Figure 6-3. *Adding the custom script to retain logs beyond the trace period*

3. Replace the contents of the script with the following code.

```
import com.sap.gateway.ip.core.customdev.util. Message;
import java.util. HashMap;
def Message processData(Message message)
{
    def body = message.getBody(java.lang. String) as String;
    def messageLog = messageLogFactory.getMessageLog(message);
    if(messageLog!= null)
    {
    messageLog.addAttachmentAsString("Log current Payload:", body, "text/
    plain");
    }
    return message;
}
```

4. With this Groovy script, you can see the logs in Monitor Message Processing under
    Monitor Message Processing Artifacts, as shown in Figure 6-4. See Chapter 3,
    Section 3.2.5, to learn more about the entire integration flow.

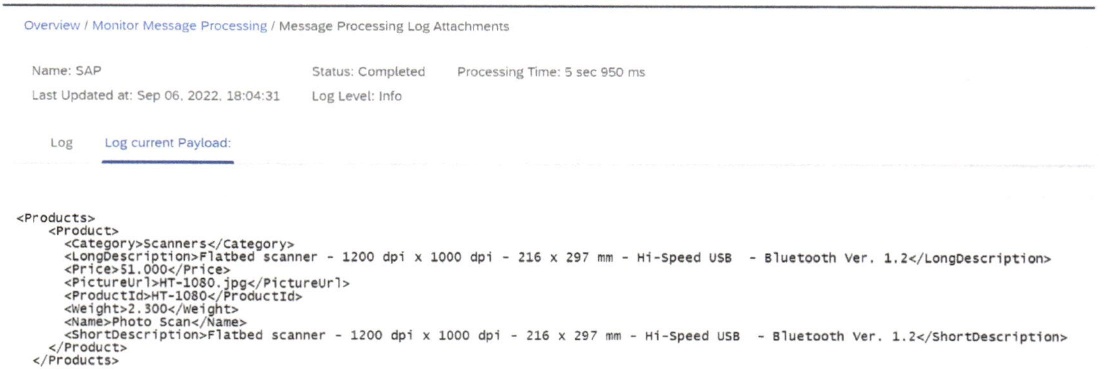

Overview / Monitor Message Processing / Message Processing Log Attachments

Name: SAP   Status: Completed   Processing Time: 5 sec 950 ms
Last Updated at: Sep 06, 2022, 18:04:31   Log Level: Info

Log   Log current Payload:

```
<Products>
    <Product>
        <Category>Scanners</Category>
        <LongDescription>Flatbed scanner - 1200 dpi x 1000 dpi - 216 x 297 mm - Hi-Speed USB  - Bluetooth Ver. 1.2</LongDescription>
        <Price>51.000</Price>
        <PictureUrl>HT-1080.jpg</PictureUrl>
        <ProductId>HT-1080</ProductId>
        <Weight>2.300</Weight>
        <Name>Photo Scan</Name>
        <ShortDescription>Flatbed scanner - 1200 dpi x 1000 dpi - 216 x 297 mm - Hi-Speed USB  - Bluetooth Ver. 1.2</ShortDescription>
    </Product>
</Products>
```

Figure 6-4. Log configuration

You learned about the message logs and saw the step to print the data into the payload. In the next section, you learn about the message processing logs.

6.1.2 Message Processing Logs: View Content

A log header and log stages are contained in the message processing log. The log header contains more general information. The log steps show particular processing events and stages as well as other information, depending on the messages reported. For example, there might be an EscalationEvent, in which case a message processing log alert will be generated.

The attributes listed in Table 6-6 are displayed in the log header.

Table 6-6. *Message Processing Log Header*

Property	Description
StartTime	Start of message processing.
StopTime	End of message processing.
OverallStatus	Corresponds to the State attribute in the Message Monitoring Editor and specifies the end-to-end status of message processing.
ChildCount	Shows the current processing step's serial number.
ChildrenCounter	The total number of message processing stages is specified.
ContextName	Specified integration flow.
CorrelationId	Identifier for correlating messages.
CustomHeaderProperties	The attributes can only be introduced by declaring a script in the script API; they are displayed in the message processing log header.
ExecutedMapping	Contains the name of the performed message mapping artifact.
Id	The MPL header displays the ID if it was previously set as a header property.
IntermediateError	True if there was a message processing fault (even temporarily) or if it took more than a minute to process a message.
MessageGuid	Key used to uniquely identify the message in the database.
MessageType	If the message type has already been declared as a header property, it shown in the MPL header.
Node	Hostname of the message-processing node.
ReceiverId	The receiver ID is displayed in the MPL header if it has already been defined as a header property.
SenderId	The MPL header displays the sender ID if it has already been set as a header property.
LocalSuccessor Id	Established during aggregation, and the aggregated MPL is used as the successor in the incoming message processing logs.
LocalPredecessor Id	If either the SAP MessageProcessingLogID or the SAPPredecessorMessageProcessingLogID message header is present and set during the creation of a new MPL.

You can see the data in Table 6-6 in Figure 6-5.

Log   Log current Payload:

```
Message Processing Log:
    StartTime           = Tue Mar 28 15:23:33.088 UTC 2023
    StopTime            = Tue Mar 28 15:23:33.495 UTC 2023
    OverallStatus       = COMPLETED
    MessageGuid         = AGQjBnWbHF7IsgOVLTYvwVR7uvo5
    LogLevel            = INFO
    LogLevelExternal    = NONE
    ArchivingLogAttachments= false
    ArchivingPersistedMessages= false
    ArchivingReceiverChannelMessages= false
    ArchivingSenderChannelMessages= false
    ChildCount          = 0
    ChildrenCounter     = 4
    ContextName         = JDBC
    CorrelationId       = AGQjBnW6S_mJbANMye52N4NSIm8u
    IntermediateError   = false
    Node                = 0
    OriginComponentName = CPI_1ca54e24trial
    PreviousComponentName= CPI_1ca54e24trial
    TransactionId       = 2eb6fab3b77b4762b78db436a8f866d1
```

Figure 6-5. Message processing logs

The message processing log steps are listed in Table 6-7.

Table 6-7. Message Processing Logs

Property	Description
Branch	Shows that the subsequent stages were handled in the same branch. The steps from one submessage are grouped together inside one branch if a Split or Multicast Step is employed.
ModelStepId	This ID is used to establish the relationship between the MPL entry and the modeled step (in the integration flow).
StepId	Identification of the relevant integration flow phase. The Camel framework is in charge of allocating it.

You have seen the different functions and the different attributes associated with message processing logs. In the next section, you will learn about the log level setup.

6.1.3 Log Level Setup

The message processing log's log level describes the level of detail at which the log collects information. You can select the log level for a specific integration flow scenario in the Web-based Monitor's Manage Integration Content tab. Choose your integration flow in the Manage Integration Content section, and then select the log level for this integration flow in the Log Configuration section.

You can select from the available log levels shown in Table 6-8.

Table 6-8. *Log Level*

Log Level	Description
None	When processing messages, no data is captured, and no data is displayed when data monitoring is enabled.
Info	When a message is processed, the basics are logged. The header is constantly visible.
Error	Only unsuccessful message executions are recorded in the message processing log. The latest 50 message steps' worth of information are stored and shown in great detail in the message processing log. If you have no interest in logging finished messages, use this log level. Attachments to the message processing log are not kept.
Debug	The message processing log keeps thorough records of all the message processing procedures. The Debug expiry time is set for 25 hours.
Trace	The message processing log keeps track of the message content as well as every stage of the message processing process in great detail. After a given amount of time, the trace function becomes inactive (default value: 10 minutes). The log level returns to the initial setting after expiry. Additionally, the substance of recorded messages is only kept for so long (default value: 1 hour).

In Monitor Message Processing, you have successfully learned about the message logs, message processing logs, and the log level setup. In the next section, you learn about the management integration content and look at the different tiles associated with it.

6.2 Managing Integration Content

The Manage Integration Content section gives a summary of the integration content artifacts, such as integration flows, that have been deployed on the tenant.

The amount of deployed integration content artifacts for a specific kind (such as integration flow) and with a specific state is shown in a collection of tiles in the management integration content section (for example, Started). Each tile can be set up to filter according to a distinct artifact category, state, or even both. Place the cursor on a tile, then select Edit from the context menu to do this.

Clicking a tile will launch the integrated content monitor for that group of artifacts (as determined by the tile filter), as shown in Figure 6-6.

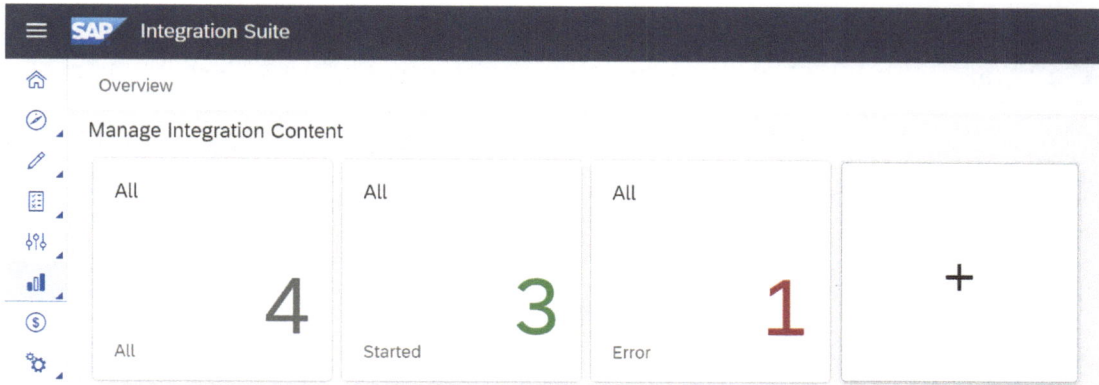

Figure 6-6. *The Manage Integration Content tile*

You can customize the Manage Integration Content by clicking the (+) on the right side, as shown in Figure 6-7. You can filter according to the Status and All Artifacts, which you will learn about in this section.

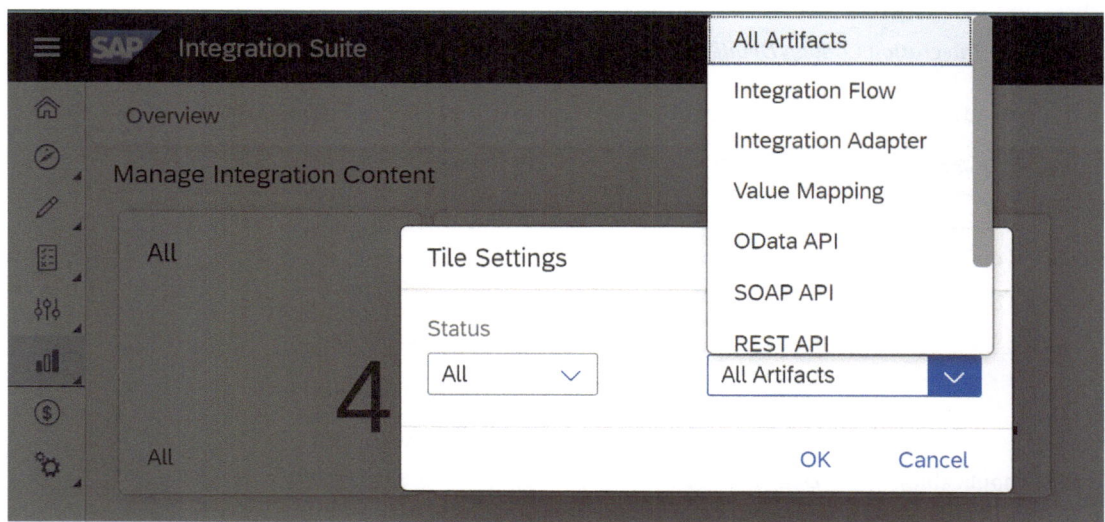

Figure 6-7. *Customize the Manage Integration Content tile*

The Management Integration Content contains the different tiles that you learn about in the following sections.

6.2.1 Integration Content Detail

Integration Content Details provides detailed information about integration content artifacts, such as integration flows, message mappings, and adapters.

You can easily check out the details available in the header box in Figure 6-8.

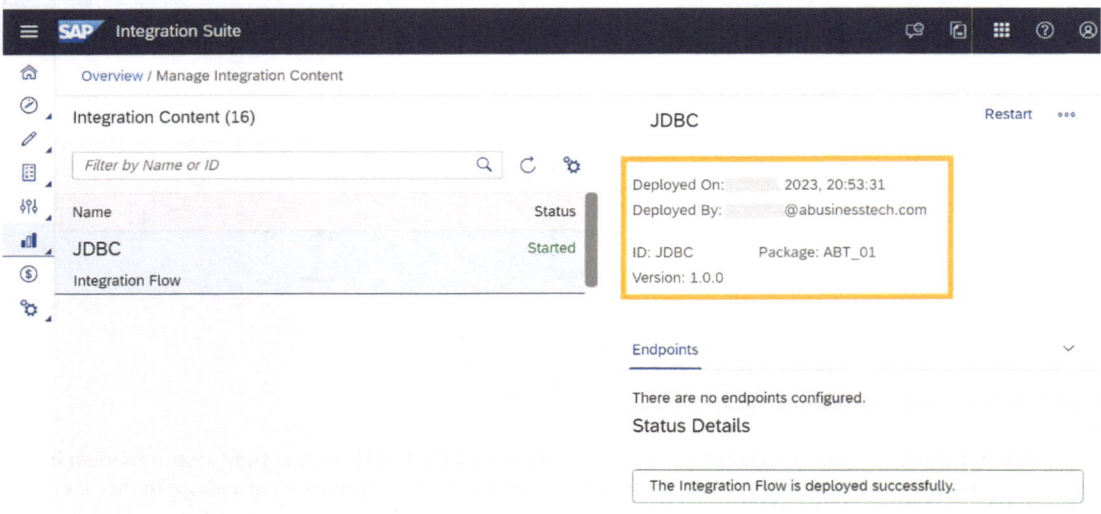

Figure 6-8. *Integration Content Detail*

The header box contains details regarding the chosen artifact, as listed in Table 6-9.

Table 6-9. *Information Content Detail*

Information	Description
Deployed on	Date of the artifact's deployment.
Deployed by	The person who used the item.
Version	For instance, the integration flow version of the item.
ID	Name of the package from which the item was deployed.
Package	If the material is configure-only indicated.
Mode (if applicable)	Date of the artifact's deployment.

Additionally, the header area offers the features (the features offered vary depending on the kind of selected artifact) listed in Table 6-10.

Table 6-10. *Integration Content Details*

Information	Description
Restart	Offers the option to restart an artifact.
Undeploy	Gives you the option to undeploy a current integration flow.
Download	Enables the item to be downloaded to your computer.
Endpoints	Displays for a transmitting application the URLs of the services exposed by the artifact. By selecting the Copy option, you can copy the URL.
Status Details	Displays the artifact's state in relation to its use in runtime (for example, if it is starting). Additionally, displays the polling Information for all active message polling adapters, including SFTP and mail adapters. The view offers the pertinent data regarding: • The adapter details. • The exact time and date when the most recent survey was conducted. • Whether or not further polls are scheduled (Additional polls scheduled/No additional polls scheduled). • Whether the poll was successful or not. • A thorough error message if one is provided.
Log Configuration	The log level is set to info when you first deploy an artifact.

Artifact Details

Figure 6-9 shows the integration flow artifact details.

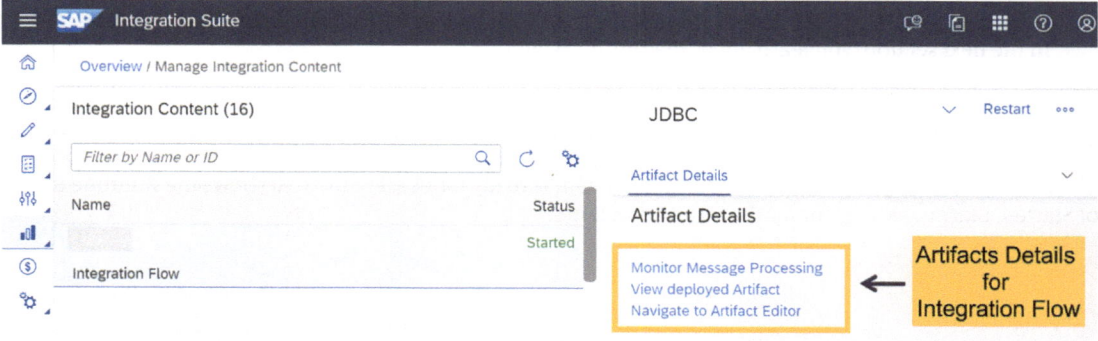

Figure 6-9. *Artifact details*

Information is provided based on the type of item that has been chosen, as listed in Table 6-11.

Table 6-11. *Artifact Details*

Artifact Type	You Can...
Integration flows	Look at the Monitor Message Processing process.
	Look at the deployed artifact.
	Open the Artifact Editor.
Value mapping	Look at the deployed artifact.
	Open the Artifact Editor.
SOAP API	Look at the Monitor Message Processing process.
	Look at the deployed artifact.
	Open the Artifact Editor.
REST API	Reach the Monitor Message Processing process.
	Watch the deployed artifact.
	Access the Artifact Editor.
OData API	Look at the Monitor Message Processing process.
	Open the Artifact Editor.
Integration Adapter	Access the vendor.
Script Collection	Look at the deployed artifact.
	Open the Artifact Editor.
Message Mapping	Look at the deployed artifact.
	Access the Artifact Editor.

In the next section, you learn about the runtime status.

6.2.2 Runtime Status

The runtime status specifies whether a deployed item is ready for use. Figure 6-10 shows the Runtime Status of Started, Starting, and Error for the integration flows.

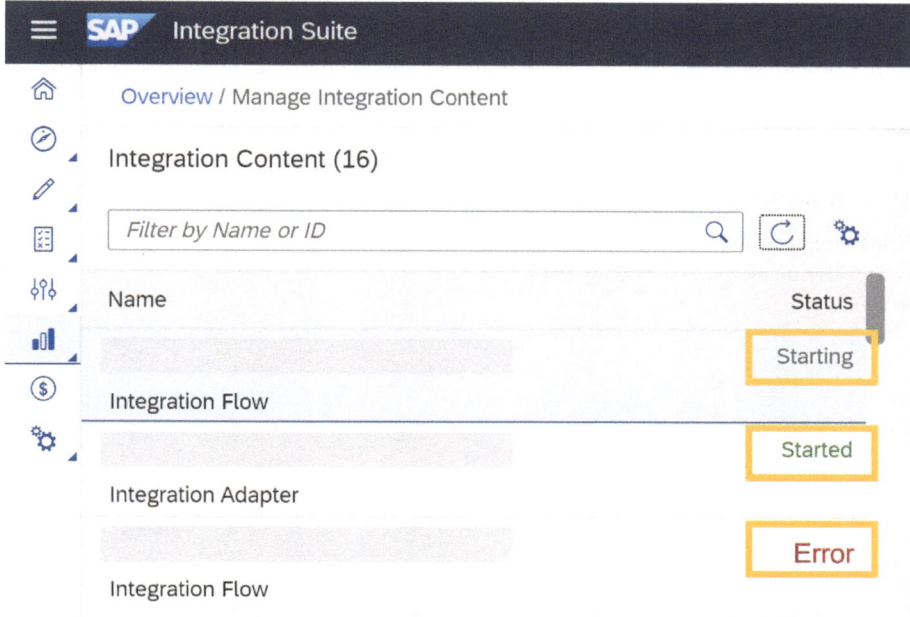

Figure 6-10. Runtime status

The status values listed in Table 6-12 are all conceivable.

Table 6-12. Runtime Status Values

Status	Description
Started	The artifact is prepared for use.
	The state of an artifact with this classification is complete.
Error	The user must pay attention to the artifact (administrator).
	A piece of art with this status has reached its completion.
Starting	The artifact is starting to be deployed on the worker node(s) if it has already been deployed.
	This condition designates an artifact that is in a transitional state.
Stopping	On the worker node(s), the artifact is currently being halted, and if stopped, it will soon be undeployed.
	This condition designates an artifact that is in a transitional state.
	The artifact's deployment status is stopping now. This condition prevents an artifact from being downloaded, undeployed, or restarted.

In the next section, you learn the iflow endpoint view.

6.2.3 iflow Endpoint View

The Manage Integration Content section contains a visual representation of the adapter endpoints. The endpoint definitions of some adapters are extracted with the most recent release of the integration service.

To reach the endpoints, follow these steps:

1. Suppose you have created the integration flow with the inbound communication of HTTPS connecting the Sender and Start message.

2. In the Processing tab of HTTPS, you provide the address. This address is the last part of the endpoint, as shown in Figure 6-11.

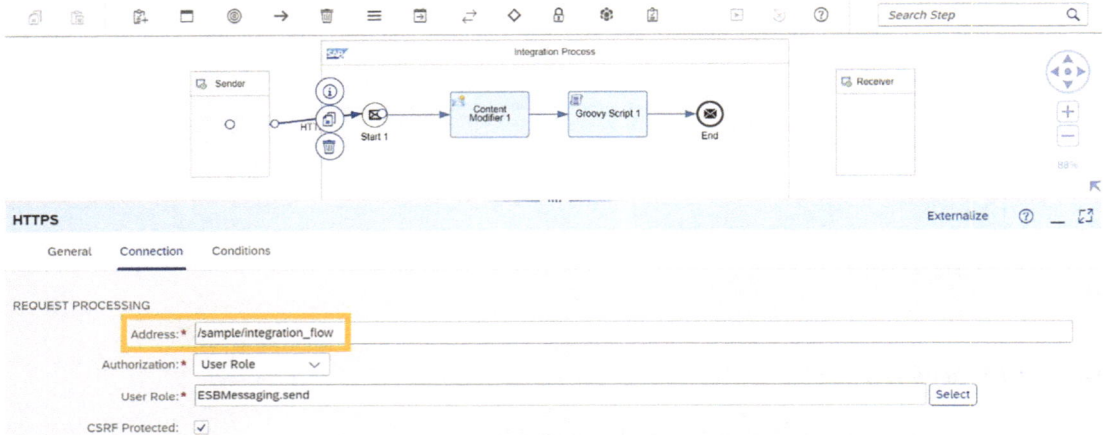

Figure 6-11. *Configuration HTTPS*

3. You have saved and deployed the flow successfully.

4. To obtain the endpoints, navigate to the Monitor screen of the Cloud Integration and open the Message Integration Content tile.

5. Pick an integration flow from the list by going to Manage Integration Content, then select Endpoints in the detailed view, as shown in Figure 6-12.

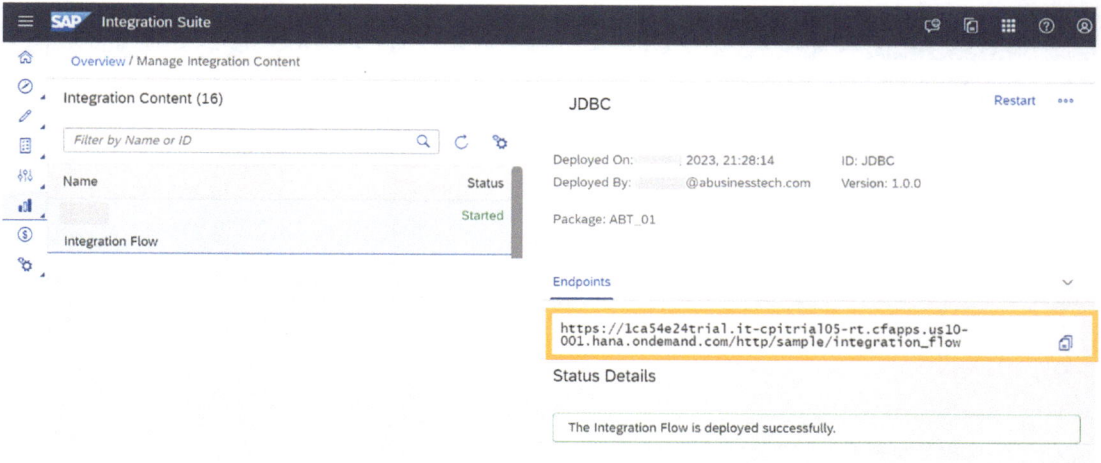

Figure 6-12. *Endpoint of an iflow*

You learned about the different attributes related to the management integration content. In the management integration content area, you can monitor the messages filtered based on Started, Error, and All. You can also add more tiles there and customize the Manage Integration Content tile according to the user.

In Chapters 3 and 4, you learned about the maintenance of user credentials and keystores. The Manage Security, SAP Cloud Integration area allows you to manage all-security material, keystore, data stores, and so on. Next, you look briefly into the Manage Security material and see the different tiles associated with it.

6.3 Managing Security

You can manage different types of security content (such user credentials and keystore entries) and run outbound connectivity checks using the Manage Security module, as shown in Figure 6-13.

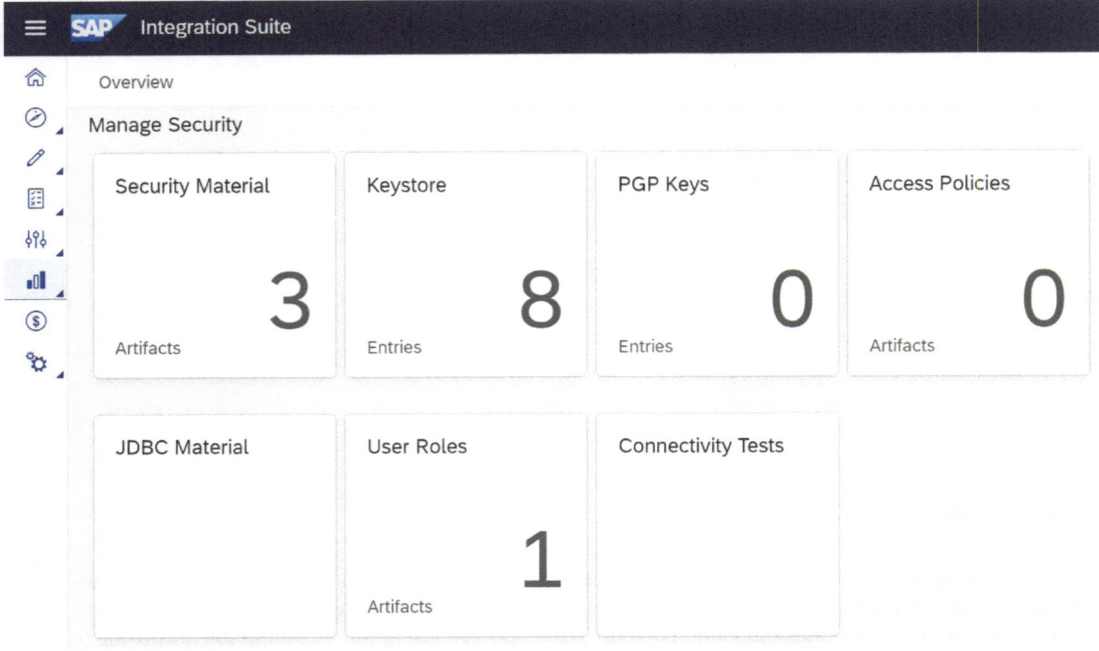

Figure 6-13. *Manage Security tile*

The Manage Security Material area contains the different tiles, and the following sections look briefly at each tile and explain the operations associated with them.

6.3.1 Create Security Material

An overview of security-related artifacts can be found in the Manage Security Material section.

The following steps enable you to access the Manage Security Material section:

1. Navigate to the Security Material tile.

2. Security artifacts are shown together with the tile's state in each case.

Figure 6-14 shows the Security Material details listed in Table 6-13.

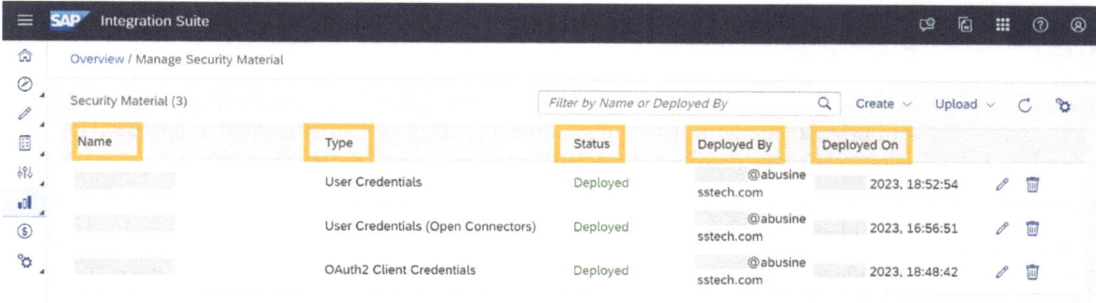

Figure 6-14. *Security material*

Table 6-13 is a list of the security materials.

Table 6-13. *Manage Security Material Home Page Details*

Name	Display Name of the Artifact
Type	Possible values: • User Credentials • OAuth2 Credentials • OAuth2 Authorization Code • Secure Parameter • Known Hosts (SSH) • PGP Public Keyring • PGP Secret Keyring
Status	This status shows whether an artifact has been successfully deployed on a tenant or whether it still requires authorization. Possible values: • Stored • Deployed • Error • Unauthorized (for OAuth2 Authorization Code)
Deployed By	User who deployed the artifact.
Deployed On	Time when the artifact was deployed.

The Actions tab, which is available on the home page of Manage Security Material, is explained in Table 6-14.

Table 6-14. *Action Tab Details*

Action	Description
Create	User credentials, OAuth2 client credentials, OAuth2 SAML bearer assertion, OAuth2 authorization code, and secure parameter artifacts can all be created or deployed by selecting Create and the relevant artifact type.
Authorize	Only OAuth2 authorization codes are supported; therefore, choose the existing artifact in the table and click Authorize.
Edit	For changes and redeployment support only, select the existing artifact in the table and choose Edit.
Download	Select the item in the table and choose Download to download an artifact (only supported for Known Hosts).
Delete	An artifact can also be eliminated (supported for all artifact types except Keystore).

The Create Security Material area contains functions in which you can perform various tasks. You learn briefly about those functions in the following sections.

Create User Security Material Artifact

To establish a connection using username token authentication or basic authentication, for instance, username and password authentication, you must provide the relevant characteristics. Figure 6-15 shows the Create User Credentials in managing security.

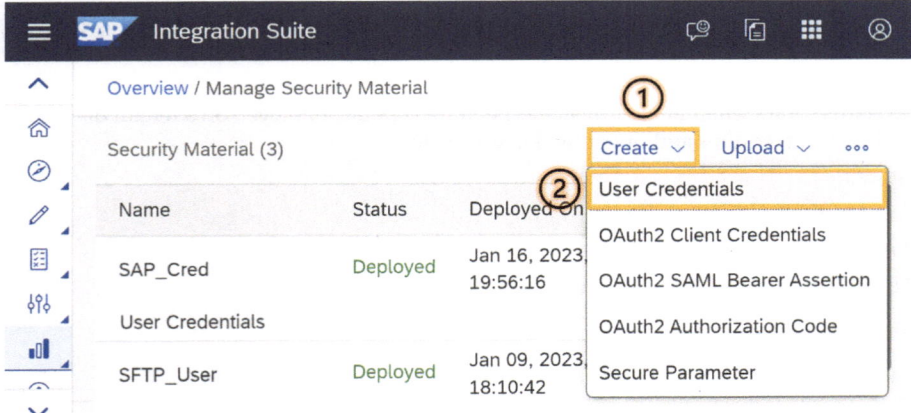

Figure 6-15. *Creating user credentials in the Manage Security Material area*

Procedure:

1. Choose Security Material by clicking the tile under Manage Security Material.

2. Choose Create ➤ User Credentials.

3. Specify the properties listed in Table 6-15 and then click Deploy, as shown in Figure 6-16.

Table 6-15. *Create User Credential Details*

Option	Description
Name	Name of the artifact.
Description	Description of the artifact.
Type	With this, you can configure a specific system to connect to your integration flow artifact.
	If that you choose SuccessFactors, enter the Business ID.
	Determine the Organization and Element for Open Connections
User	The user who calls the receiver system.
Password	Password used to verify the user.

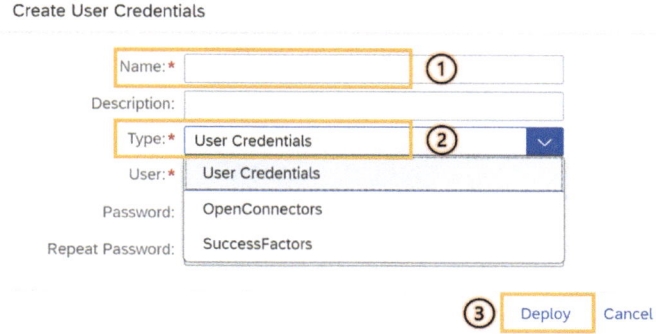

Figure 6-16. *The Create User Credentials area*

You have now created the User Security Material Artifact in the Security Material area. Next, you see4 how to create the OAuth2 Client Credentials Artifacts in the next section.

OAuth2 Client Credentials Artifact

OAuth 2.0 is widely used by web servers for authorization. You must deploy an OAuth2 Credentials artifact using the following steps if you want to connect to a system that employs OAuth 2.0 authentication.

Procedure:

1. Choose Security Material by clicking the tile under Manage Security Material.

2. Choose Create ➤ OAuth2 Client Credentials.

3. Specify the properties listed in Table 6-16. The fields with red asterisks are the mandatory fields, and the rest are the optional fields, as shown in Figure 6-17.

Table 6-16. *Create OAuth2 Credential Details*

Attribute	Description
Name	Name of the credential
Token Service URL	URL of the access token issuing the OAuth2 authorization server
Client ID	The client's ID to which you are connected
Client Secret	The client's secret key to which you are connected

Create OAuth2 Credentials

Name: *	ABT_OAuth
Grant Type:	Client Credentials
Description:	
Token Service URL: *	https://1ca54e24trial.authentication.us10.hana.ond...
Client ID: *	sb-1c7ba24f-0693-4118-a4d5-41cd2926f3ca!b139...
Client Secret: *	•••...
Client Authentication: *	Send as Request Header
Scope:	
Content Type:	application/json
Resource:	
Audience:	

Deploy   Cancel

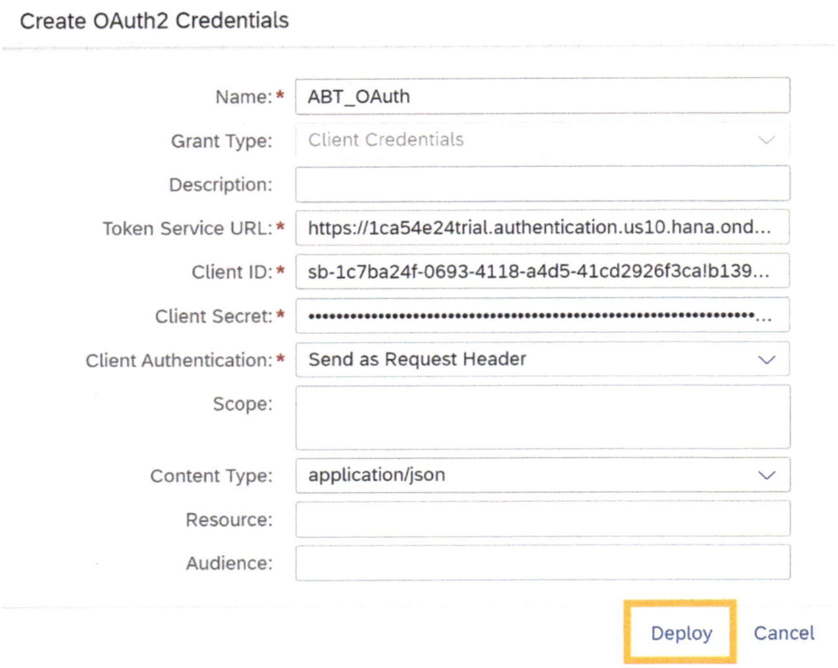

Figure 6-17. *The Create OAuth2 Credentials area*

The client ID and client secret can be found in the SAP BTP Cockpit. You have created the Service Instance along with the Secret key (see Chapter 2, Section 2.9). This service key contains the Client ID, Client Secret, and Token URL.

Open the Service Key, and you will find the details shown in Figure 6-18.

Credentials

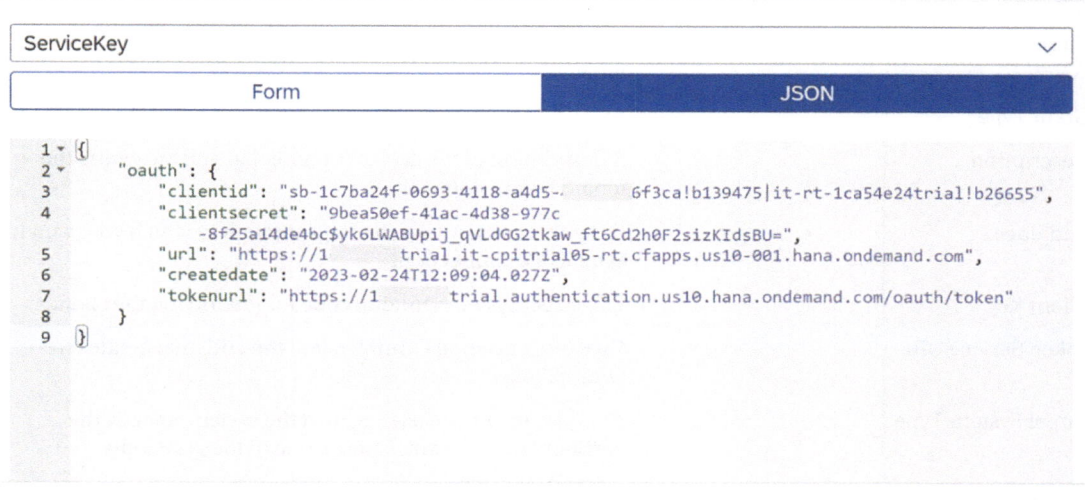

Figure 6-18. Credentials

4.   Select Deploy.

You have now created the OAuth2 Client Credentials in the Security Material area. Next, you learn how to create the OAuth2 SAML Bearer Assertion.

OAuth2 SAML Bearer Assertion

OAuth 2.0 is widely used by web servers for authorization. You must deploy an OAuth2 Credentials artifact using the following steps if you want to connect to a system that employs OAuth 2.0 authentication.

Procedure:

1.   Choose Security Material by clicking the tile under Manage Security Material.

2.   Choose Create ➤ OAuth2 SAML Bearer Assertion.

3.   Specify the properties in Table 6-17. The fields with red asterisks are the mandatory fields, and the rest are the optional fields, as shown in Figure 6-19.

Table 6-17. *Create OAuth2 SAML Bearer Assertion Details*

Attribute	Description
Name	Give the artifact its own name before installing it on the tenant.
Grant Type	An immutable `OAuth2SAMLBearerAssertion` is the grant type.
Description	A description of the artifact's name that you are giving the tenant before deployment.
Audience	Enter the hostname of the target system to which you want to establish a connection.
Client Key	The target system's unique code for identifying the client.
Token Service URL	Give the registered OAuth2 client the URL that creates the OAuth2 token.
Target System Type	To authenticate the user against the system, specify the pertinent host system. Choose one of these systems: • SuccessFactors • SAP BTP (Neo) • SAP BTP (CF)
Company ID (SuccessFactors)	Specify the company ID of your SuccessFactors instance.
User ID (Only if SuccessFactors is chosen as the target system type)	Choose one of the following: • Principal Propagation • Key Pair Common Name (CN)
Additional Properties (Only if you choose Principal Propagation for the user ID and SuccessFactors as the destination system type)	Keep the following value for each environment's userId Source key field:
Key Pair Alias (Only if you choose Key Pair Common Name (CN) for the user ID and SuccessFactors for the destination system type)	Provide the name of the alias that you specified in the requirements.
Token Service User (Only if the target system type is SAP BTP)	To access the URL for the token service, you need a username.
Token Service Password (Only if you choose SAP BTP as the destination system type)	You need a password to access the URL for the token service.

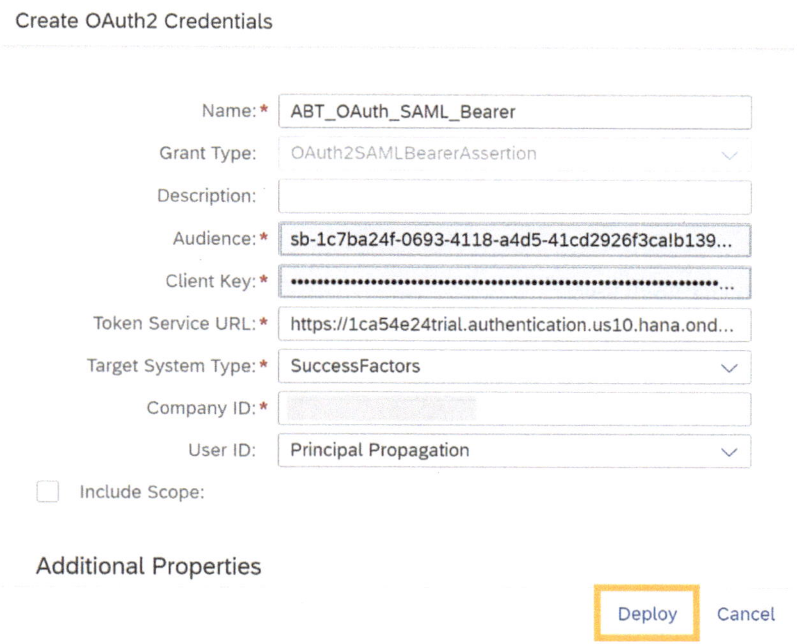

Create OAuth2 Credentials

Name: *	ABT_OAuth_SAML_Bearer
Grant Type:	OAuth2SAMLBearerAssertion ⌄
Description:	
Audience: *	sb-1c7ba24f-0693-4118-a4d5-41cd2926f3calb139...
Client Key: *	•••...
Token Service URL: *	https://1ca54e24trial.authentication.us10.hana.ond...
Target System Type: *	SuccessFactors ⌄
Company ID: *	
User ID:	Principal Propagation ⌄

☐ Include Scope:

Additional Properties

Deploy   Cancel

Figure 6-19. *Create OAuth2 SAML Assertion Credentials*

4.   Select Deploy.

You have seen how to create the OAuth2 SAML Bearer Assertion in the Security Material area. In the next section, you learn how to create the OAuth2 Authorization Code.

OAuth2 Authorization Code

Several web servers can use OAuth 2.0 to request permissions. A token service is used as a middle step to authenticate you if you utilize an OAuth2 Authorization Code. The client can obtain an access token by using the OAuth2 Authorization Code. Your login details are never disclosed to the client.

Procedure:

1.   Choose Security Material by clicking the tile under Manage Security Material.

2.   Choose Create ➤ OAuth2 Authorization Code.

3.   The provider should be selected as Generic.

4.   Specify the properties shown in Table 6-18. The fields with red asterisks are the mandatory fields, and the rest are the optional fields, as shown in Figure 6-20.

Table 6-18. *Create OAuth2 Authorization Code Details When the Provider Is Set to Generic*

Attribute	Description
Name	Name of the artifact you want to utilize with the tenant.
Provider	In the space provided, type the platform provider name for the OAuth2 client.
Authorization URL	To give the OAuth client access to a user's resources, you must specify the Authorization URL.
Token Service URL	Provide the URL that generates the OAuth2 access and refresh tokens to the registered OAuth2 client and the given user.
Client ID	The ID of the client you want to connect to.
Client Secret	The client's private key, which you have access to.

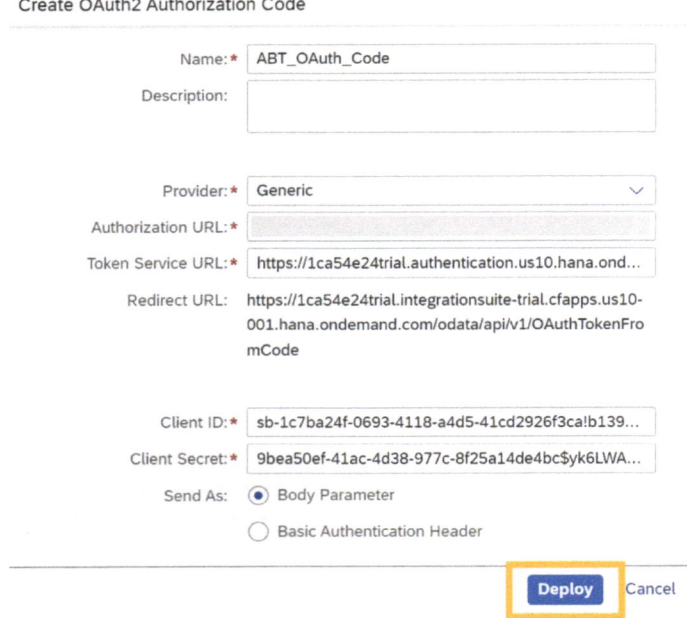

Figure 6-20. *Create an OAuth2 authorization code when the provider is set to Generic*

    5.   Select Deploy.

    6.   When the provider is selected as Microsoft 365, as shown in Figure 6-21, the details are specified in Table 6-19.

Create OAuth2 Authorization Code

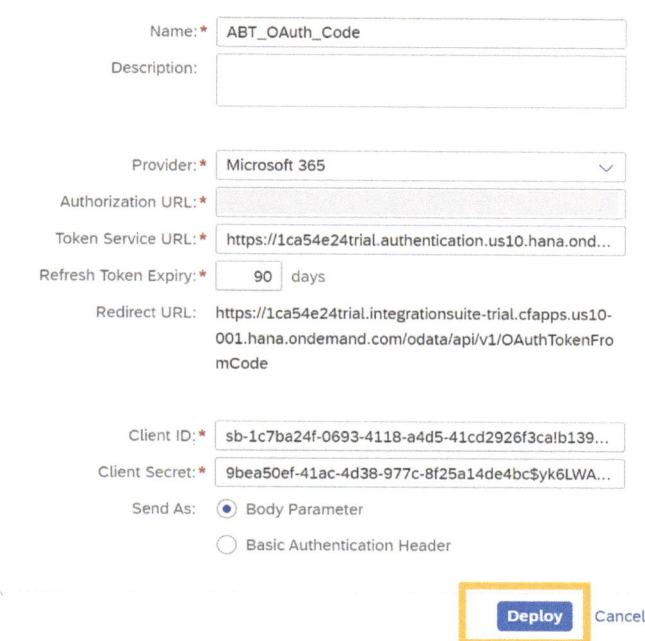

Name:*	ABT_OAuth_Code
Description:	
Provider:*	Microsoft 365 ⌄
Authorization URL:*	
Token Service URL:*	https://1ca54e24trial.authentication.us10.hana.ond...
Refresh Token Expiry:*	90 days
Redirect URL:	https://1ca54e24trial.integrationsuite-trial.cfapps.us10-001.hana.ondemand.com/odata/api/v1/OAuthTokenFromCode
Client ID:*	sb-1c7ba24f-0693-4118-a4d5-41cd2926f3ca!b139...
Client Secret:*	9bea50ef-41ac-4d38-977c-8f25a14de4bc$yk6LWA...
Send As:	● Body Parameter ○ Basic Authentication Header

Deploy Cancel

Figure 6-21. *The CreateOAuth2 authorization code when the provider is Microsoft 365*

Table 6-19. *The Create OAuth2 Authorization Code Details When the Provider Is Microsoft 365*

Attribute	Description
Name	The artifact's name that you want to use with the tenant.
Provider	Enter the name of the OAuth2 client's platform's provider in the box provided.
Authorization URL	Specify the authorization URL to grant the OAuth client access to the user's resources.
Token Service URL	Give the registered OAuth2 client and the supplied user the URL that creates the OAuth2 access and refresh tokens.
Refresh Token Expiry	Provide a deadline by which the refresh token must expire.
Client ID	The client's ID that you want to connect to.
Client Secret	The client's secret key that you are connected to.
Send As	• Body Parameter—When calling the Authorization URL or Token Service URL, send the Client ID and Client Secret in the request body. • Basic Auth Header—It is necessary to send the Client ID and Client Secret via Basic Authorization whenever a call to the Authorization URL or Token Service URL is performed.
User Name	The user's name whose resources the OAuth2 client can access.
Scope	OAuth2 scopes guard against unauthorized access to resources.

7. Select Deploy.

You have successfully created the OAuth2 Authorization Code in the Security Material area. Next, you learn how to create the Secure parameter in the Security Material area.

Secure Parameter Artifact

You use the Secure Parameter artifact to distribute private information, such as for custom adapters.

Procedure:

1. Choose Security Material by clicking the tile under Manage Security Material.

2. Choose Create ➤ Secure Parameter.

3. Specify the properties listed in Table 6-20. The fields with red asterisks are the mandatory fields, and the rest are the optional fields, as shown in Figure 6-22.

Table 6-20. Secure Parameter Details

Attribute	Description
Name	The artifact's name that you want to use with the tenant.
Secure Parameter	Enter the attribute's secret value.
Repeat Secure Parameter	Specify the attribute's confidential value once more.

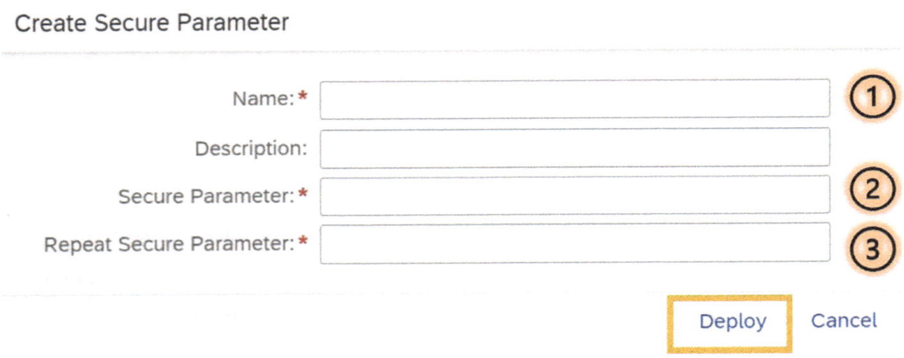

Figure 6-22. The Create Secure Parameter option

4. Select Deploy.

You have now created the Secure Parameter in the Security Material area. Next, you learn about the SSH Known Host Files.

Deploy SSH Known Host File

You can upload the artifacts in the Manage Security Material area. In the Upload section, you can only upload the SSH known host artifact file.

When setting up secure connectivity using the SSH file transfer protocol, this artifact type defines the known host file that is used (SFTP).

Procedure:

1. Choose Security Material by clicking the tile under Manage Security Material.

2. Select Upload Known Hosts (SSH).

3. Go to your computer's recognized hosts file, as shown in Figure 6-23.

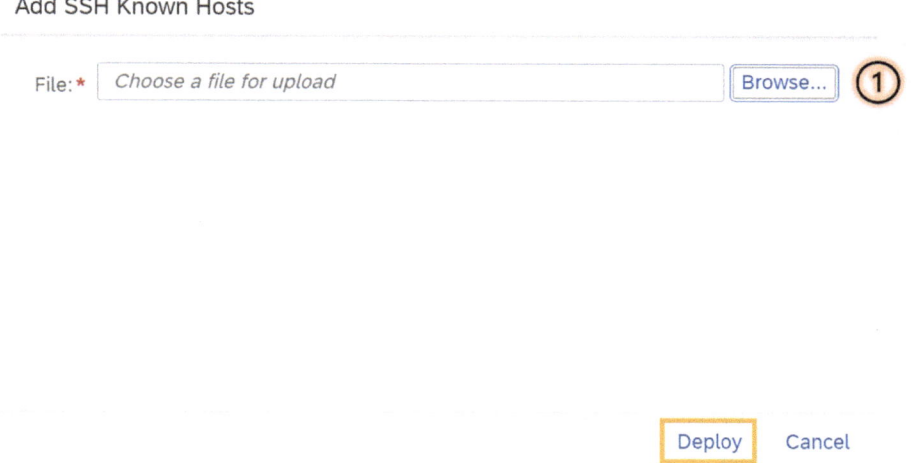

Figure 6-23. Adding an SSH known host file

4. Select Deploy.

In the Manage Security area, you learned how to create the security material and the different artifacts. In the next section, you learn about another tile that falls under Manage Security, Keystore Entries.

6.3.2 Manage Keystore Entries

A tenant administrator can manage the tenant keystore and its entries, including X.509 certificates and key pairs, using the Keystore Monitor.

Open the Monitor section of SAP Cloud Integration. Under Manage Security, choose the second tile of Keystore, and you will land in the home page, as shown in Figure 6-24. The various options to manage the keystore are listed in Table 6-21.

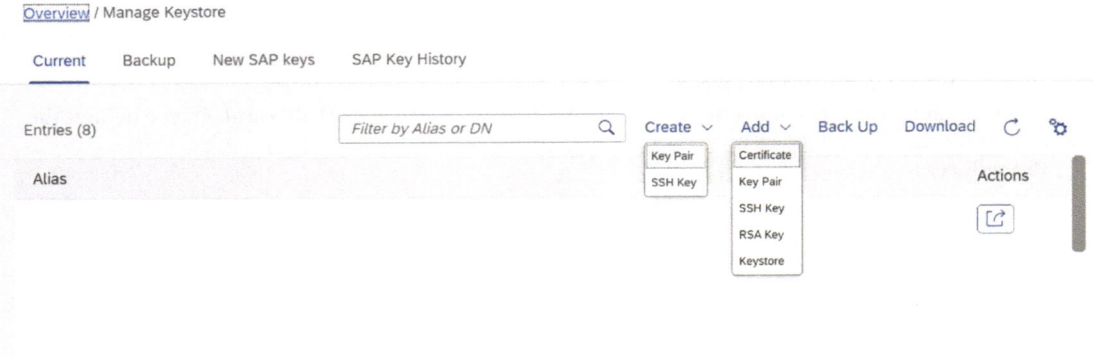

Figure 6-24. *The Manage Keystore area*

Table 6-21. *Manage Keystore Dashboard Details*

Available Options	Description
Create	Select Create and enter the artifact's details to create a key pair or an SSH key pair. Complete the required fields.
Add	Choose Add to add a keystore or upload an entry to the keystore.
	To an already-existing keystore, you can upload specific entries. You can either preserve the current entries or rewrite them in the latter scenario.
	A lock icon denotes keystore entries that belong to SAP. They are not editable or erasable. You can include the following things:
	• Certificate
	• Key Pair
	• SSH Key
	• RSA Key
	• Keystore
Download	Choose Download to download a keystore's public content or a specific keystore entry.
Back Up	Check out this article for information on backing up keystore entries.
Download Back Up	Choose Back Up to download backed-up keystore entries.
Delete	After choosing an artifact in the table, navigate to Monitor Handling Security Material and select Delete.
Restore	You can restore the value for future reference.
Update	You can update any entry made in the Manage Keystore area.

Table 6-21 lists the choices related to the Keystore Monitor.

You learned about managing keystore entities and the attributes and functions associated with them. In the next section, you learn about the different functions you can perform in Manage Keystore. You will learn about the creation of the keystore entries.

Create Keystore Entries

The keystore entry is a security feature in SAP Cloud Integration that enables you to securely store and manage digital certificates, keys, and trust configurations used for authentication, encryption, and decryption of messages and data. In the keystore, you can create the artifacts shown in Figure 6-25.

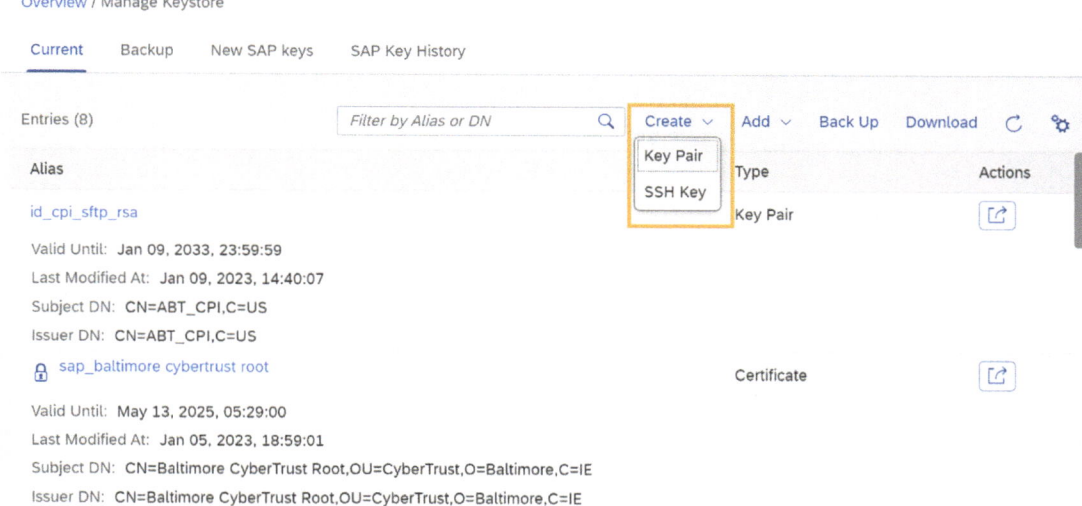

Figure 6-25. *Creating keystore entries*

Procedure:

1. Select Manage Security Keystore from the Operations view.

2. Depending on your requirements, choose Generate SSH Key or Create Key Pair from the Current tab.

3. Select the key pair or SSH key according to your needs, as shown in Figure 6-26.

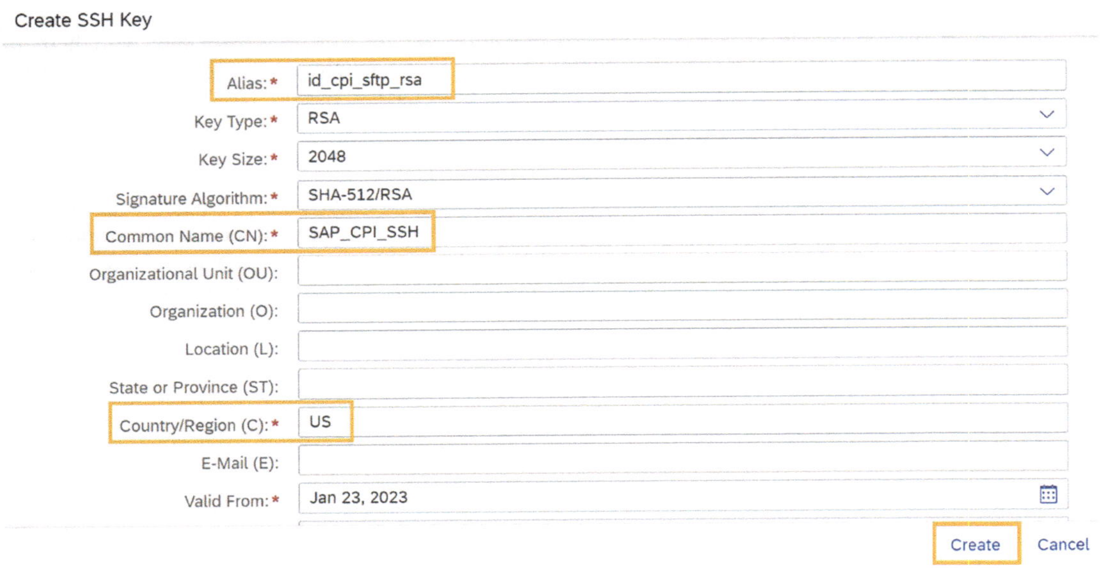

Create SSH Key

Figure 6-26. *Creating an SSH key*

4. Specify the fields shown in Table 6-22.

Table 6-22. *Create Keystore Entry Details*

Attribute	Details
Alias	A different name for the artifact you want to make.
Key Type	The available options are: • RSA • EC • DSA (only for key pair)
Key Size	Choose a key-bit value from the list of options.
Signature Algorithm	Choose a hashing algorithm from the list.
Common Name (CN)	Give the technical user a recognizable name. To deploy an OAuth2 credential, you use this common name.
Country/Region (C)	Type in the nation's two-letter ISO code.

5. Select Create.

You have successfully created the keystore entries in the Keystore. In the next section, you learn how to add keystore entries.

Add Keystore Entries

You can import the following artifacts into a keystore, as shown in Figure 6-27.

- Certificate
- Key Pair
- SSH Key
- KEYSTOREMANAGEMENT/RSA_KEY
- Keystore

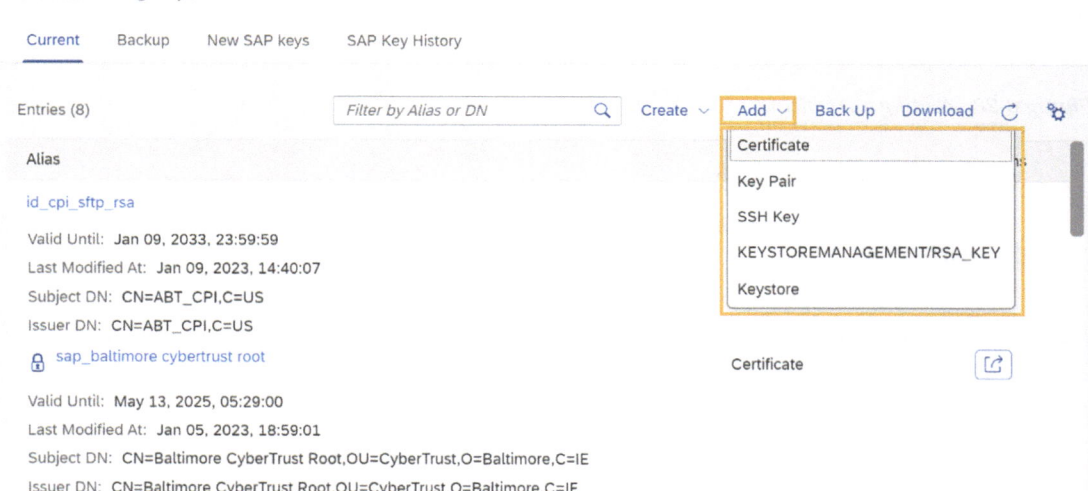

Figure 6-27. *Adding keystore entries*

You will now learn about each entry that can be added to the keystore. The following section starts with the certificates.

Certificate

In SAP Cloud Integration, a certificate is a digital document that verifies the identity of a user or system and is used for secure communication between different systems.

1. In the Keystore tab under Manage Security, select Add, then Certificate, as shown in Figure 6-28.

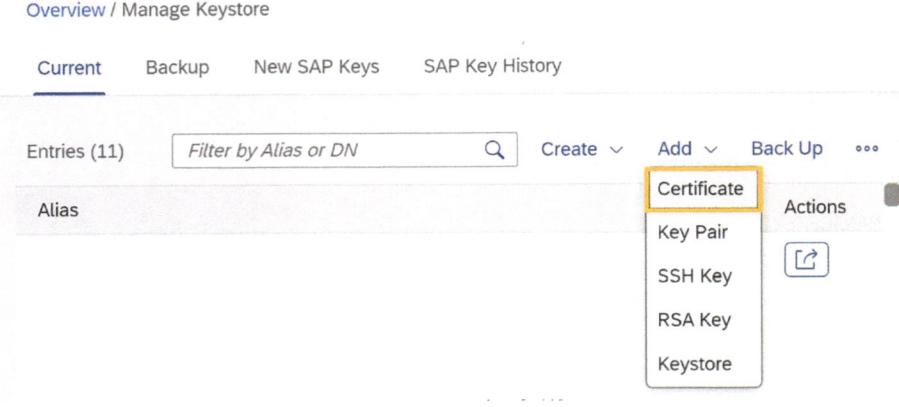

Figure 6-28. *Adding a certificate*

2. Add the alias (Alias field).

3. Select the certificate from your local drive by choosing Browse, as shown in Figure 6-29.

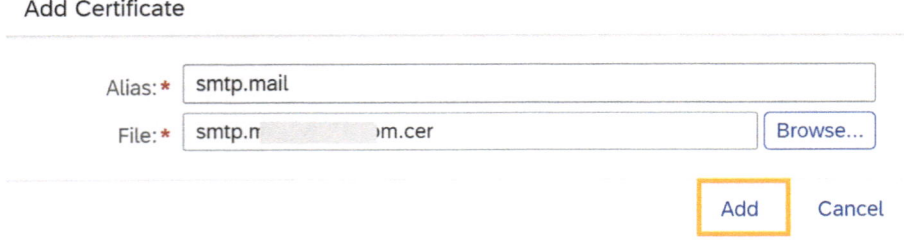

Figure 6-29. *Adding a certificate*

4. Click Add.

You have successfully added the certificate to the keystore. Next, you learn how to add the key pair to the keystore.

Key Pair

A key pair is a set of cryptographic keys used for secure communication between systems. A key pair consists of two related keys—a public key and a private key—that are mathematically linked. The private key is kept secret and is used to encrypt messages, while the public key is shared with other systems and is used to decrypt messages.

1. In the Keystore tab under Manage Security, select Add Key Pair from the Current tab, as shown in Figure 6-30.

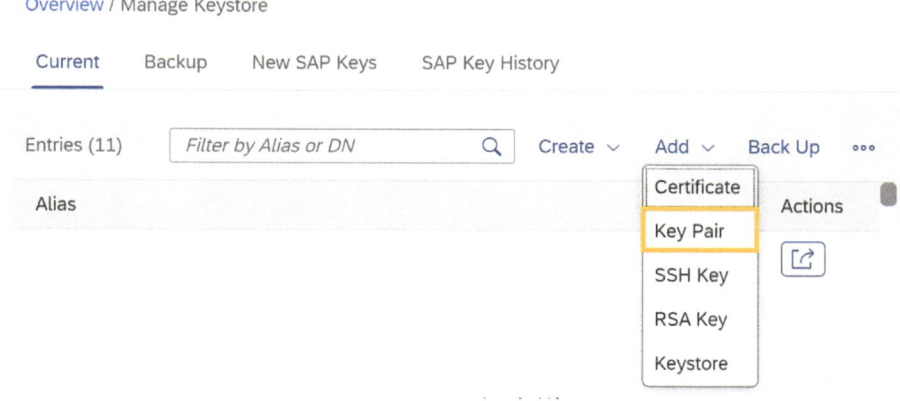

Figure 6-30. *Adding a key pair*

> 2. Provide the alias (Alias field). Go to your PC and choose the file. Include the password connected to the private key, as shown in Figure 6-31.

Figure 6-31. *Adding a key pair*

> 3. Select Add.

You have now added the key pair to the keystore. In the next section, you see how to add the keystore to the keystore entries.

Keystore

A secure repository known as a keystore is utilized by SAP Cloud Integration to store digital certificates and private keys for encryption and authentication.

> 1. In the Keystore tab under Manage Security, choose Add ➤ Keystore from the Current tab, as shown in Figure 6-32.

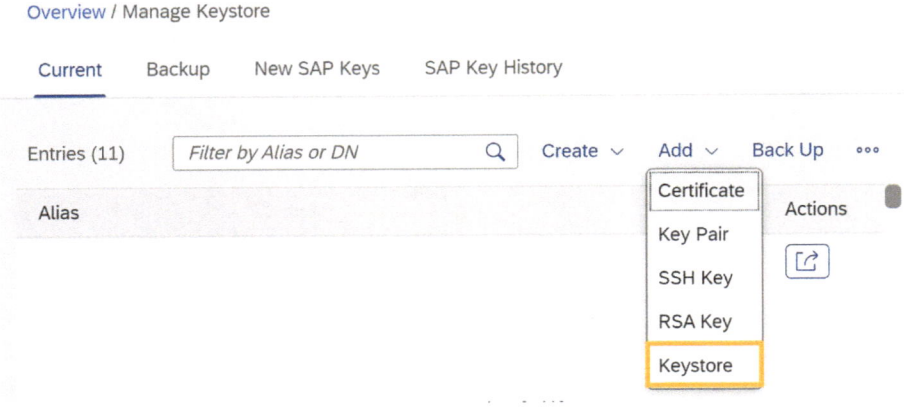

Figure 6-32. *Adding a keystore*

2. Choose Browse and pick the keystore on your local drive, as shown in Figure 6-33.

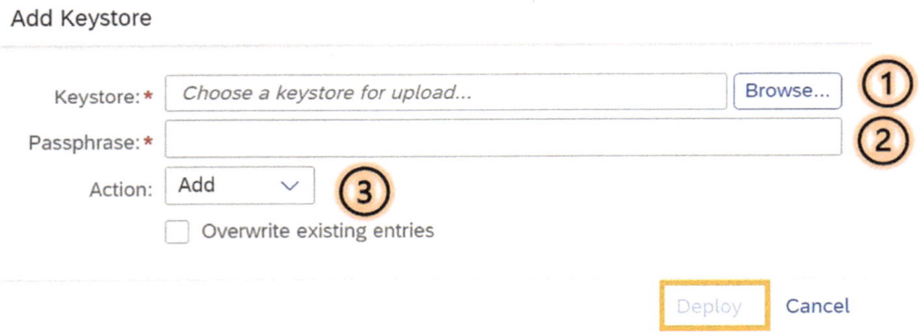

Add Keystore

Keystore:*	*Choose a keystore for upload...* Browse... ①
Passphrase:*	②
Action:	Add ▽ ③
	☐ Overwrite existing entries

Deploy   Cancel

Figure 6-33. *Adding the local keystore*

3. Add the password of the keystore.

4. In the Action field, select the Add or Replace option. Select Deploy.

You have successfully added a keystore to the keystore entries. In the next section, you learn about the additional keystore operations in the keystore entries.

Additional Keystore Operations

In Keystore, you can perform the various actions after creating a particular alias.
You can perform the following tasks, as shown in Figure 6-34:

- Rename

- Update

- Update Signing Response

- Update SSH Key

- Download Certificate

- Download Certificate Chain

- Download Public OpenSSH Key

- Download Root Certificate

- Download Signing Request

- Delete

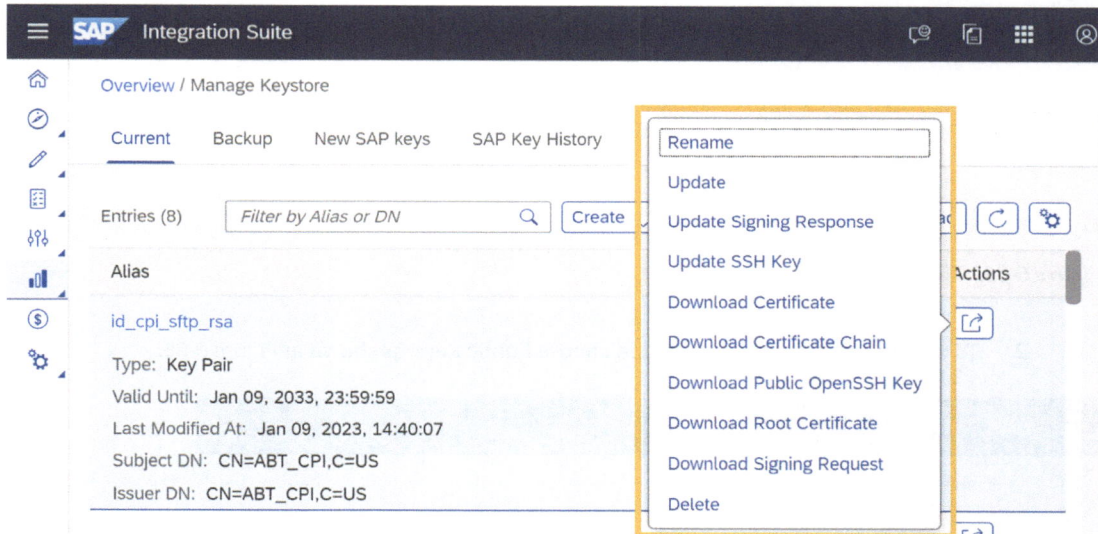

Figure 6-34. *Various actions performed with a particular alias*

You have briefly learned about the Manage KeyStore entries in the Manage Security area and found that you can upload, create, and add different security materials there. In the next section, you learn about another tile, called PGP Keys.

6.3.3 Manage PGP Keys

A tenant administrator can control the public and private PGP keys using the PGP Keys monitor.

When managing PGP keys, you can perform the various options in which you can deploy the PGP public and secret keyring. In the next section, you will learn about the PGP Public Keyring.

Deploy PGP Public Keyring

This artifact holds the tenant's public key and can be used by the tenant to encrypt or validate messages using the **Pretty Good Privacy (PGP)** standard.

1.   Select the PGP Keys tile, as shown in Figure 6-35.

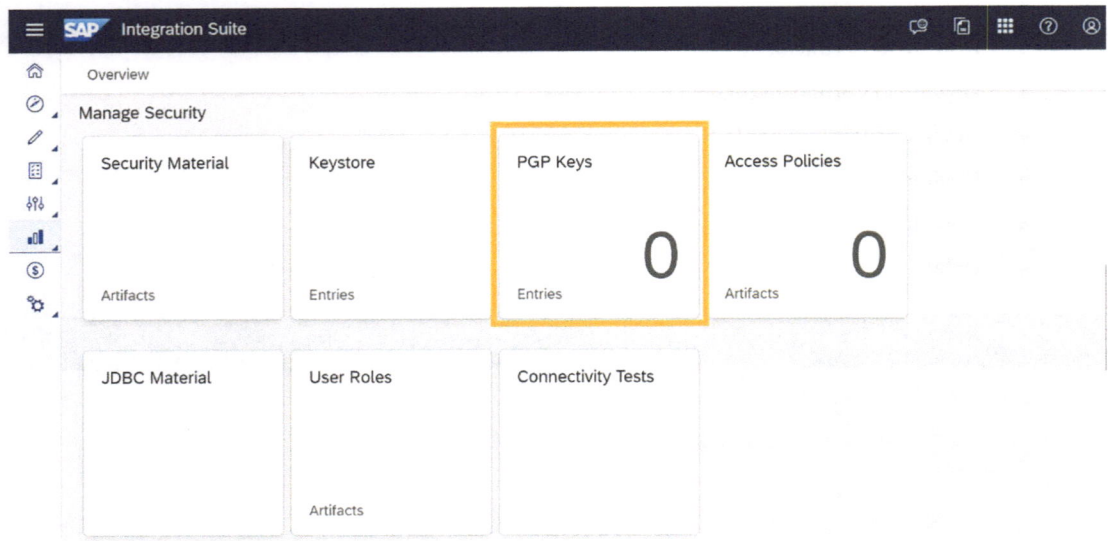

Figure 6-35. *The PGP Keys tile*

2.   To upload an artifact, go to Add and choose Public Keys, as shown in Figure 6-36.

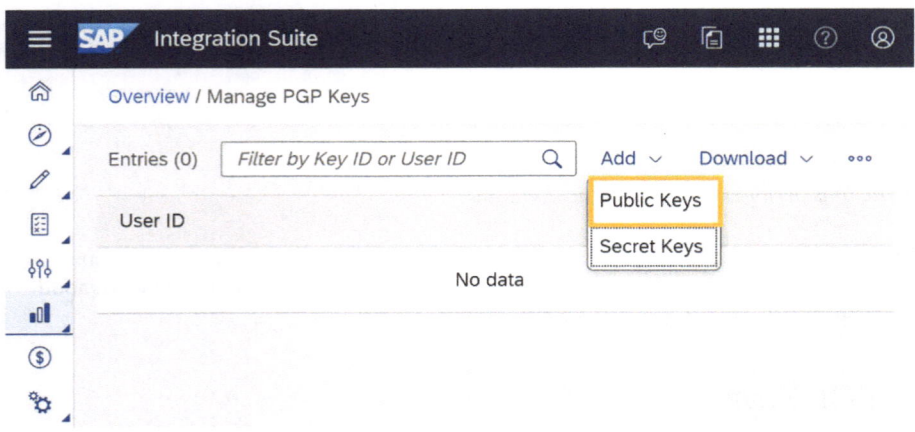

Figure 6-36. *Adding public keys*

3.   Upload the public key file from your local computer, as shown in Figure 6-37.

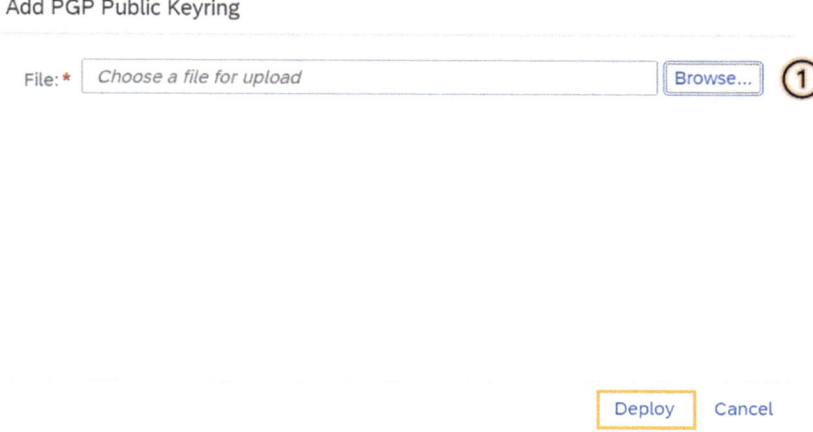

Figure 6-37. *Adding a PGP Public Keyring*

4.   To deploy your artifact, click Deploy.

You can easily download the public key by choosing Download ➤ Public Keys.

You have successfully deployed the PGP Public Keyring. In the next section, you see how to deploy the PGP Secret Keyring.

Deploy PGP Secret Keyring

Deploying a PGP Secret Keyring in SAP Cloud Integration allows you to securely manage and use PGP keys for encryption and decryption of data and messages in integration scenarios.

1.   Select the PGP Keys tile, as shown in Figure 6-38.

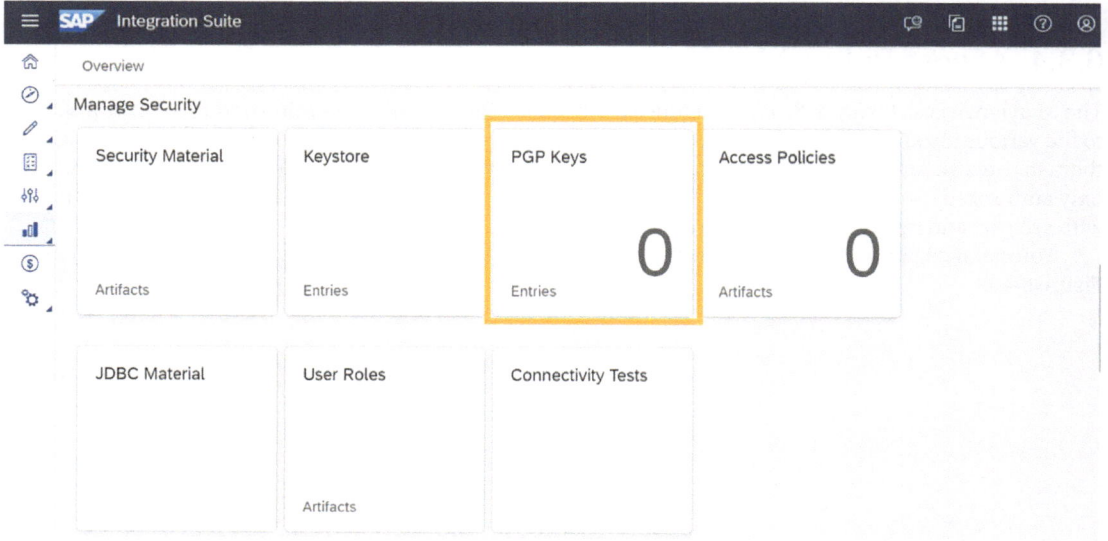

Figure 6-38. *The PGP Keys tile*

2. To upload an artifact, go to Add and choose Secret Keys, as shown in Figure 6-39.

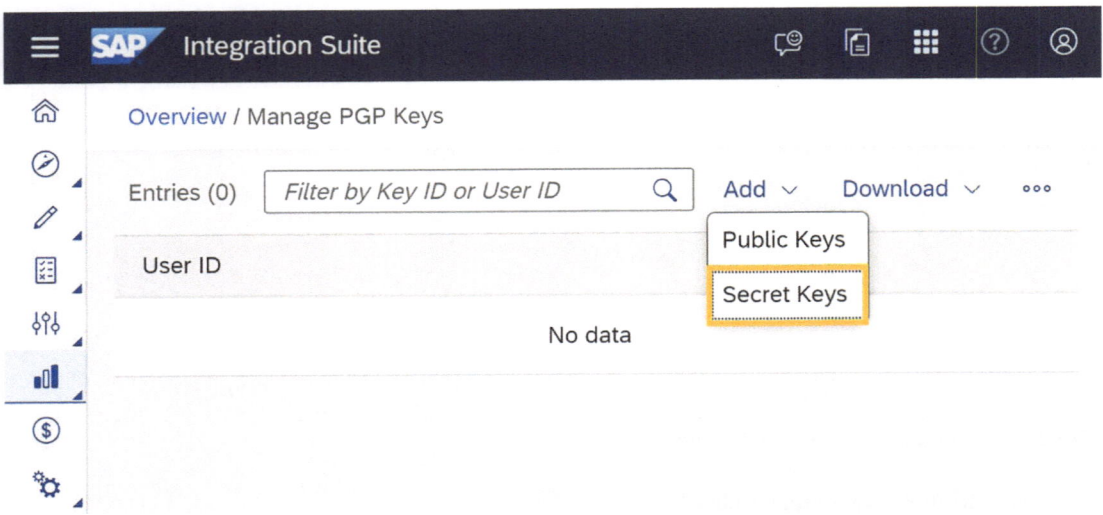

Figure 6-39. *Adding secret keys*

3. Enter the secret key passphrase and look for the secret keyring file on your computer.

4. To deploy your artifact, click Deploy.

You can easily download the secret keyring by choosing Download ➤ Secret Keyring.

You have successfully deployed and downloaded the secret keyring in PGP keys. You have learned about the different options available for performing the different operations in PGP keys. In the next section, you see the different tiles of management security, called access policy management.

6.3.4 Access Policy Management

The Manage Access Policy in SAP Cloud Integration is a feature that allows administrators to control access to the various resources and functionalities within the platform. It allows them to define roles and assign them to users and to specify which actions each role is allowed to perform. This feature is used to ensure that only authorized users can access sensitive data and perform specific actions and help maintain compliance with security and regulatory requirements.

You can navigate to the access policies by choosing Manage Security ➤ Access Policies, as shown in Figure 6-40.

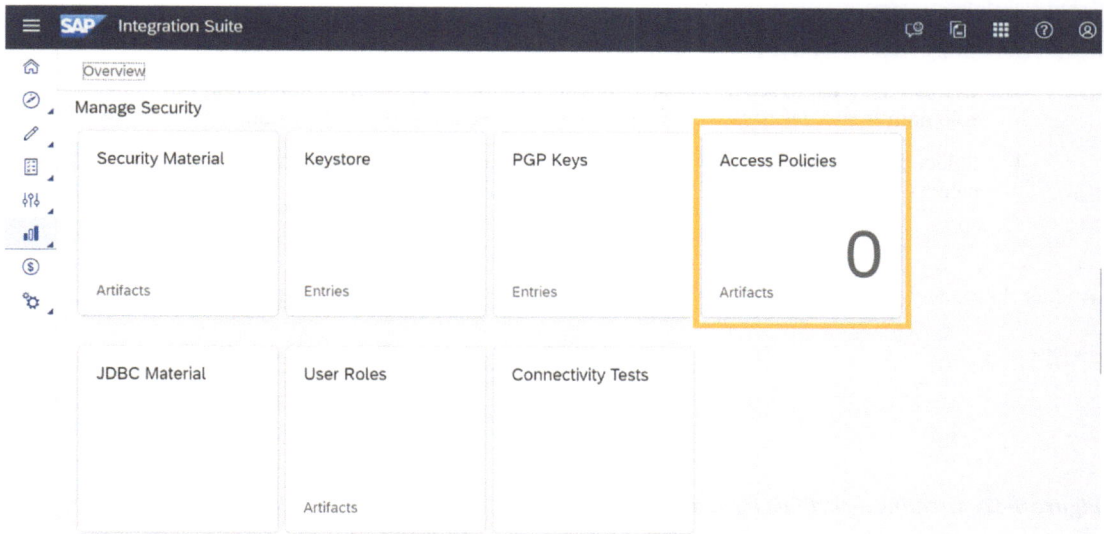

Figure 6-40. *Access policies*

In the next section, you learn how to create an access policy in SAP Cloud Integration.

Create an Access Policy

In SAP Cloud Integration, access policies are used to manage user access to integration artifacts and resources within the integration platform. Access policies define who can perform specific actions on integration artifacts, such as creating, reading, updating, or deleting them.

To create access policies in SAP Cloud Integration, follow these steps:

1.   Select the Access Policies tile under the Manage Security section. The current list of access policies is shown in Figure 6-41.

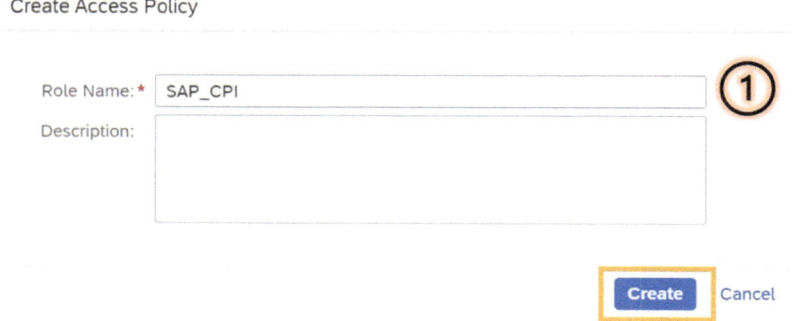

Figure 6-41. *Creating an access policy*

2. To create a new access policy, choose Create New Access Policy (+) and input the role name and description.

3. Refresh the page to see the most recent access policy. By running a search in the role names and descriptions, you can filter the list and sort it by role name.

4. Select the access policy you want to add artifact references to from the list, then select + (Create Artifact Reference), as shown in Figure 6-42.

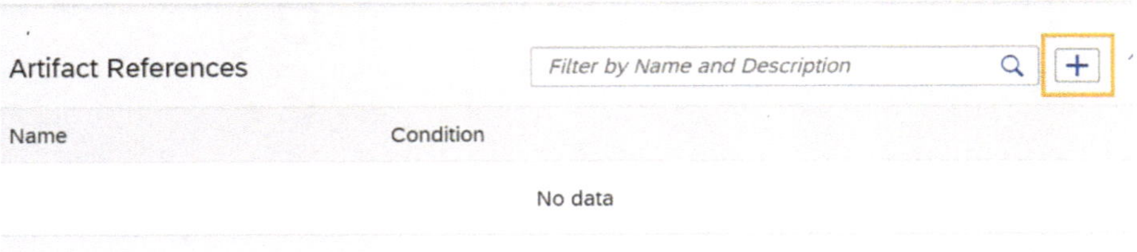

Figure 6-42. *Creating an artifact reference*

5. Provide the following mandatory details to create the artifact reference, as shown in Figure 6-43.

- **Name**—Enter name of the artifact.
- **Value**—Enter the ID or name of the integration artifact if you select equals.

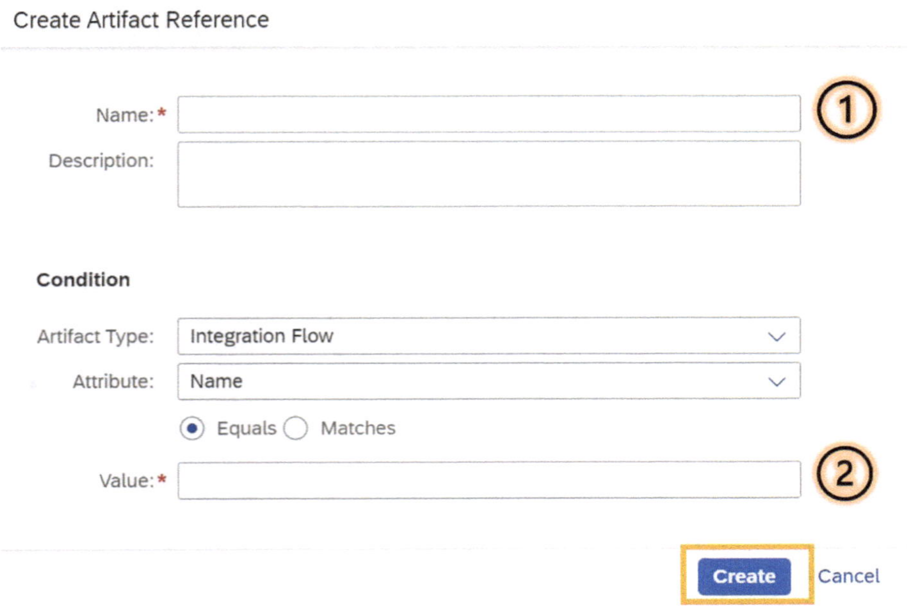

Figure 6-43. *Creating an artifact reference*

6. Click Create.

You have seen the different operations and learned about access policy management in the Management Security area. In the next section, you see another tile of Manage Security, called JDBC Material Management.

6.3.5 JDBC Material Management: Practical Example

Manage JDBC Material in SAP Cloud Integration is a feature that allows administrators to create and manage connections to JDBC (Java Database Connectivity) data sources. This feature allows users to define the connection settings, such as the JDBC driver, the connection URL, and the credentials needed to access the data source. Once JDBC material is created, it can be used in various integration flows to read and write data from the data source. This feature allows for the integration of on-premise data sources with the cloud, making it easy for customers to bring their existing data into their cloud integration scenarios.

Figure 6-44 shows the JDBC Material tile under Manage Security in SAP Cloud Integration.

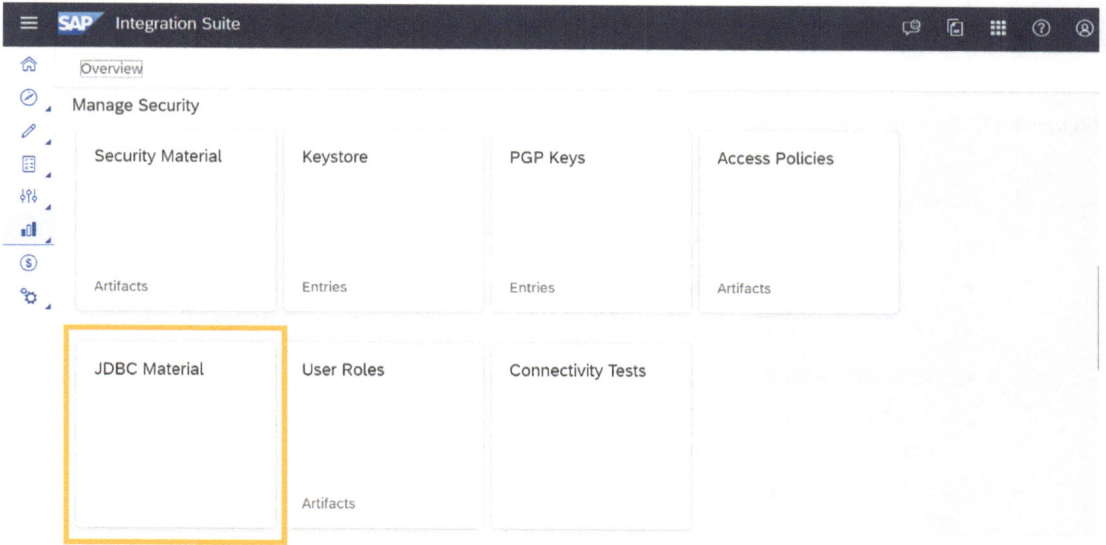

Figure 6-44. *The JDBC Material tile*

In the JDBC Material tile, you can perform many operations, which you learn about in the following sections.

Manage JDBC Data Sources

While interacting with a database, you can construct and maintain a cluster of artifact connections using JDBC Data Sources (DB). Each data source contains information on the kind of database it uses as well as the configuration options that are unique to that database.

1. Select the JDBC Material tile in the Monitor view, as shown in Figure 6-44.

2. Take one of the following actions (Add or Edit) after selecting the JDBC Data Sources tab, as shown in Figure 6-45.

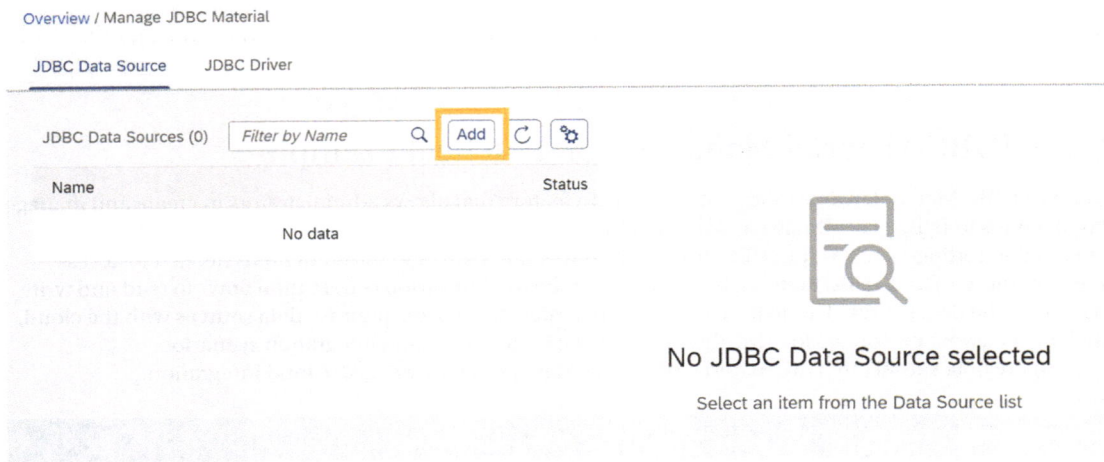

Figure 6-45. *Adding the JDBC data source*

3. Specify the following mandatory fields:

    ● Name—Define the name of the data source.

    ● Database Type—Choose a supported database type from the list.

    ● User—Type in the username for the target database.

    ● Password—Type in the password for the target database.

    ● JDBC URL—Provide the JDBC driver the database connection URL needs to connect to an on-premises or cloud database.

4. Select Deploy, as shown in Figure 6-46.

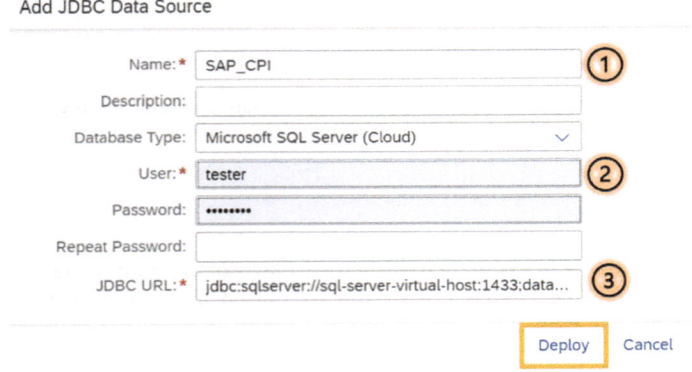

Figure 6-46. *Adding the JDBC data source*

JDBC Driver Configuration

In SAP Cloud Integration, "configuring JDBC drivers" refers to the process of setting up and registering the necessary driver files and configurations to connect to a JDBC data source. This process typically involves uploading the JDBC driver files to the platform and then providing the necessary information, such as the driver class name and the connection URL, to establish a successful connection.

Follow these steps to deploy the JDBC driver:

1. To upload a driver from your file system, select the JDBC Driver option and then choose Add, as shown in Figure 6-47.

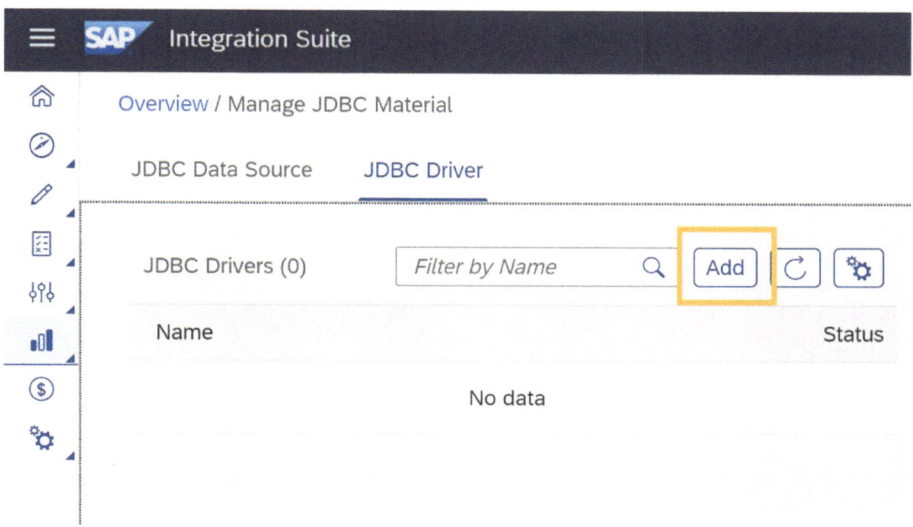

Figure 6-47. *Adding a JDBC driver*

2. Provide the following details:

   • Database Type—Pick one of the supported database types.

   • JDBC Driver File—Browse and choose the file from the system.

   This JDBC driver file is provided by the database vendor website, for example, Microsoft, Oracle, and so on.

3. Click Deploy, as shown in Figure 6-48.

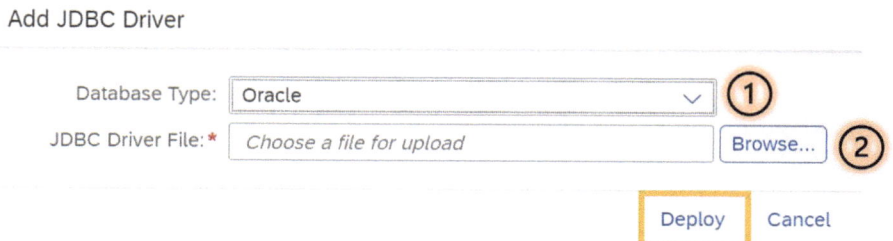

Figure 6-48. *Adding the JDBC driver*

You have briefly discussed JDBC Material and the different operations associated with it. In the next section, you see another tile of managing Security, called User Roles.

6.3.6 User Roles

A tenant administrator can manage user roles through the User Role Monitor, and these roles can then be used for inbound authorization of an integration flow.

Under the Monitor view's Manage Security section, select the User Roles tile. The tenant administrator's and SAP's configured user roles are shown in Figure 6-49.

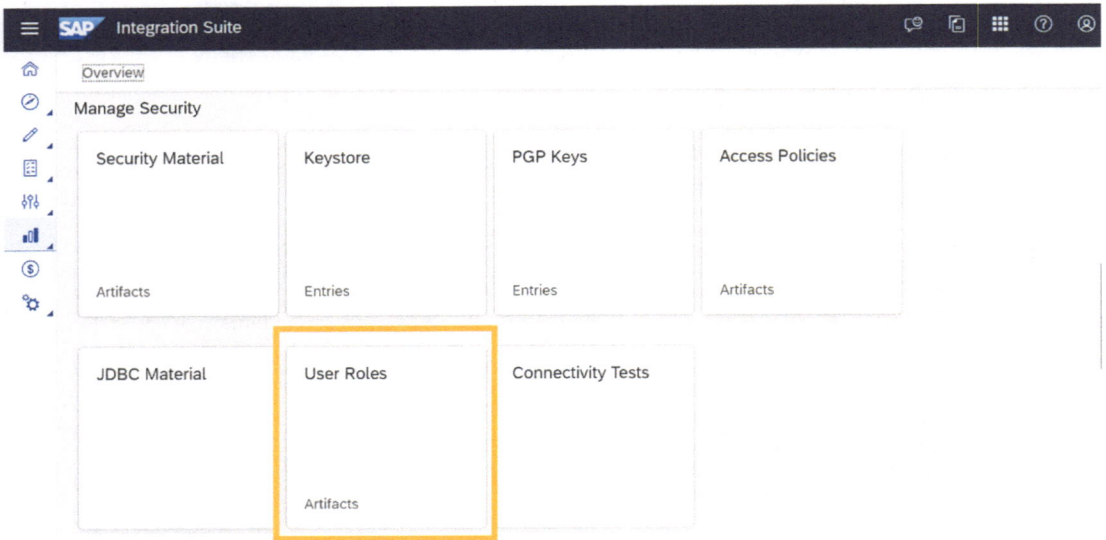

Figure 6-49. *The User Roles tile*

The actions in Figure 6-50 can be performed when entering the Users tile under Manage Security.

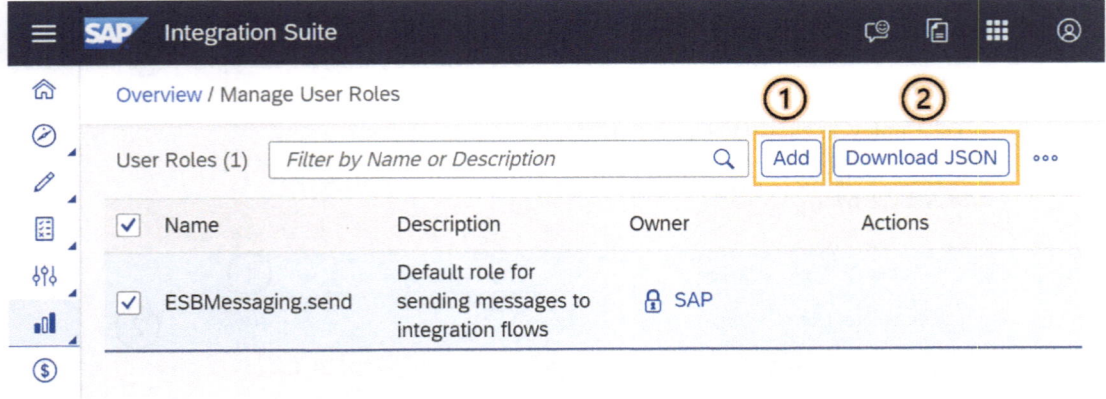

Figure 6-50. *User Role actions*

1. **Add**—You must enter the role name and role description in the input box that displays when you click the button. After that, only the role description can be changed.

2. **Download to JSON**—The JSON file for the selected user roles is downloaded. You must import the JSON file in the Process Integration Runtime tile to create a service instance in the Cloud Foundry environment that is associated with a user role.

You can add different users and assign them different roles. In the next section, you see another tile of management security, in which you can test the connectivity from SAP or non-SAP applications.

6.3.7 Connectivity Test

You can check the receiver system's connectivity. Choose the Connectivity Tests tile from the Overview page's Manage Security section to access the tests. By choosing the corresponding tab, you can test different connectivity kinds.

A Request Form and a Send button are available on each tab. The answer displays a message along with other connectivity-specific data. Navigate to the Manage Security area, as shown in Figure 6-51.

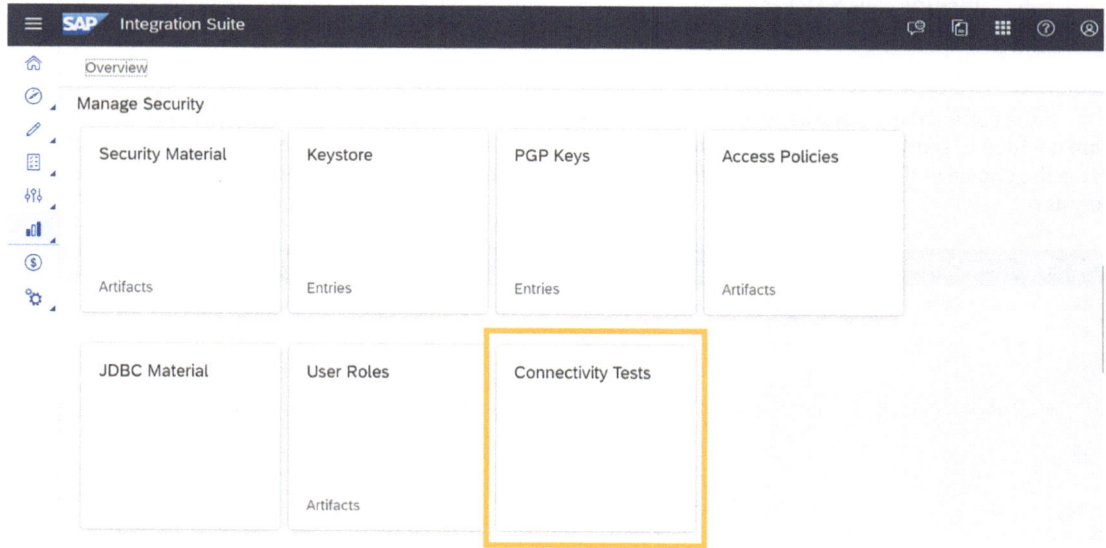

Figure 6-51. *The Connectivity Tests tile*

Different options are available from the connectivity test. In the next sections, you learn about the different connectivity tests you can perform within this tile.

TLS Connectivity

The test tool verifies the following once you select a TLS connection:

- If the tenant can reach the receiver (host).

- If the deployed keystore has the keys needed for the given authentication method during the TLS handshake.

The following settings must be entered to conduct the TLS connection test:

1. **Host**—Hostname of the receiver.

2. **Port No**—The port for outgoing communication. The common port is 553.

3. **Authenticate with Client Certificate (optional)**—Choose the mutual authentication option if the receiver (server) has to confirm the client during the TLS handshake.

4. **Alias**—Include the alias for the key pair that corresponds to client certificate authentication. (This is only needed when authentication with client credential is selected.)

5. **Include New SAP Key**—You can run the connectivity test using the new key pair from the New SAP Keys keystore by selecting Include New SAP Key.

6. **Validate Server Certificate**—Enables verification of the server certificate.

If the connectivity test was successful, as shown in Figure 6-52, information about the various checks you decided to send along with the request is provided. The server certificates are then shown, and you have the choice of downloading the certificate chain or adding the root CA certificate straight to the tenant keystore.

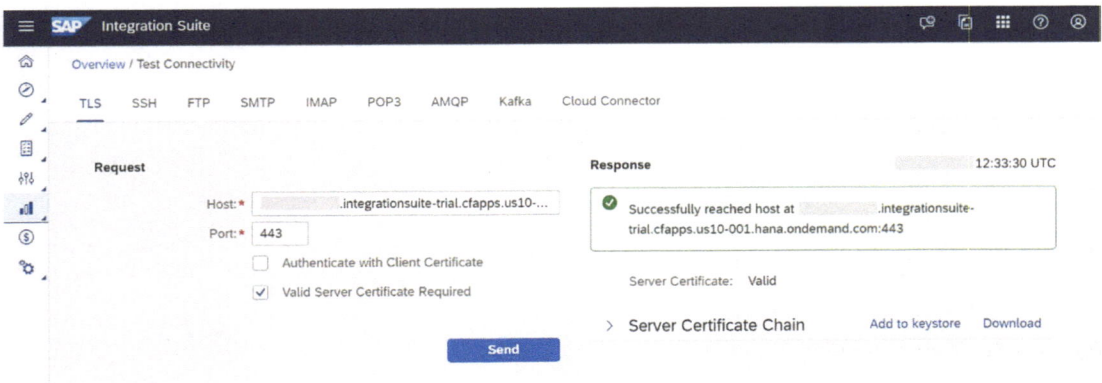

Figure 6-52. *TLS connectivity*

SSH Connectivity

After selecting the SSH connection type, the test tool checks whether the SSH outbound connection connects to the associated SFTP server.

The following are verified by the test according to the selected authentication:

- If the tenant can access the server (host).

- If the tenant's configured known hosts file has the SSH server's certificate.

Procedure:

1. Select connectivity tests from the management security section, as shown in Figure 6-53.

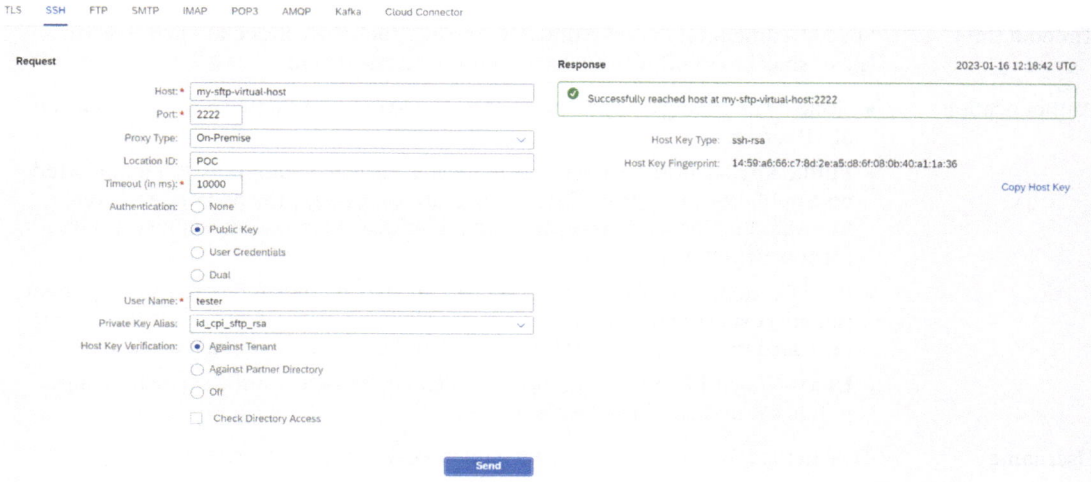

Figure 6-53. *The SSH Connectivity Test*

2. Choose SSH.

3. Define the test's characteristics listed in Table 6-23.

Table 6-23. *Attributes for an SSH Connectivity Test*

Attribute	Description
Host	Enter the hostname of the receiver.
Port	Enter the port number to be used in outbound communication.
Proxy Type	Select the proxy type from the dropdown list: • Internet • On-premises
Location ID (When proxy type is on-premises)	Enter the location ID you established for this instance in the destination settings on the cloud side in order to connect to a Cloud Connector instance connected to your account.
Timeout (ms)	Specify a timeout (in milliseconds) following which the connection to the server (host) shall be cut off. The 10.000 ms number is the default.
Authentication	• **None**—The calling component does not need to be authenticated when using an SFTP server. • **Public Key**—The SFTP server authenticates the caller component's tenant based on a public key. If chosen, the system analyzes the key provided under Private Key Alias and the username provided under User Name to determine the tenant's identity with the SFTP server. • **User Credentials**—The SFTP server verifies the identity of the calling component (tenant) using the username and password. For this configuration setting to work, the username and password must be defined in a User Credential artifact. • **Dual**—The SFTP server confirms the caller component's authenticity by using a public key and user credentials (tenant).
Username	Type in the ID of the user that the tenant uses to access the SFTP server.
Private Key Alias	A name used to distinguish the private key from the keystore that is used to connect to the SFTP server.
Credential Name	The tenant's artifact's name for user credentials.
Host Key Verification	Verify the host key: • Against tenant • Against partner directory • Off
Check Directory Access	If you want to check the destination directory's access, choose this option. The default directory is your home directory if the Directory option is left blank.

FTP Connectivity

To confirm the parameters needed by the FTP adapter, FTP connectivity tests were run, as shown in Figure 6-54.

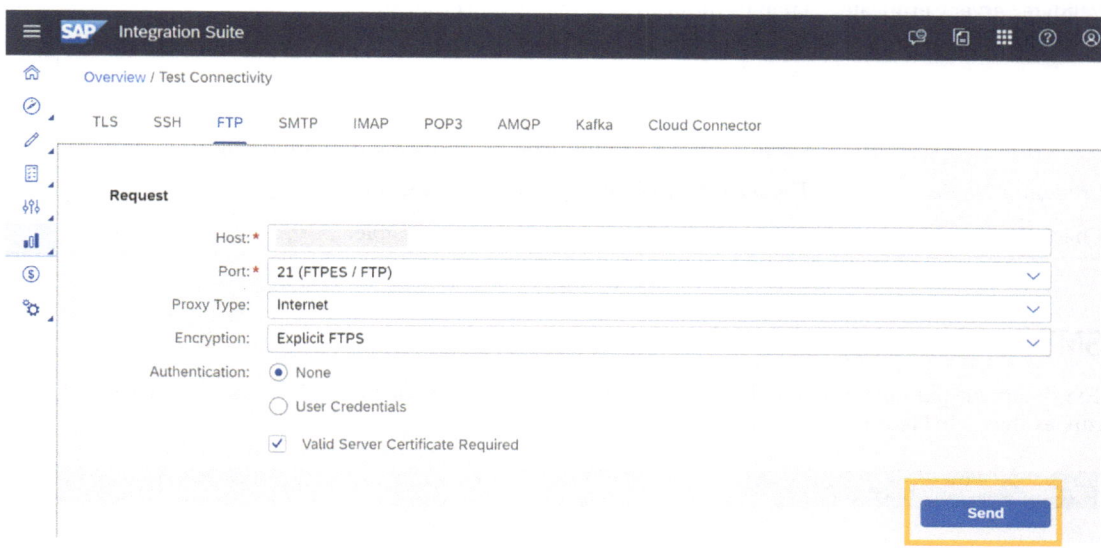

Figure 6-54. *FTP Connectivity Tests*

Provide the details listed in Table 6-24.

Table 6-24. *Attributes for the FTP Connectivity Test*

Attribute	Description
Host	Type the receiver's hostname here.
Port	Enter the port number to be used in outbound communication.
Proxy Type	Choose Internet or on-premises from the dropdown box for the proxy type.
Location ID (When proxy type is on-premises)	Enter the location ID you established for this instance in the destination settings on the cloud side to connect to a Cloud Connector instance connected to your account.
Encryption	Choose the encryption method: Implicit FTPS, Simple FTP, and Explicit FTPS (none)
Authentication	Select the authentication method that will be used to connect to the server. Potential values include: • None • User Credentials

(*continued*)

Table 6-24. (*continued*)

Attribute	Description
Credential Name	Select the deployed credential's name to be used for authentication.
Validate Server Certificate	Enables verification of the server certificate.
	When the default value of the Valid Server Certificate option is chosen, the following tests are performed:
	• If the client connects to a server that has a certificate from that server.
	• If an instance that the client trusts signed the certificate.
Credential Name	The tenant's artifact's name for user credentials.
Check Directory Access	You can determine if a directory is accessible.

SMTP Connectivity

To validate the parameters needed to configure the receiver mail adapter, SMTP connectivity tests can be run, as shown in Figure 6-55.

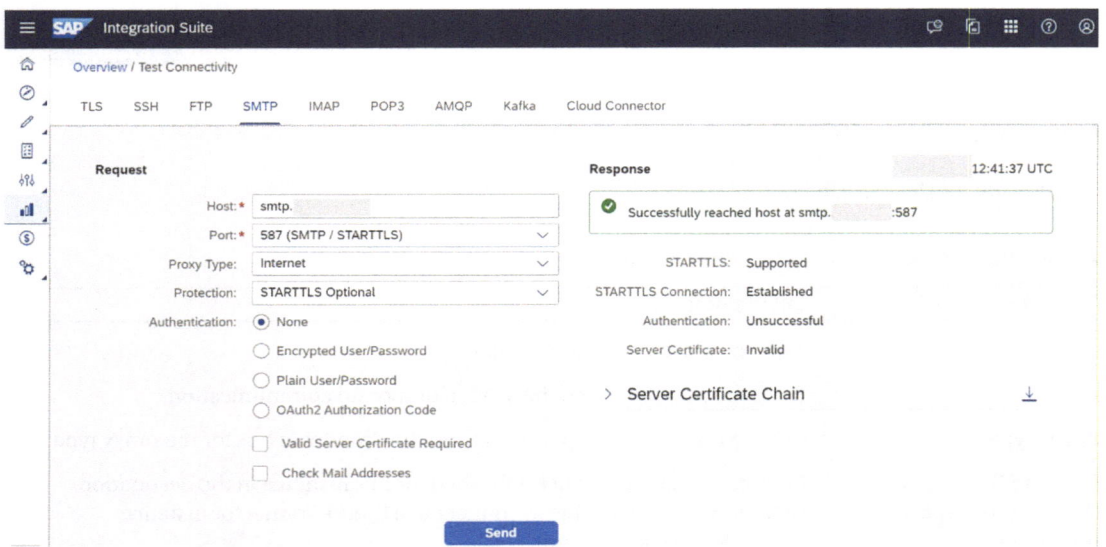

Figure 6-55. *The SMTP Connectivity Test area*

Provide the details listed in Table 6-25 to perform the SMTP test.

Table 6-25. *Attributes for the SMTP Connectivity Test*

Attribute	Description
Host	Type the receiver's hostname here.
Port	Enter the port number to be used in outbound communication.
Proxy Type	From the dropdown list, choose the proxy type: • Internet • On-premises
Location ID (When proxy type is on-premises)	To connect to a Cloud Connector instance linked to your account, enter the location ID you created for this instance in the destination settings on the cloud side.
Protection	Select the protection type: • None • SMTPS • STARTTLS Mandatory • STARTTLS Optional
Authentication	Select the authentication method that will be used to connect to the server. Potential values include: • None • Encrypted User/Password • Plain User/Password • OAuth2 Authorization Code
Credential Name	The tenant's artifact's name for user credentials.
Validate Server Certificate	Makes it possible to verify the server certificate (only when Protection is not off). The following tests are carried out when you choose Validate Server Certificate, which is the default setting: • If the client connects to a server that has a certificate from that server. • If an instance that the client trusts signed the certificate.
Check Mail Address	Enter the sender or receiver email address.

IMAP Connectivity

Select the IMAP (Internet Message Access Protocol) connection, as shown in Figure 6-56.

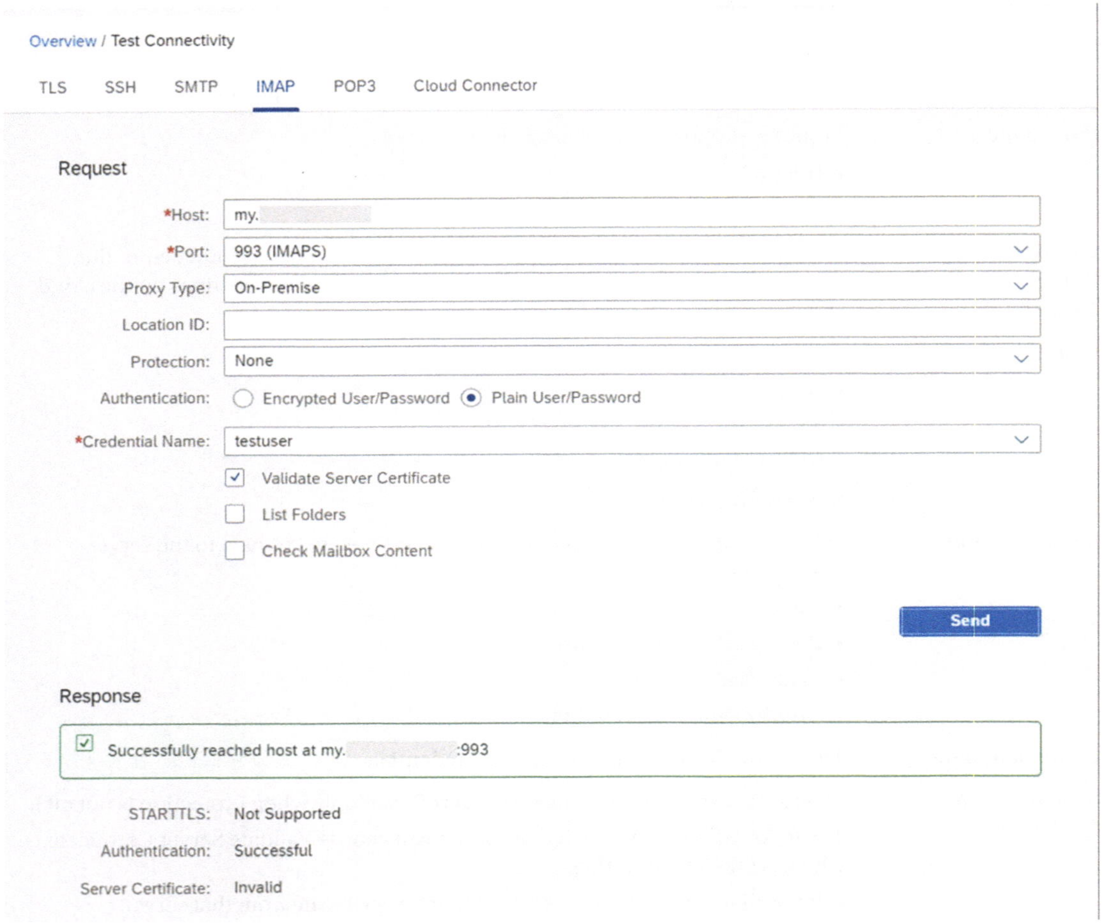

Figure 6-56. *IMAP connectivity test*

Provide the details listed in Table 6-26.

Table 6-26. Attributes for the IMAP Connectivity Test

Attribute	Description
Host	Type the receiver's hostname here.
Port	Enter the port number to be used in outbound communication.
Proxy Type	From the dropdown list, choose the proxy type: • Internet • On-premises
Location ID (When proxy type is on-premises)	To connect to a Cloud Connector instance linked to your account, enter the location ID you created for this instance in the destination settings on the cloud side.
Protection	Select the protection type: • None • IMAPS • STARTTLS Mandatory • STARTTLS Optional
Authentication	Select the authentication method that will be used to connect to the server. Potential values include: • Encrypted User/Password • Plain User/Password
Credential Name	The tenant's artifact's name for user credentials.
Validate Server Certificate	Only allows the server certificate to be verified when Protection is not off. When you select the Validate Server Certificate option, which is the default setting, the following tests are run: • If the client connects to a server that has a certificate from that server. • If an instance that the client trusts signed the certificate.
Folder	You can specify a folder if you choose Check Mailbox Content.

AMQP Connectivity

The test tool determines whether the connection was successful if you choose the Advanced Message Queuing Protocol (AMQP), as shown in Figure 6-57.

Figure 6-57. *AMQP connectivity*

This test is performed when the SAP event mesh is set up in the tenant.
You must enter the settings listed in Table 6-27 to run the AMQP connectivity test:

Table 6-27. *Attributes for the AMQP Connectivity Test*

Attribute	Description
Transport Protocol	A transport protocol selection menu will appear. You have a choice of • TCP • WebSocket
Host	Type the receiver's hostname here.
Port	From the dropdown menu, choose a transport protocol. You have a selection of • 5671 for TCP • 553 for WebSocket
Path	A WebSocket protocol characteristic.
Proxy Type	From the dropdown list, choose the proxy type: • Internet • On-premises
Location ID (When proxy type is on-premises)	To connect to a Cloud Connector instance linked to your account, enter the location ID you created for this instance in the destination settings on the cloud side.
Connect with TLS	Enable the TLS-based secure connection (Transport Layer Security). Checkbox is checked by default.
Validate Server Certificate	Only allows the server certificate to be verified when Protection is not off. When you select the Validate Server Certificate option, which is the default setting, the following tests are run: • If the client connects to a server that has a certificate from that server. • If an instance that the client trusts signed the certificate.

POP3 Connectivity

The test tool verifies the following when POP3 (Post Office Protocol) is selected, shown in Figure 6-58.

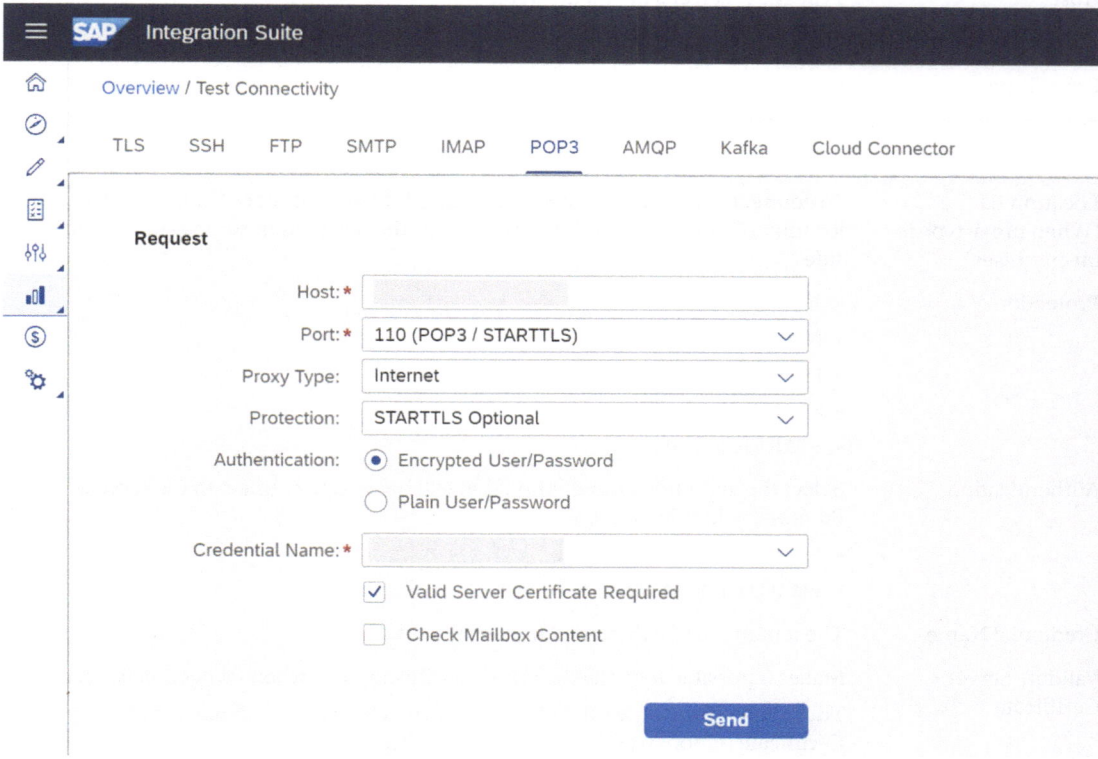

Figure 6-58. *POP3 connectivity*

Provide the details listed in Table 6-28.

Table 6-28. *Attributes for POP3 Connectivity Test*

Attribute	Description
Host	Type the receiver's hostname here.
Port	Enter the port number to be used in outbound communication.
Proxy Type	From the dropdown list, choose the proxy type: • Internet • On-premises
Location ID (When proxy type is on-premises)	To connect to a Cloud Connector instance linked to your account, enter the location ID you created for this instance in the destination settings on the cloud side.
Protection	Select the Protection type • None • POP3S • STARTTLS Mandatory • STARTTLS Optional
Authentication	Select the authentication method that will be used to connect to the server. Potential values include: • Encrypted User/Password • Plain User/Password
Credential Name	The tenant's artifact's name for user credentials.
Validate Server Certificate	Makes it possible to verify the server certificate (only when Protection is not off). The following tests are carried out when you choose the Validate Server Certificate option, which is the default setting: • If the client connects to a server that has a certificate from that server. • If an instance that the client trusts signed the certificate.
Check Mailbox Content	The mailbox is checked when you select Check Mailbox Content, and the number of emails in the inbox is shown.

Kafka Connectivity

When you select the Kafka connectivity test, the test tool determines whether the connection was successful, as shown in Figure 6-59.

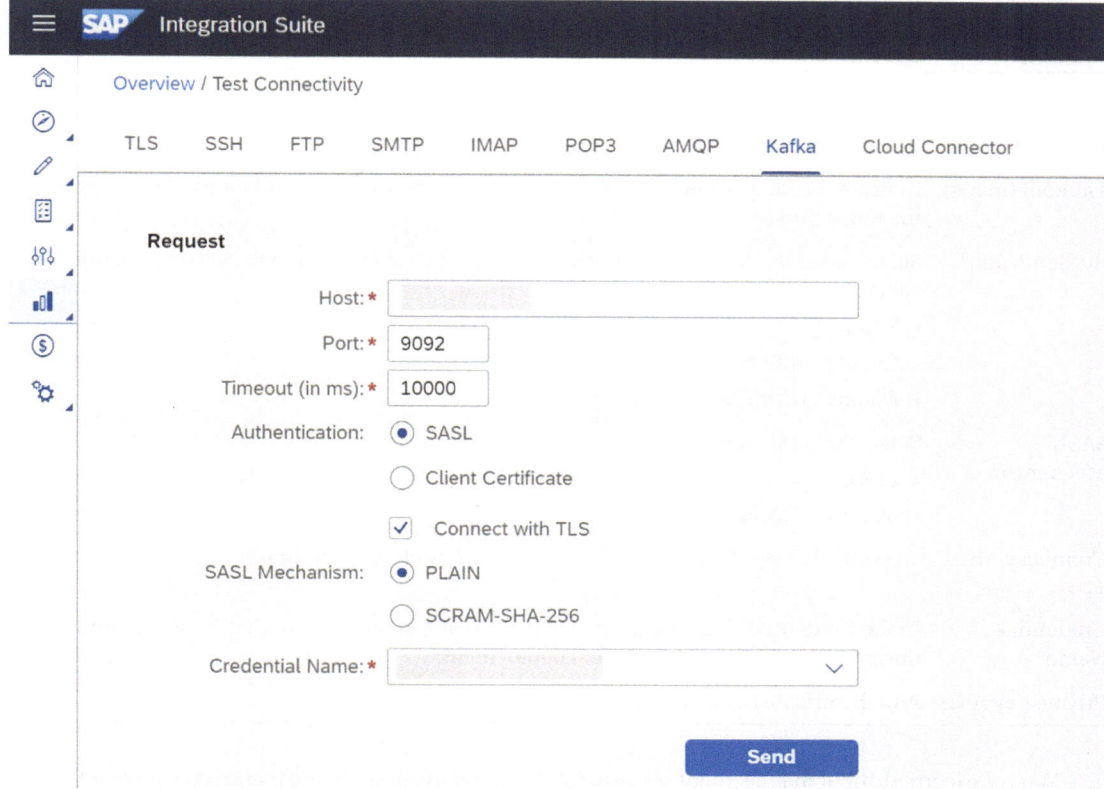

Figure 6-59. *Kafka connectivity test*

The settings must be entered as listed in Table 6-29 to conduct the Kafka connectivity test.

Table 6-29. *Attributes for the Kafka Connectivity Test*

Attribute	Description
Host	Type the receiver's hostname here.
Port	Add the port that will be used for outbound communications.
Timeout (in ms)	Indicate the longest period of time the client will be forced to wait for a response from the Kafka broker.
Authentication	Select the authentication method that will be used to connect to the server. Potential values include: • SASL • Client Certificate • Connect with TLS
SASL Mechanism	Select the SASL mechanism: • PLAIN • SCRAM-SHA-256
Connect with TLS	To switch between PLAIN and SCRAM-SHA-256, choose this option.
Credential Name	In the Credential Name field, enter the username and password that were provided during the tenant's user credential deployment.
Private Key Alias	Add the private key alias here.

You have learned about management security in SAP Cloud Integration and identified various tiles associated with it. In the next section, you learn about the management stores, where you manage the different variables, data stores, message queries, and number ranges. You also learn about each tile in this area.

6.4 Managing Stores

You can manage different temporary data storage on your tenant through the Manage Stores feature.

Data stores are used to store and retrieve data during integration procedures, and SAP Cloud Integration offers a number of choices for managing them. These data stores can be created, configured, and managed by users utilizing the web interface for SAP Cloud Integration. Users of the platform can specify the properties of the data source, such as login information, connection information, and other relevant variables. Additionally, the platform uses encryption and decryption processes to offer a safe place to store private information, such as passwords and authentication credentials. Users can ensure seamless integration processes and dependable data transmission between multiple systems and applications by managing data storage efficiently. Figure 6-60 shows the managed stores in SAP Cloud Integration.

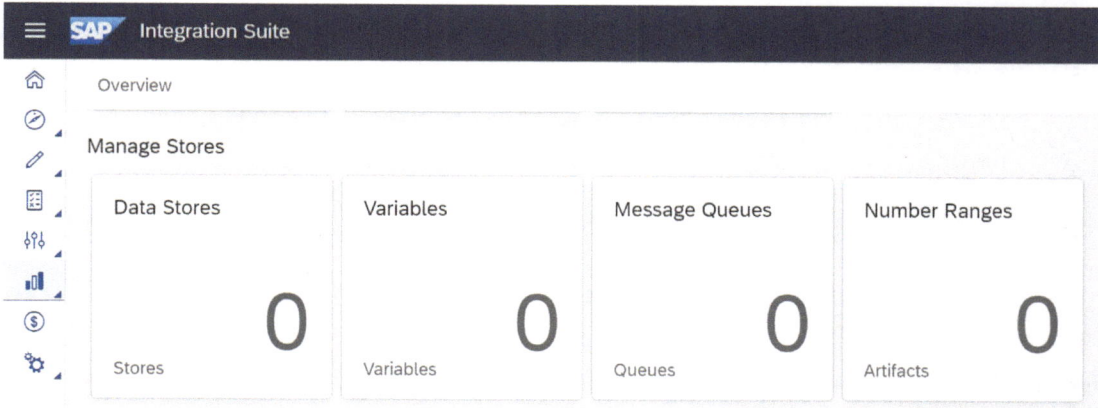

Figure 6-60. *The Manage Stores tile*

You learn about the tiles associated with management stores in the upcoming sections.

6.4.1 Manage Data Stores

With SAP Cloud Integration, managing data stores requires setting up links to data sources that hold the data that integration flows need. Databases, file systems, or other systems that store data can be used as these data stores. Figure 6-61 shows the management data stores in SAP Cloud Integration.

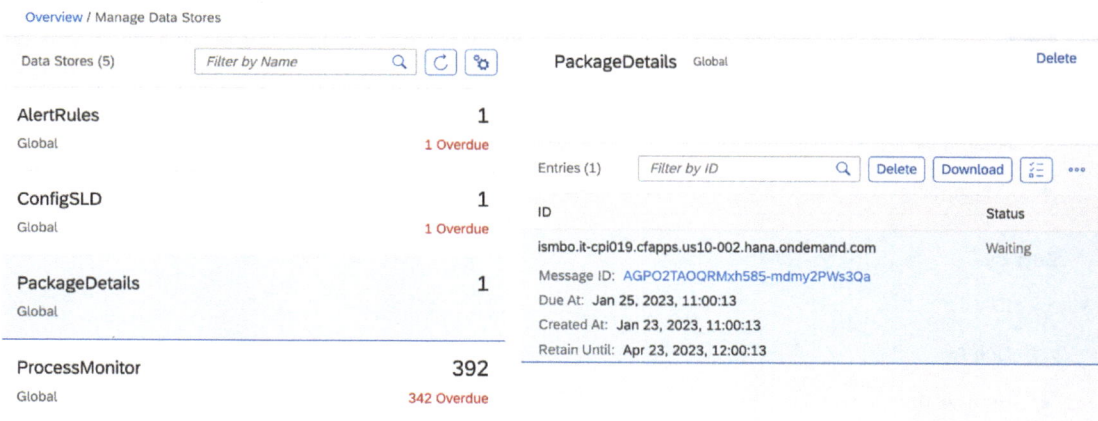

Figure 6-61. *Managing data stores*

1. Select the Data Stores tile to launch the Manage Data Stores pane, as shown in Figure 6-62.

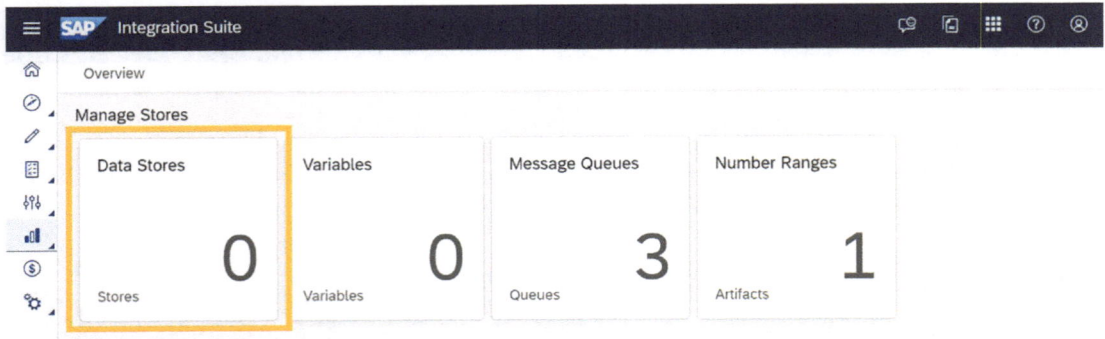

Figure 6-62. *Data stores*

2. The Manage Data Stores left pane displays the data store's name.

3. You can access the following features by going to the top of the list of data stores, as shown in Figure 6-63.

- Filter By Name

- Sort By

- Filter By

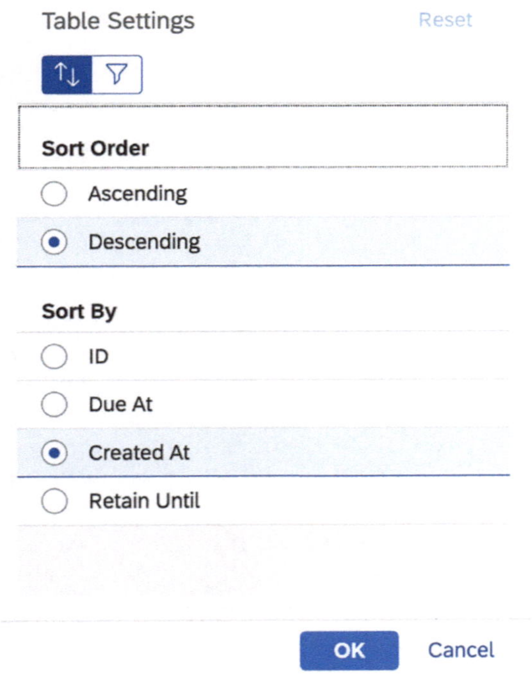

Table Settings   Reset

↑↓   ▽

Sort Order

○ Ascending

◉ Descending

Sort By

○ ID

○ Due At

◉ Created At

○ Retain Until

OK   Cancel

Figure 6-63. *Table settings of data stores*

4. Information regarding the data store selected on the left side of the screen is shown on the right side of the screen in Figure 6-64.

- ID

- Message ID

- Status

- Due At

- Created At

- Retain Unit

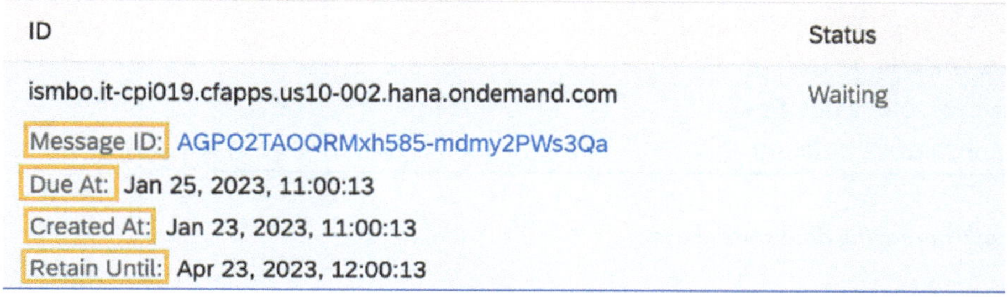

ID	Status
ismbo.it-cpi019.cfapps.us10-002.hana.ondemand.com	Waiting

Message ID: AGPO2TAOQRMxh585-mdmy2PWs3Qa

Due At: Jan 25, 2023, 11:00:13

Created At: Jan 23, 2023, 11:00:13

Retain Until: Apr 23, 2023, 12:00:13

Figure 6-64. Data store details

5. To manage the entries, select one of the following choices from the header above the table, as shown in Figure 6-65.

- Filter By ID

- Delete

- Download

- Reload Content

- Table Settings

- Multiselect Mode

PackageDetails   Global   **Delete**

Entries (1)   [Filter by ID   🔍]   [**Delete**] [**Download**]   [☑]   ○○○

☑	ID	Status
☑	ismbo.it-cpi019.cfapps.us10-002.hana.ondemand.com	Waiting

   Message ID: AGPO2TAOQRMxh585-mdmy2PWs3Qa

   Due At: Jan 25, 2023, 11:00:13

   Created At: Jan 23, 2023, 11:00:13

   Retain Until: Apr 23, 2023, 12:00:13

Figure 6-65. *Manually managing data store entries*

In the next section, you learn about the Manage Variables tile.

6.4.2 Manage Variables

You can monitor the variables utilized in integration processes by using the Variables view. Choose the Variables tile from the Manage Store menu, as shown in Figure 6-66.

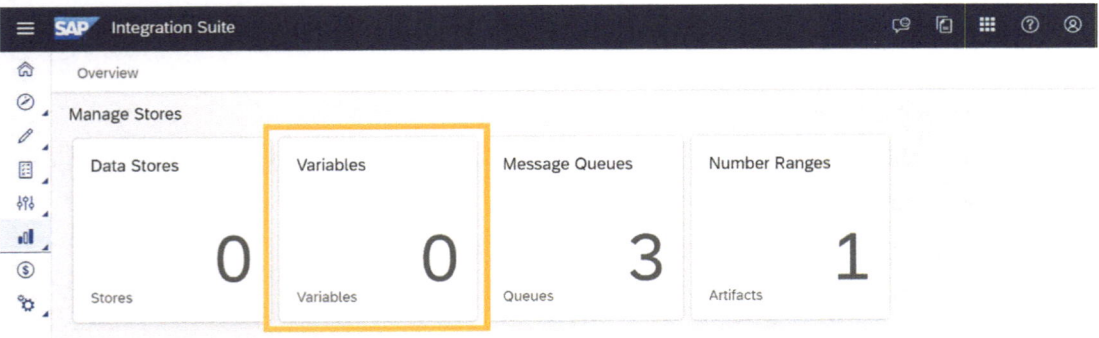

Figure 6-66. *The Variables tile*

You receive a summary of the current variables along with the following characteristics:

- Name
- Visibility
- Integration Flow
- Updated At

360

- Retain Unit

- Actions

You can view the content of a variable by clicking its name in a table. The content of the variable cannot be displayed, and a message is displayed if it is not declared as a string value. You can download the variable by choosing Download. If you choose to store the variable, the system creates n .ZIP file that contains the header properties file, as shown in Figure 6-67.

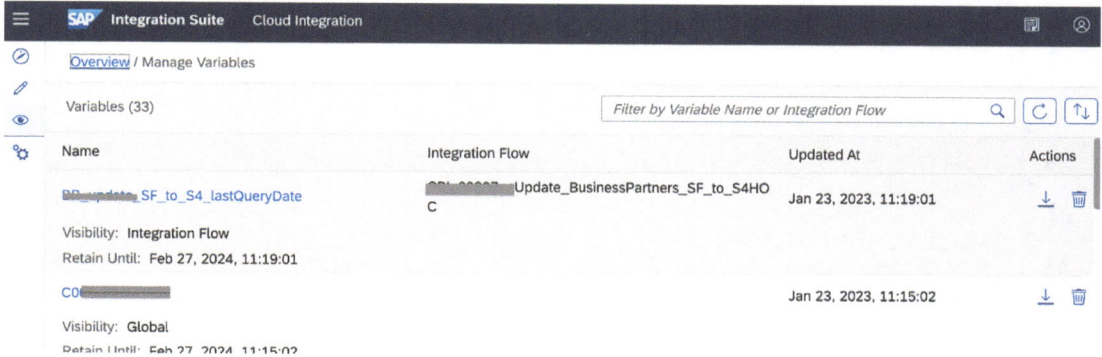

Figure 6-67. *Manage variables*

In the next section, you learn about the Manage Message Queues tile.

6.4.3 Manage Message Queues

Several adapters let you queue up messages for later delivery. Active queues for a tenant are visible to you. Select Monitor to launch the queue-monitoring application, as shown in Figure 6-68.

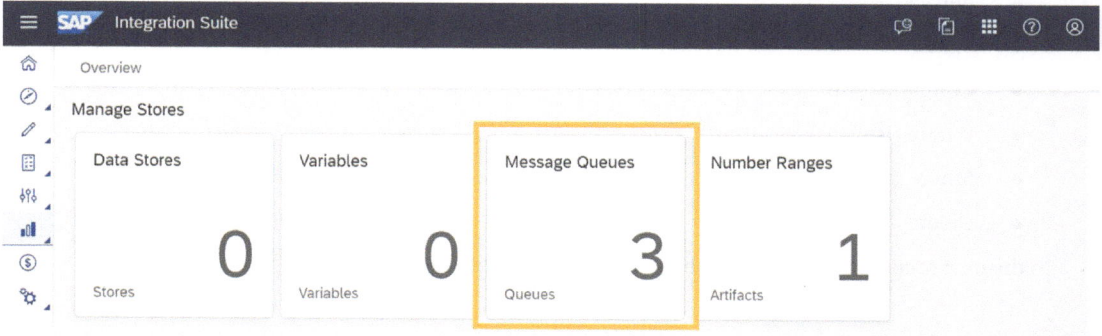

Figure 6-68. *The Message Queues tile*

The Message Queues tile will appear if your tenant has one or more active queues (under Manage Stores).

To filter and sort the table's contents, select Table Settings. By defining an attribute and setting whether the entries should be sorted for that attribute in ascending or descending order, you can choose how the table entries are organized. Additionally, you can filter the table entries based on a variety of factors.

361

By entering a portion of a queue's name in the search, you can filter out particular queues, as shown in Figure 6-69.

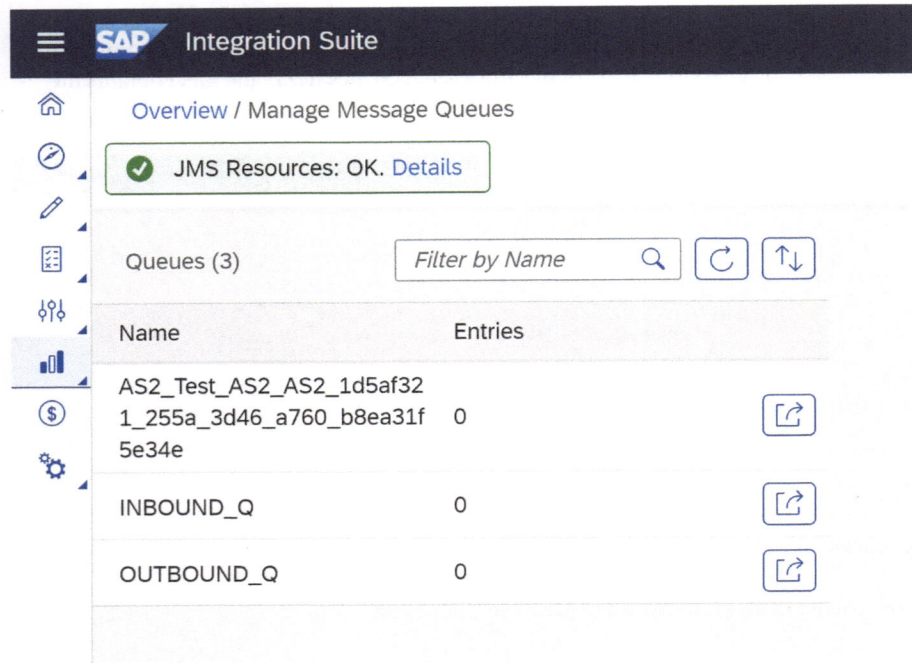

Figure 6-69. *The Manage Message Queue area*

To carry out the following actions, choose a queue and click Actions:

- Retry
- Status
- Configure Size
- Where Used
- Move
- Delete

In the next section, you examine the management number ranges.

6.4.4 Manage Number Ranges

The topic is a general review of artifacts connected to number ranges. To examine the artifacts with the corresponding status, select the Number Ranges tile in the Manage Stores section, as shown in Figure 6-70.

Figure 6-70. *The Number Ranges tile*

Each document that is sent out for EDI processing must have a special interchange number applied to it. You can use the Number Range Object to add such an exchange number. The Interchange Control Reference for EDIFACT messages has a length of 1 to 9 digits.

A table presents a list of number ranges. The following qualities for each artifact are shown:

- Name
- Minimum Value
- Maximum Value
- Net Value
- Field Length
- Rotate
- Deployed By
- Deployed On

Here are the actions, as shown in Figure 6-71:

- Select Add to create or deploy a new artifact.
- To edit and redeploy an existing item, select the artifact in the table and then choose Edit.
- To remove an artifact from deployment, pick it in the table and then choose Undeploy.

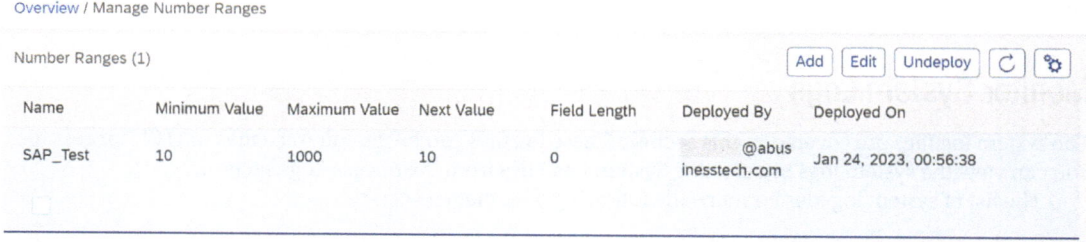

Figure 6-71. *The Manage Number Ranges area*

You have learned about the management stores and seen the different tiles associated with them. In the next section, you will learn about the access logs.

6.5 Using Access Logs

For monitoring and debugging integration activities, SAP Cloud Integration's access logs are essential. The platform records different kinds of access logs, such as system access logs and user access logs, which provide information about user behavior and system performance. The logs keep track of details including but not limited to user login attempts, successful and unsuccessful logins, and system faults. These logs can be accessed by users using the platform's various APIs or the web interface for SAP Cloud Integration. Access logs can be used to monitor user behavior and identify issues with integration procedures, such as connection drops, data errors, and system crashes. Users can spot potential security lapses, unauthorized access attempts, and other anomalies that can affect system performance and data by reviewing access logs.

The access logs are available in the Neo and Cloud Foundry environments with different functionalities. You will learn about the different access logs associated with different environments in the following sections.

6.5.1 Access Logs in the Neo Environment

The access logs section and system log files can be used to evaluate errors that occur during incoming HTTP processing and track audit logs generated as a result of system adjustments.

Monitor Audit Log

The audit log contains details about system modifications. Examples of these events include the deployment of an integration flow and a configuration change. Select the Audit Log tile under Manage Security to view the audit log.

You can modify the filter's Time Range setting to alter how the messages are displayed. The following preset time ranges are available for selection:

- All
- Past Hours
- Past 25 Hours
- Past Week
- Custom

You can access the information and sort the items in the audit log list by object name, user, or source. Moreover, you can sort the audit log list by Time, Action, Object Kind, or Object Name.

Monitor System Logs

The system log files are covered in this section. These log files can be default trace files or HTTP access files. You can view the system logs by selecting System Log Files from the Access Logs section.

The list of system log files contains the following data that you can retrieve:

- Name
- Log Type

- Updated At

- Size

- Actions

The log files can be filtered by name, and the list can be sorted by name, updated at, or file size. Either download the desired file or obtain the URL for the desired file.

You learned about the access logs in the Neo environment and looked at the audit and system logs associated with the Neo environment. In the next section, you learn about the access logs in the Cloud Foundry environment.

6.5.2 Access Logs in the Cloud Foundry Environment

You can keep track of and examine errors that occur during inbound HTTP processing using the access logs section, as shown in Figure 6-72.

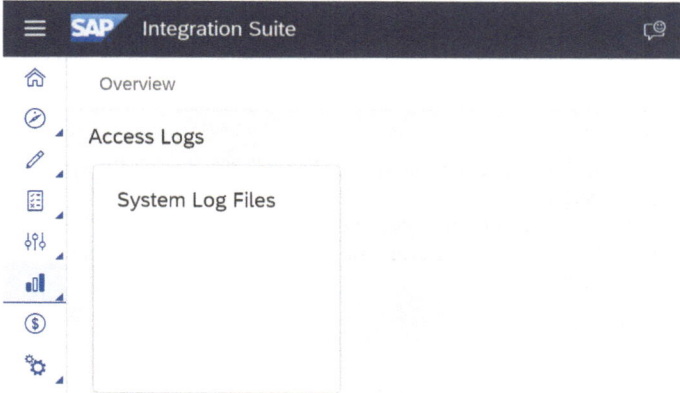

Figure 6-72. *Access logs*

Monitor System Log Files

Information on system log files can be found in this section. HTTP access files and standard trace files can be found in these log files. You can view the system log files by choosing System Log Files from the Access Logs section.

The following details are accessible from the list of system log files, as shown in Figure 6-73:

- Name

- Log Type

- Updated At

- Entries

- Action

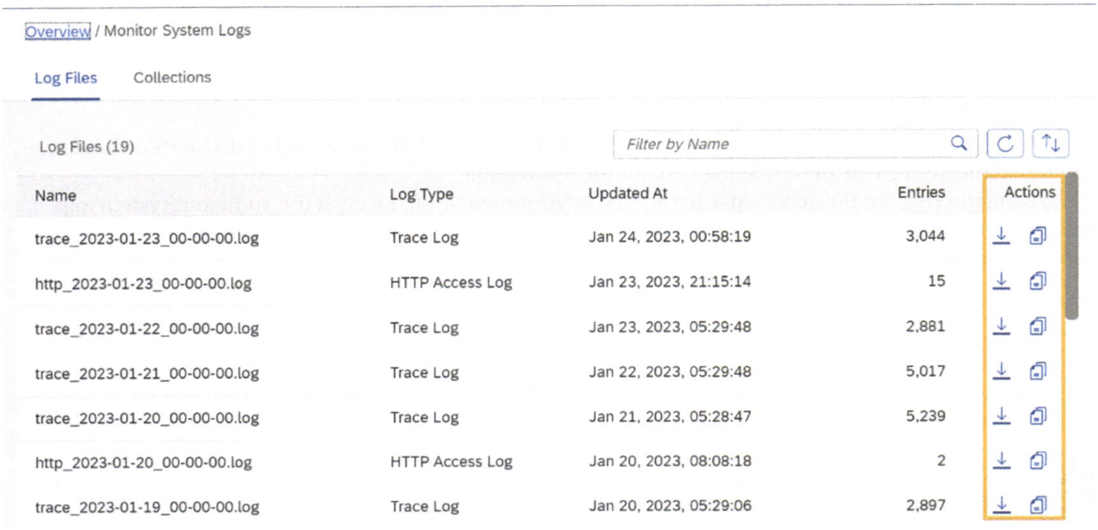

Figure 6-73. *Monitor system log files*

The log files can be searched by name, and the list can be arranged by name, updated at, or file size. Get the URL for the required file or download the desired file.

You learned about the access logs for the Neo and Cloud Foundry environments. You also saw the monitor system log files for both environments. The next section covers the process of managing locks.

6.6 Managing Locks

Information on system log files can be found in this section. Either HTTP access files or standard trace files can be found in these log files, as shown in Figure 6-74.

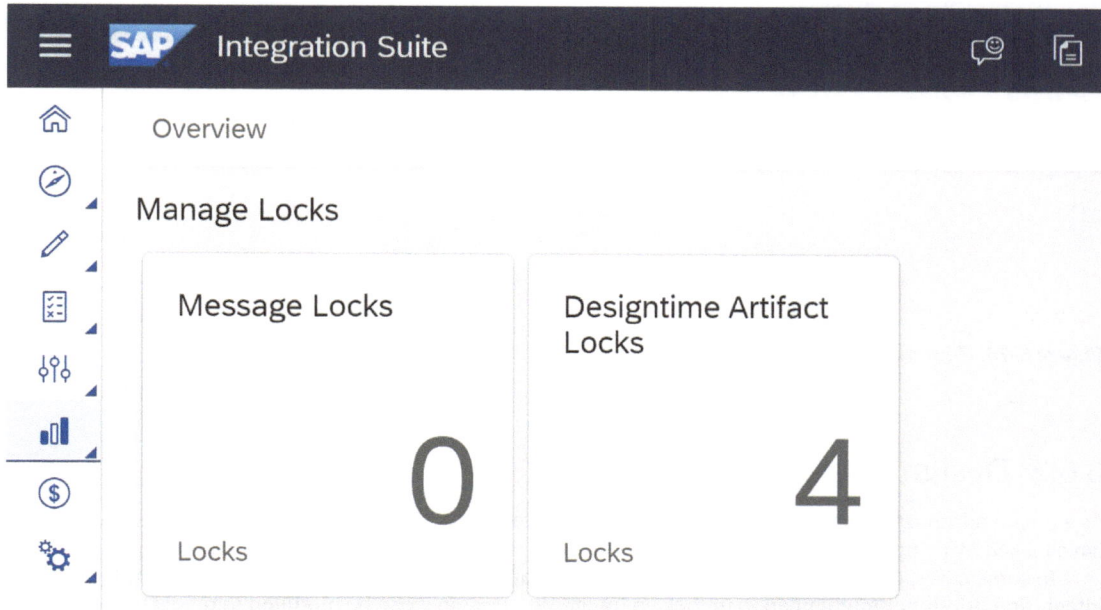

Figure 6-74. *Manage locks*

The management blocks are also associated with two tiles, which you learn about in the following sections.

6.6.1 Message Locks

You can view and manage lock entries created in this section to prevent the same message from being processed more than once concurrently.

When you choose the Manage Locks tile in the Monitor program and then click the Message Locks tile, a list of locks is displayed. The following details are shown for every lock entry:

- Component

- Source

- Entry

- Created At

- Expires At

Additionally, you can look up table entries (search field). Select settings to sort and filter the table's contents. Select the entry and select Release to release the lock entry and restart message processing.

Figure 6-75 shows the Message Locks dashboard containing all these options.

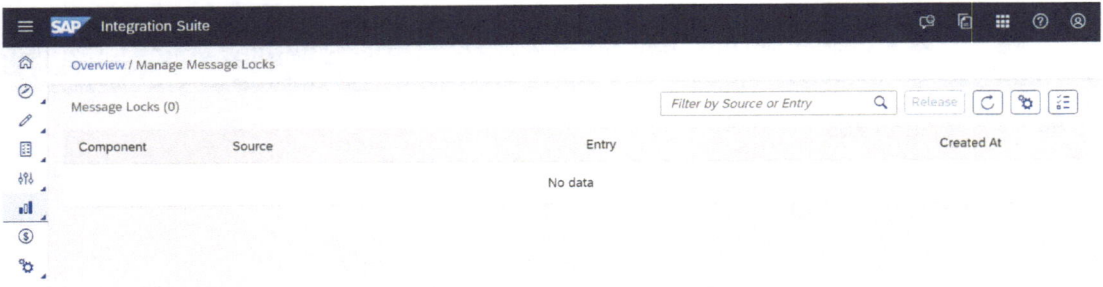

Figure 6-75. *Message locks*

6.6.2 Designtime Artifact Locks

As a tenant administrator, you can inspect and unlock integration artifacts and packages that tenants' users have locked by using this self-service capability.

In some cases, users can lock an artifact at the time of design and fail to unlock it. For example, an integration developer might need to unlock an integration flow so it can be modified to reflect the updated business logic. If the user is unavailable for any reason and there is a critical business need to change the artifact, tenant administrators can release the artifact. The tenant administrators can unlock an artifact at any moment.

Choose Monitor Designtime Artifact Locks to see a list of locked designtime artifacts. All users who are tenants have access to the locked artifacts. Only the tenant administrators can unlock the artifacts using the (unlock) icon. Once you confirm the unlock action, the artifact is unlocked and removed from the list.

The following details are shown for each locked artifact in Figure 6-76:

- Name
- Type
- Package
- Locked At
- Locked By

Figure 6-76. *Designtime artifact locks*

6.7 Summary

Users can view the status and performance of their integration processes in real time on the SAP Cloud Integration monitor screen. The screen shows a variety of data, including system alarms, error rates, message flow, and message processing times. Users can alter the display of relevant information on the monitor screen to suit their individual requirements and tastes. The platform's different filtering and sorting features let users concentrate on particular integration flows or messages. The monitor panel also enables users to take instant actions, such as restarting integration flows, configuring alarms, or resending and cancelling communications.

This chapter dug deep into the Monitoring part of SAP Cloud Integration. You learned that you can easily monitor your integration flows and Cloud Integration enviornment. The next chapter examines the security aspects of SAP Cloud Integration in the Cloud Foundry and Neo environments.

CHAPTER 7

■ ■ ■

SAP Cloud Integration: Security

In today's digital era, security is a major concern of businesses as they move their operations to the cloud. With the ever-evolving cybersecurity landscape, it is essential for organizations to ensure that their data and systems are secure and protected.

This chapter examines the security aspect of SAP Cloud Integration in both the Cloud Foundry and the Neo environments. It delves into the security measures that are in place for the Cloud Foundry environment. The chapter also discusses Cloud Foundry's certificate management, technical environment, identity access management, data storage, protection, and security. It examines the various Cloud Foundry data storage types, including message logs, contents, storage, and retention, as well as the malware scanner.

The chapter explores the security measures for the Neo environment as well. It discusses identity access control, data storage security, and data flow for cloud integration Neo. Together with other security information, such as UI security and remote API security, it also examines the different categories of stored data in Neo as well as particular data sets.

7.1 Security Cloud Foundry Environment

Cloud-based SAP Cloud Integration technology, which integrates many systems and apps to optimize business processes, places a high priority on security. Strong security measures are provided by SAP Cloud Integration to safeguard enterprises' systems and data. These controls include, among others, identity access management, data storage security, data flow protection, message logging storage and retention, user interface security, and remote API security. Organizations can securely integrate their systems and applications while ensuring the confidentiality, integrity, and availability of their data, thanks to the security safeguards in place for SAP Cloud Integration in both the Cloud Foundry and Neo environments.

Security in Cloud Foundry in SAP Cloud Integration refers to the various measures that are in place to protect the integrity, availability, and confidentiality of data and systems. This can include measures such as authentication and authorization, encryption, and network security. Additionally, SAP Cloud Integration provides built-in security features, such as role-based access control, secure communication between services, and integration with external identity providers. These measures help ensure that only authorized users and systems can access the data and services in the Cloud Foundry environment and that the data is transmitted and stored in a secure manner.

Transport Level Security

The actions made to safeguard communication between two systems or applications at the transport layer are referred to as transport layer security in the SAP Integration Suite. The transport layer, which is the third layer in the OSI paradigm, is in charge of guaranteeing reliable end-to-end communication between two systems. Table 7-1 shows the transport level security protocols.

© Jaspreet Bagga 2023
J. Bagga, *A Practical Guide to SAP Integration Suite*, https://doi.org/10.1007/978-1-4842-9337-9_7

Table 7-1. *Transport Level Security Protocols*

Transport Protocol	Description
SFTP	This protocol is supported by the SFTP transmitter and receiver adapters.
	SSH uses a symmetric key length of at least 128 bits to protect FTP transfer. The default length of the SAP asymmetric key is 2048 bits.
	When asymmetric key pairs are used, SFTP also ensures that the participants are utilizing only authorized public keys.
HTTP(S)	This protocol is supported by all adapters that allow communication over HTTPS, including the Soap Adapter, IDoc Adapter, and the HTTP Adapter.
	Moreover, receiver adapters provide main propagation based on the SAP Cloud Connector.
	Simple authentication options include using client certificates, user credentials, or OAuth, depending on the sender or receiver adapter being used.
SMTP	Email communication is supported using these protocols.
POP3	The STARTTLS extended operation supports transport encryption.
IMAP	You can send the username and password to the email server in plain text or encrypted.

Message-Level Security

The methods used to safeguard a message's payload at the application layer are referred to as message-level security in the SAP Integration Suite. The application layer, the seventh layer in the OSI model, is in charge of managing services that are specific to the running program. You can see the message-level security standards in Table 7-2.

Table 7-2. *Message Level Security Standards*

Standards	Security Features
PKCS#7/CMS Enveloped Data and Signed Data	Encrypting and decrypting messaage content. Payload authentication and signature.
PKCS#7/CMS Enveloped and Signed Data	Payload encryption, decryption, signature, and verification.
Pretty Good Privacy—PGP	Payload encrypting, decrypting, signing, and verifying. Encrypting and decrypting message content.
XML Signature	Verification and signature of the payload.
WS-Security	Authenticating the SOAP body with a signature.

The next section covers certificate management.

7.1.1 Certificate Management

Depending on the transport- and message-level security solutions you choose, a tenant can need to maintain and deploy a variety of security artifacts:

- **Certificates in X.509**—Used for transport-level security using TLS as well as message-level security utilizing PKCS#7, WS-Security, and XML Digital Signature. They are stored in a Java Keystore.

- **A PGP key**—Used for message-level security with Open PGP.

- **Files for known hosts**—Transport-level SFTP security is necessary. SFTP keys are also stored in a Java Keystore.

The next section explains the technical landscape of the SAP Cloud Foundry environment of SAP Cloud Integration and the different roles associated with cloud integration.

7.1.2 Technical Landscape and Identity Access Management

In this section, you learn about the security aspects of the SAP Integration Suite and the security measures you can take to protect client data when it is communicated over the system while an integration scenario is being executed.

Customers who use SAP Cloud Integration concur that a large portion of their own and their clients' sensitive data is handled and kept in a noncustomer infrastructure.

The main job of an integration platform is to serve as a central point for messages that could contain confidential client information. First and foremost, these messages must be kept private and protected from unauthorized access.

Technically speaking, the cloud-based clustered and containerized integration platform is how the platform is constructed. Messages handled by the platform's integration flow from various clients are handled by distinct sections (referred as *tenants*).

Tenants handling integration flows from distinct clients are tightly segregated from one another in terms of CPU, data storage, and user access.

Figure 7-1 shows the high-level architecture of the SAP Integration Suite.

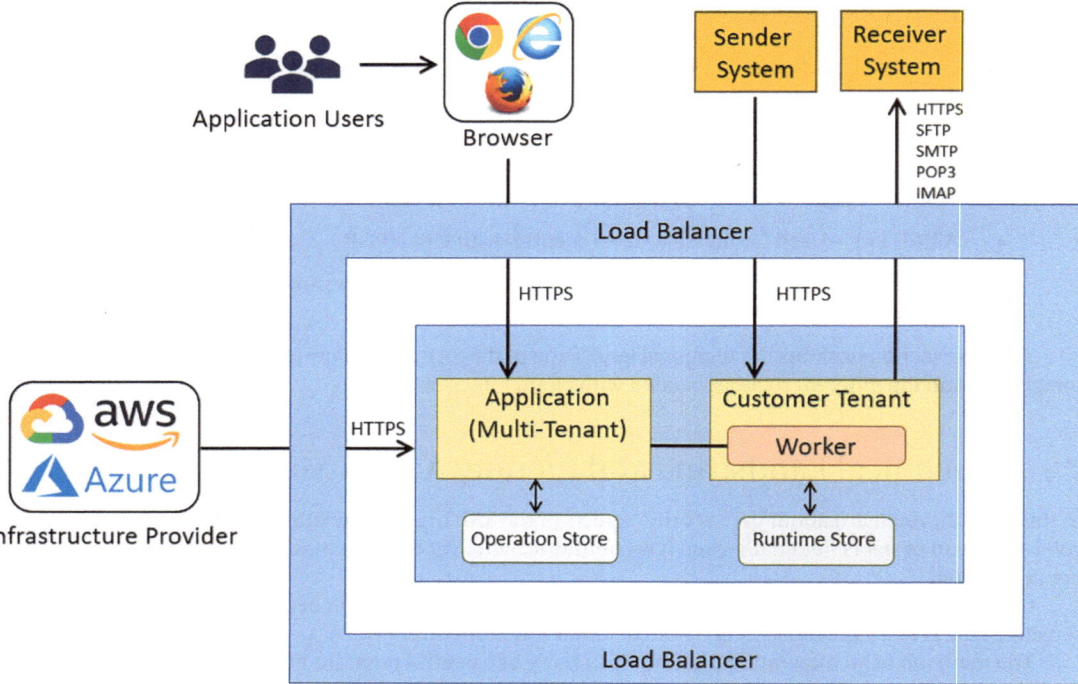

Figure 7-1. *Architecture of the SAP Integration Suite*

Identity Access Management

An identity provider verifies the identity of dialog users who access the platform. ID Service for SAP Identity Service is employed by default. The management of IDs and their lifecycles is centralized via SAP ID Service.

Access to the platform's dedicated features is restricted and secured by authorization checks. To handle the authorizations of dialog users, a number of authorization groups are offered. Based on a persona, an authorization group establishes a set of specific permissions pertaining to the activities that take place over the course of an integration project.

When managing users through the SAP BTP Cockpit, as illustrated in Figure 7-2, you can assign account users several predefined roles. Based on the major responsibilities associated with integration projects, these jobs are linked to certain personas that are beneficial for integration initiatives.

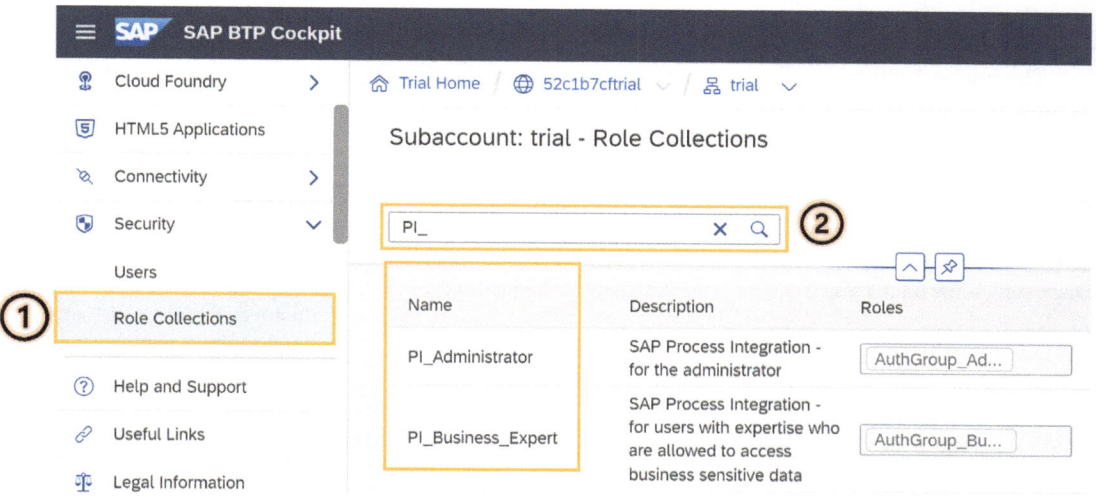

Figure 7-2. Role collections for SAP Cloud Foundry

Persona

There are a number of predefined roles that you can apply to users of the account when managing users in the SAP BTP Cockpit. These roles are connected to specific personas that are useful for integration projects based on the primary tasks connected with integration projects. Table 7-3 shows the personas describing the different role collections required in SAP Cloud Integration.

Table 7-3. Identity Access Management (Cloud Foundry)

Persona	Role Collection (Cloud Foundry)	Description
Business expert	PI_Business_Expert	Enables a business professional to perform business tasks, including looking at the payload.
Administrator	PI_Administrator	Enables the tenant administrator, also referred to as the tenant cluster administrator, to connect to a cluster and control its operations.
Integration developer	PI_Integration_ Developer	Enables a cluster connection in Integration Designer so that an integration developer can see, download, and deploy artifacts. You must be a member of this authorization group in order to access Cloud Integration.
Read-only persona	PI_Read_Only	You can communicate with a tenant and check your messages. You can download WSDL files and have read-only access to integration flow artifacts in the Monitoring section with the help of this permission group.
Partner Directory configurator	AuthGroup TenantPartnerDirectory	Enables the Partner Directory administrator to access and edit Partner Directory content.

Check out the SAP Help Portal for more details on access management for cloud integration.

Task and Permission

The responsibilities necessary to complete the various Cloud Integration-related tasks are summarized in Table 7-4. Additionally, Table 7-4 states to what extent the roles and duties relate to the key persona.

Table 7-4. *Task and Permission (Cloud Foundry)*

Area	Task	Role Template (Cloud Foundry)	Persona
Discover	View packages	CatalogPackagesRead	Integration Developer Business Expert Read-Only Persona System Developer Tenant Administrator
	View package artifacts	CatalogPackageArtifactsRead	Integration Developer Business Expert Read-Only Persona/System Developer Tenant Administrator
	Copy package to workspace	CatalogPackagesCopy	Integration Developer
Design	View packages and package artifacts	WorkspacePackagesRead	Integration Developer Business Expert Read-Only Persona/System Developer Tenant Administrator
	Create, edit, export, and remove a package and all its artifacts	WorkspacePackagesEdit	Integration Developer
	Update packages	WorkspacePackagesEdit	Integration Developer
	Configure artifacts	WorkspacePackagesConfigure	Integration Developer
	Deploy/undeploy artifacts	WorkspaceArtifactsDeploy	Integration Developer Tenant Administrator
	Export packages for transport	WorkspacePackagesTransport	N/A
	Import packages from transport	WorkspacePackagesTransport	N/A

This section explained the technical landscape and the identity access management for Cloud Foundry. In the next section, you learn about data storage, protection, and security in Cloud Foundry.

7.1.3 Data Storage, Protection, Privacy, and Security in Cloud Foundry

In data storage, customer data that is kept indefinitely is strictly isolated and divided by tenants. The physical infrastructure that various tenants use can be shared, but each tenant's data is stored in a different schema.

The customer can specify whether their data at rest is encrypted for particular use cases. It is possible to store encrypted message content. If this security measure is configured, the automatically generated encryption key is different for each tenant and is periodically renewed.

AES is used for data storage encryption, and a 256-bit key length is used. The encrypted data is not kept in the same place as the encryption key.

Data Protection and Privacy in Cloud Foundry

The integration platform processes and stores different kinds of customer data at various times. The highest level of security is provided for this data, and SAP takes special precautions to ensure it.

Data Protection and Privacy

The integration platform processes and stores different kinds of client data at various times. Sensitive data is protected to the highest standard, and SAP takes special steps to ensure this degree of security.

Governments impose legal responsibilities on businesses to safeguard privacy and data. To assist you in fulfilling these demands, there are various features and functions.

It is assumed that software operators, like SAP clients, obtain and store users' consent before collecting their personal information. A data privacy professional can determine whether the data subjects have given, withdrawn, or declined their consent at a later time.

The compiled information about a data subject makes up an information report. Such a report can be provided by a data privacy specialist, or an application can include a self-service option. Cloud integration assumes that software developers, such as SAP clients, can offer this data.

This section briefly described data storage, protection, privacy, and data security in the Cloud Foundry environment. In the next section, you learn about the types of stored data in Cloud Foundry.

7.1.4 Types of Stored Data in Cloud Foundry

While an integration scenario is running, various types of data, including message content and monitoring data, can be stored.

Due to the possibility of containing personal information, this data must be regarded as sensitive. This includes such data as:

- **Message content**—Messages processed on a runtime node frequently contain business data from an integration scenario, including sensitive customer data such as addresses, names, and financial information.

- **Monitoring data**—The steps of an integration flow's processing are recorded in the message processing log. Only individuals allocated to this tenant and those with certain permissions can view this data.

In the next section, you learn about the message logs, contents, storage, and retention in Cloud Foundry.

7.1.5 Message Logs, Contents, Storage, and Retention in Cloud Foundry

The SAP Cloud Integration architecture stores a variety of data types over the course of an integration project. Table 7-5 lists the different groups and characteristics, such as the type of storage and the length of retention.

Table 7-5. *Specific Data Sets (Cloud Foundry)*

Data	Description	Logical Storage	Classification	Retention Time
Message processing log	Information that is organized about how a message is processed		Log data	30 days
Message processing log attachments	Data recorded in a runtime message processing log	Message store	Log data Business data	30 days
Integration flow tracing information	Being aware of the message flow, the message payload, and any processing errors that took place	Trace store	Log data Business data	60 minutes
Integration content (design time)	Camel XML representation of integration flows and other design-time entities, created or modified by an integration developer. Integration flow models and value mappings	Workspace	Configuration data	Unlimited
Integration content (runtime)	Knowledge of the message flow, including the message payload, as well as any processing blunders that occurred		Configuration data	Unlimited
Information kept in a Data Store operation step	Certain integration flow steps keep track of message content, which includes details like the payload, tenant ID, message GUID, and message processing log GUID Further phases in the integration cycle use it to process messages in more depth	Data store	Business data	Developer of the integration can define it; the default value is 30 days
Information kept by the Persist step	Message data is kept in specific integration flow steps	Message store	Business data	90 days
JMS adapter stores message content	JMS message queues contain the message content	JMS queue	Business data	Developer of the integration can define; the default value is 90 days
Lock entries	To prevent the same message from being processed more than once concurrently, lock entries that are created			

The next section explains the malware scanner found in SAP Cloud Foundry.

7.1.6 Malware Scanner

Users of the tenant upload a variety of design-time files to the tenant, including integration artifacts and their resources. Such files are vulnerable to malware attacks, which threatens the tenant's security.

When a user tries to upload such files, a malware scanner from SAP Cloud Integration scans them. The scanner is not enabled by default. You can activate the choice in settings (select the Malware Scanner tab). Larger files can take slightly longer to upload when a malware scan is enabled than when it is disabled.

To enable the malware scanner:

1. Go to Settings and select Integrations.

2. Select Edit from the Malware Scanner tab, as shown in Figure 7-3.

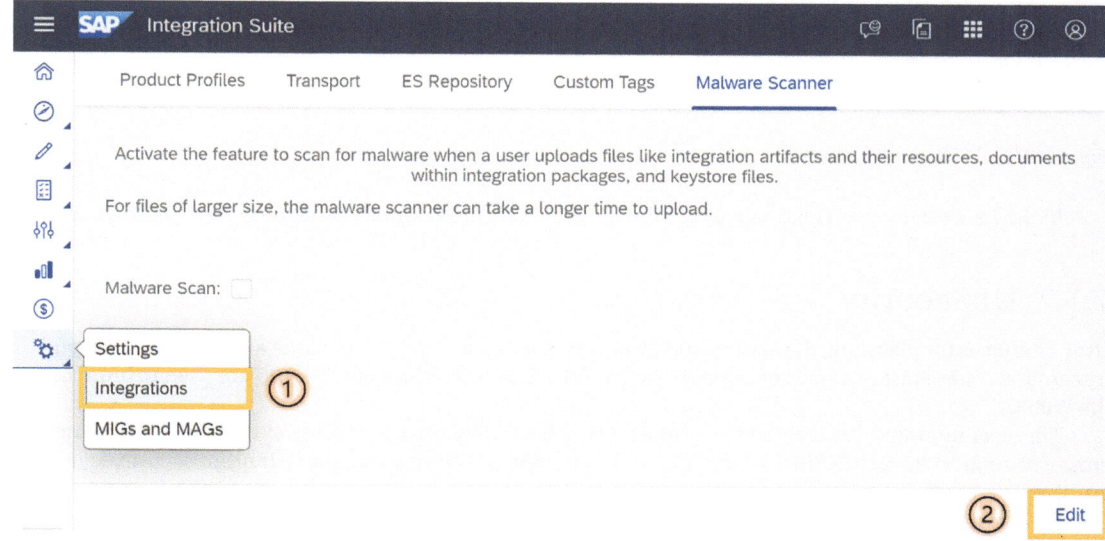

Figure 7-3. *Malware scanner in Edit mode*

3. Select Save after enabling the Malware Scan feature, as shown in Figure 7-4.

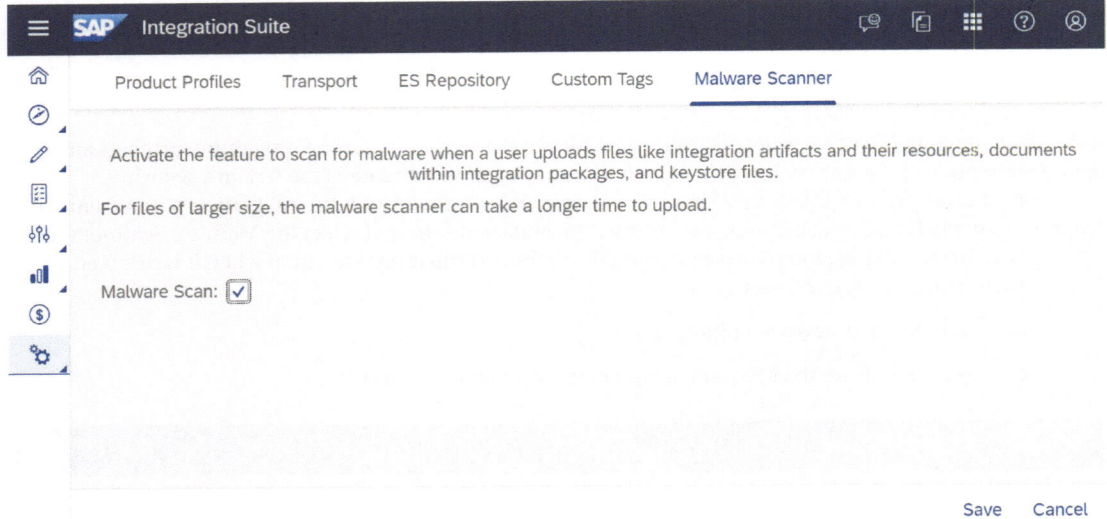

Figure 7-4. Malware scanner

In the next section, you learn about the user interface security options associated with Cloud Foundry.

7.1.7 UI Security

User interfaces for planning, deploying, and monitoring message flows in real time are provided by Cloud Integration. These tasks can be completed using a web UI. JavaScript and HTML UI5 are used to implement the Web UI.

This user interface was created to guard against vulnerabilities such as cross-site scripting (XSS) and cross-site request forgery (XSRF). Along with secure design and coding standards, built-in security features are also leveraged.

You have briefly learned about the security aspects of SAP Cloud Integration in Cloud Foundry. In the next section, you learn about the security aspects of SAP Cloud Integration in the Neo environment.

7.2 Security in the Neo Environment

The technological infrastructure consists of various technical components that can safely communicate with one another and with remote components using, for example, HTTPS or SFTP protocols. The design of user access further ensures that only users with certain permissions can access the various components of the technical infrastructure.

The primary access points and connections, as well as the high-level technical infrastructure, are depicted in Figure 7-5.

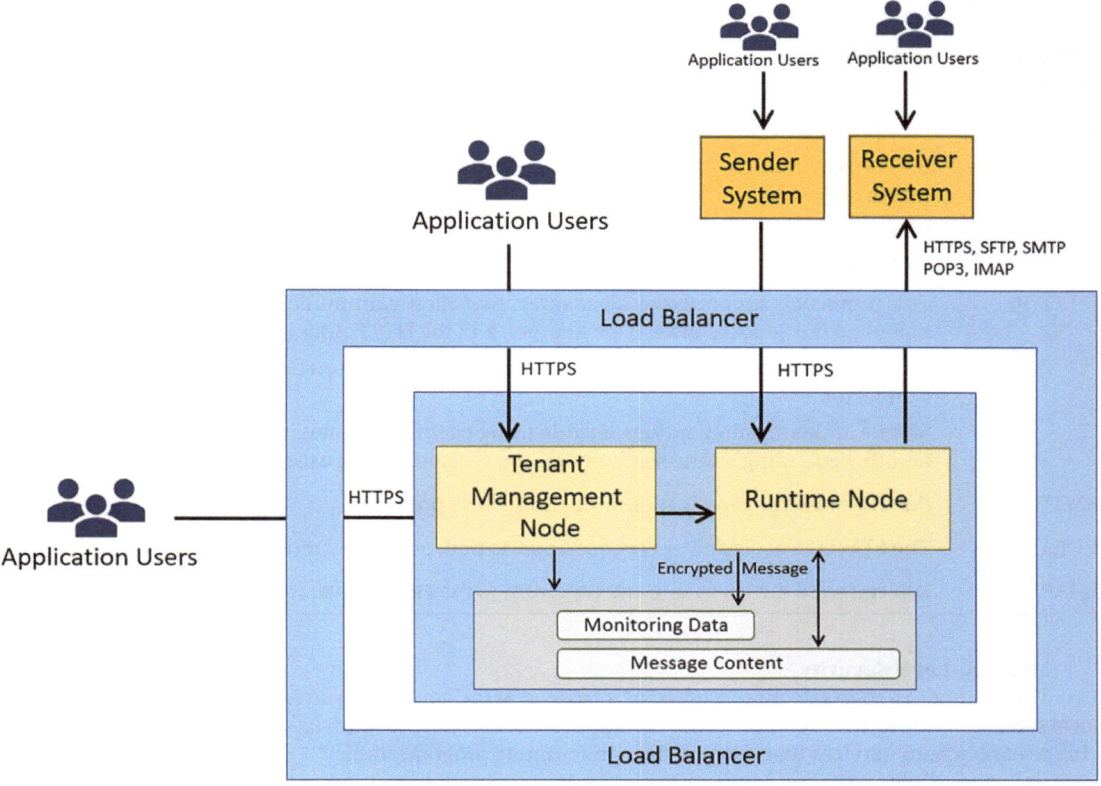

Figure 7-5. *High-level architecture of SAP cloud integration, Neo environment*

This section covers the different security aspects of SAP Cloud Integration in the Neo environment. You will see the data protection and data flow capabilities for SAP Cloud Integration in the Neo environment.

7.2.1 Data Protection and Data Flow for Cloud Integration in Neo

There are two levels of protection for data in transit. The linked remote systems communicate with one another using the established transport protocol. These protocols provide a variety of choices to safeguard the transferred data from unauthorized access. Digital signatures and encryption can be used to safeguard the content of exchanged communications in addition to transport level security.

Transport Level Security

The actions made to safeguard communication between two systems or applications at the transport layer are referred to as transport layer security in the SAP Integration Suite. The transport layer, which is the third layer in the OSI paradigm, is in charge of guaranteeing reliable end-to-end communication between two systems. Table 7-6 shows the transport level security protocols for SAP Cloud Integration in the Neo environment.

Table 7-6. *Transport Layer Security Protocols (Neo)*

Transport Protocol	Description
SFTP	This protocol is supported by the SFTP transmitter and receiver adapters.
	SSH uses a symmetric key length of at least 128 bits to protect FTP transfer. The default length of the SAP asymmetric key is 2048 bits.
	When asymmetric key pairs are used, SFTP also ensures that the participants are utilizing only authorized public keys.
HTTP(S)	This protocol is supported by all adapters that allow communication over HTTPS, including the Soap Adapter, IDoc Adapter, and the HTTP Adapter.
	Moreover, receiver adapters provide main propagation based on the SAP Cloud Connector.
	Simple authentication options include using client certificates, user credentials, and OAuth, depending on the sender or receiver adapter being used.
SMTP	Email communication is supported using these protocols.
POP3	The STARTTLS extended operation supports transport encryption.
IMAP	You can send the username and password to the email server in plain text or encrypted.

Message-Level Security

The methods used to safeguard a message's payload at the application layer are referred to as message-level security in the SAP Integration Suite. The application layer, the seventh layer in the OSI model, is in charge of managing services that are specific to the running program. Table 7-7 shows the message-level security standards for SAP Cloud Integration in the Neo environment.

Table 7-7. *Message Level Security Standards (Neo)*

Standards	Security Features
PKCS#7/CMS Enveloped Data and Signed Data	Encrypting and decrypting message content. Payload authentication and signature.
PKCS#7/CMS Enveloped and Signed Data	Payload encryption, decryption, signature, and verification.
Pretty Good Privacy—PGP	Payload encrypting, decrypting, signing, and verifying. Encrypting and decrypting message content.
XML Signature	Verification and signature of the payload.
WS-Security	Authenticating the SOAP body with a signature.

This section discussed message- and transport-level security for SAP Cloud Integration in the Neo environment. In the next section, you learn about the Identity Access Management available for SAP Cloud Integration in the Neo environment.

7.2.2 Identity Access Management

In the Neo platform, you must also assign the appropriate roles based on the user persona (administrator, developer, etc.). Use the following to assign roles to SAP Cloud Integration users in the Neo platform, as shown in Figure 7-6.

1. Navigate to the Authorization tab in the SAP BTP Neo Cockpit.

2. Click New Group.

3. Assign the roles to the Role Collection.

4. Assign the Role Collection to the user.

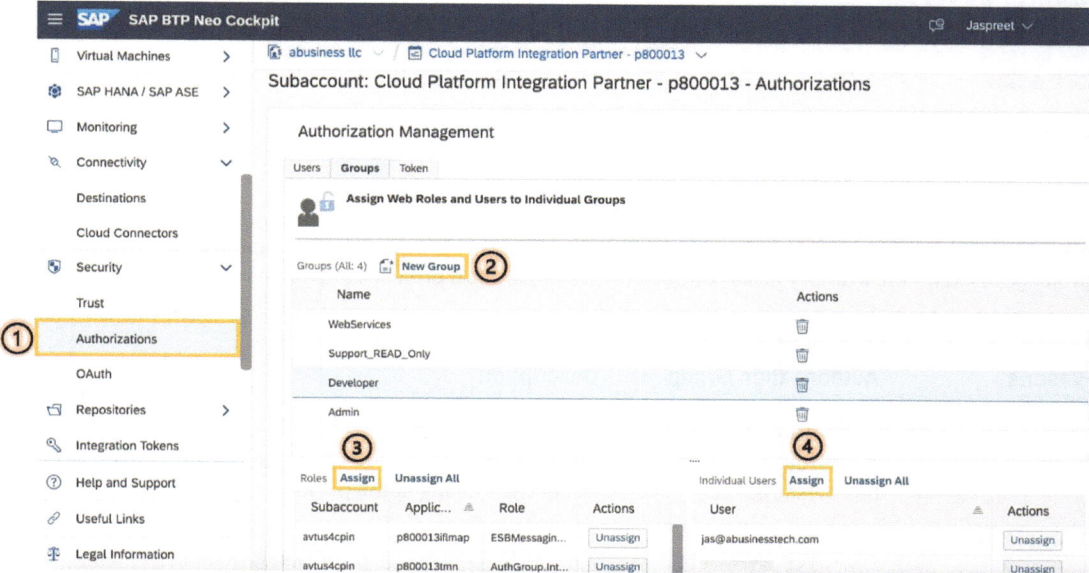

Figure 7-6. *Roles for Cloud Integration in the Neo environment*

In the Neo platform, you can create user personas (administrator, integration developer, etc.), which are specific groups as per the best practices, and then assign them roles, as shown in Figure 7-7.

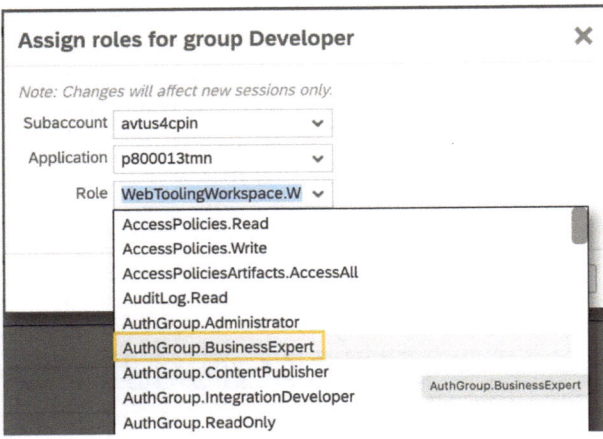

Figure 7-7. *Assigning roles for the Group Developer*

Persona

There are a number of predefined roles that you can apply to users of the account when managing users in the SAP BTP Cockpit. These roles are connected to specific personas and are useful for integration projects based on the primary tasks connected with integration projects, as shown in Table 7-8.

Table 7-8. *Identity Access Management (Neo)*

Persona	Authorization Group (Neo)	Description
Business expert	`AuthGroup.BusinessExpert`	Enables a business professional to perform business tasks, including looking at the payload.
Administrator	`AuthGroup.Administrator`	Enables the tenant administrator, also referred to as the tenant cluster administrator, to connect to a cluster and control its operations.
Integration developer	`AuthGroup.IntegrationDeveloper`	Enables a cluster connection in Integration Designer so that an integration developer can see, download, and deploy artifacts.
Read-only persona	`AuthGroup.ReadOnly`	You can communicate with a tenant and check your messages.
		You can download WSDL files and have read-only access to integration flow artifacts in the Monitoring section with the help of this permission group.
System developer	`AuthGroup.SystemDeveloper`	It enables a system developer to do tasks required for system support.
		This permission group enables read-only access to the Data Store viewer.
Partner Directory configurator	`AuthGroup.TenantPartnerDirectoryConfigurator`	Enables the Partner Directory administrator to access and edit Partner Directory content.

Task and Permission

The responsibilities necessary to complete the various Cloud Integration-related tasks are summarized in Table 7-9. Additionally, the table lists to what extent the roles and duties relate to the key persona.

Table 7-9. *Task and Permission for Neo*

Area	Task	Role (Neo)	Persona
Discover	View packages	`WebToolingCatalog.OverviewRead`	Integration Developer Business Expert Read-Only Persona System Developer Tenant Administrator
	View package artifacts	`WebToolingCatalog.OverviewRead` `WebToolingCatalog.DetailsRead`	Integration Developer Business Expert Read-Only Persona/ System Developer Tenant Administrator
	Copy package to workspace	`WebToolingCatalog.OverviewRead` `WebToolingWorkspace.Write`	Integration Developer
Design	View packages and package artifacts	`WebToolingWorkspace.Read`	Integration Developer Business Expert Read-Only Persona/ System Developer Tenant Administrator
	Create, edit, export, and remove a package and all its artifacts	`WebToolingWorkspace.Read` `WebToolingWorkspace.Write`	Integration Developer
	Update packages	`WebToolingWorkspace.Read` `WebToolingWorkspace.Write`	Integration Developer

	Configure artifacts	`WebToolingWorkspace.Read`
		`WebTooling.`
		`IntegrationFlowConfigure`
			Integration Developer

	Deploy/undeploy artifacts	`WebToolingWorkspace.Read`
		`NodeManager.read`
		`GenerationAndBuild.`
		`generationandbuildcontent`
		`NodeManager.deploycontent`
			Integration Developer
			Tenant Administrator

	Export packages for transport	`WebToolingWorkspace.Read`
		`TransportModule.read`
		`TransportModule.write`
			N/A

	Import packages from transport	`WebToolingWorkspace.Read`
		`TransportModule.read`
		`TransportModule.write`
			N/A

This section briefly discussed identity access management and listed the different personas and the tasks and permissions required in SAP Cloud Integration. In the next section, you learn about data storage security.

7.2.3 Data Storage Security

Customer information may be retained during certain stages of message processing. However, any customer data retained for an indefinite period must be securely separated and organized according to each individual customer. While multiple tenants can share the same physical infrastructure, the data of each tenant must be stored in a separate structure or format.

The customer can specify whether their data at rest is encrypted for particular use cases.

It is possible to store encrypted message content. If this security measure is configured, the automatically generated encryption key is different for each tenant and is periodically renewed. This section briefly discussed data storage security. In the next section, you learn about the types of data stored in the Neo environment.

7.2.4 Types of Stored Data

While an integration scenario is running, various types of data, including message content and monitoring data, can be stored.

Due to the possibility of containing personal information, this data must be regarded as sensitive. This includes the following examples:

- **Message content**—Since messages processed on a runtime node frequently include business data from an integration scenario, it is possible to find sensitive consumer data there, such as addresses, names, and financial information.

- **Monitoring data**—The steps of an integration flow's processing are recorded in the message processing log. Only individuals allocated to this tenant and those with certain permissions can view this data.

You have now reviewed the types of data stored in the Neo environment. In the next section, you learn about the specific data sets associated with SAP Cloud iNtegration in the Neo environment.

7.2.5 Specific Data Sets

The SAP Cloud Integration architecture stores a variety of data types over the course of an integration project.

Table 7-10 lists the different groups and characteristics, such as the type of storage and the length of retention. Backup is available for 14 days for all the data types available in this table.

Table 7-10. *Specific Data Sets (Neo)*

Data	Description	Logical Storage	Classification	Retention Time
Message processing log	Information that is organized about how a message is processed		Log data	30 days
Message processing log attachments	Data recorded in a runtime message processing log	Message store	Log data Business data	30 days
Audit log	Details about activities such data read-only accesses and changes to the system configuration		Log data	14 days
System log	Information about mistakes made while processing HTTP inbound		Log data	7 days
Integration flow tracing data	Being aware of the message flow, the message payload, and any processing errors that took place	Trace Store	Log data Business data	60 minutes
Integration content (design time)	Integration developer-created or altered design-time items, like as integration flows, are represented in Camel XML	Workspace	Configuration data	Unlimited
Integration content (runtime)	Knowledge of the message flow, including the message payload, as well as any processing blunders that occurred		Configuration data	Unlimited
Data store by Data Store operations	Certain integration flow steps keep track of message content, which includes details like the payload, tenant ID, message GUID, and message processing log GUID Further phases in the integration cycle use it to process messages in more depth	Data store	Business data	Developer of the integration can define it; the default value is 30 days
Data store by Persist step	Message data is kept in specific integration flow steps	Message store	Business data	90 days
Message content store by JMS adapter	JMS message queues contain the message content	JMS queue	Business data	Developer of the integration can define it; the default value is 90 days.
Lock entries	To prevent the same message from being processed more than once concurrently, lock entries are created (in the in-progress repository) (for example, by different runtime nodes)			

This section discussed the specific data sets associated with the Neo environment, whereby you learned about message logs, contents, storage, and retention time. In the next section, you learn about user interface security related to the Neo environment in SAP Cloud Integration.

7.2.6 UI Security

User interfaces for designing, deploying, and monitoring message flows in real time are provided by Cloud Integration. To carry out these tasks, a web tool (Web UI) is accessible. JavaScript and HTML are used to implement the Web UI (UI5).

Cross-site scripting (XSS) and cross-site request forgery (CSRF) are vulnerabilities that this user interface is designed to guard against (XSRF). Together with secure design and coding principles, these technologies' built-in security features are used.

7.2.7 Remote API Security

Cloud integration functionalities can be accessed using application programming interfaces and APIs. The OData API is secured via Basic or OAuth authentication.

To protect the API from CSRF attacks, modification operations such as POST and DELETE should be performed in conjunction with session-based authentication and client-side CSRF processing.

7.3 Summary

This chapter covered the security features of SAP's Cloud Foundry and Neo environments. First, the chapter concentrated on Cloud Foundry, which included issues such as certificate management, technical landscape, identity access management, data storage, protection, and security, as well as different types of stored data, message logs, malware scanner, and UI security. The Neo environment of SAP Cloud Integration was also discussed, and the topics covered included data flow and protection, identity access control, data storage security, specific data sets, and other security information, such as UI security and remote API security. This was the final concluding chapter of A Practical Guide to SAP Integration Suite - Cloud Integration, but an important one as Security of the data, IT assets, and intellectual property is one of the most critical requirement of businesses.

Index